A Planetary Economy

Fraser Murison Smith

A Planetary Economy

palgrave
macmillan

Fraser Murison Smith
Economist
San Rafael, CA, USA

ISBN 978-3-030-49295-3 ISBN 978-3-030-49296-0 (eBook)
https://doi.org/10.1007/978-3-030-49296-0

This Palgrave Macmillan imprint is published by the registered company Springer Nature Switzerland AG.
The registered company address is: Gewerbestrasse 11, 6330 Cham, Switzerland

Dedicated to the children of the world and all who come after.

And to my wife, Pilar, whose beauty and love, strength and patience, intelligence and insight, and inspiring motherhood are a constant joy, and the foundations for this book.

Foreword

I distinctly remember being obsessed about environmental systems when I participated at a recycling center for Earth Day 1970. My young awareness already realized that the natural systems were seriously out of balance, the air was toxic to people and rivers were on fire. People participating in that first Earth Day were not out to celebrate the planet but to protest the assault of the capitalist economic system on our ultimate life support, the primary context that connects us to every other thing. But we were also there to demonstrate that there was a much better way of doing 'business' on Earth.

That quest for a better way followed me throughout college and into the emerging service industry of 'corporate environmental management', featuring such organizational development fads as 'total quality environmental management' and 'sustainable development'. My solution was to integrate existing regulatory, economic and civic systems into new ways to improve compliance on the part of businesses and provide critical environmental feedback to their customers. The result was the first comprehensive Green Business certification program in the nation.

But there was a persistent problem. Systems thinking was fairly rare in both the business and governmental sectors, which honored granularity and specialization. Most attempts at systems integration would be blocked by a cadre of mid-level managers. There seemed no real way to get everyone on the same page within a common context for change to gain traction. That is until I met Fraser Murison Smith.

Fraser was one of the first employees at a software start-up in 1997 by the name of Datafusion, working as a systems integrator. I signed on as the company's environmental consultant, working with Fraser to develop knowledge maps: galaxies of nodes and links within a recognizable, data-rich context, undergirded by a robust content catalogue. It was electrifying to be the first infocartographers on Earth!

Since then, I have had the privilege of sharing many visual contexts with Fraser. Better yet, I have found a person who can converse with anybody in a shared understanding of complex, integrated systems without the need for continual explanation. Think about that for a minute: maximum knowledge transmission with minimum conflict.

A Planetary Economy is about all these things: protecting the planet, our primary context; creating an economy that can deliver prosperity and a high quality of life for all beings; and developing a viable future without waiting for new technology or a major scientific advancement. Fraser elegantly shows it just requires a change of perspective wired with critical feedbacks to transform the current life-destroying economic context in which we participate to one that is life-respecting. We can have our planet and live on it too.

Viewcraft, Inc., Sebastopol, CA John Garn
March 2020

Contents

About the Author

Fraser Murison Smith is an energy specialist in public utilities, formerly an information systems consultant and award-winning cleantech entrepreneur. After completing a PhD in Theoretical Ecology at Oxford University, he was a postdoctoral fellow at Stanford University in ecological economics. He has published papers on fisheries, biodiversity and economic development, as well as two books, *Environmental Sustainability: Practical Global Implications* (1997) and *Economics of a Crowded Planet* (2019). Murison Smith and his wife, a healthcare technology leader, share their home in Northern California with two wonderful children and a canoe and tent on standby for spontaneous forays into the surrounding mountains, rivers and lakes.

Abbreviations

GDP	Gross Domestic Product
GGP	Gross Global Product
GNP	Gross National Product
IMF	International Monetary Fund
LCOE	Levelized Cost of Electricity
NPP	Net Primary Production
OECD	Organization for Economic Co-operation and Development
SEEA	System of Environmental-Economic Accounting
SMC	Sovereign Money Creation
SNA	System of National Accounts
UN	United Nations

List of Figures

List of Tables

Introduction

Nature keeps us safe.
—Orion Murison Smith, age 5

… a simple yet profound insight, summing up in four words what scientists have been trying to tell policymakers for years. Might his grandchildren, by the end of the twenty-first century, be enacting policies and developing technologies to maximize the longevity of the human economy? In the hope that they will, this book is for today's children, their children and the generations to come. The impetus for this book is generative.

Unlike many writings on the problems of the present day and how to solve them, this book describes a society—an economy, specifically—which does not yet exist. It asks the simple question, how would a future economy look if one started with a blank sheet of paper, requiring the economy to persist in stable alignment with natural systems, guaranteeing prosperity for all? How would it work? What kinds of norms and values would prevail? What kinds of institutions would be needed? What kinds of controls would be in place? In proposing answers to these questions, this book works backward from this imagined future to the present day to ask how we might get there.

The effort is not an exercise in speculation, nor in utopianism. Fred Polak, writing in 1961, identified the practical importance of having a concrete perception of the future: "there is a great deal of historical evidence to suggest that a society which loses its identity with posterity and which loses its positive image of the future loses also its capacity to deal with present problems, and soon falls apart."[1]

As a graduate student studying species extinction rates[2] in the early 1990s, it was clear to me that the loss of biodiversity in today's world is not really a biological problem but a social one. Years later, upon learning that snow leopards are hunted for their pelts, my daughter asked, "why? Money's not important. Nature is important." Ignorance of this simple fact is the essence of this social problem. Although the current spike in species extinctions could become a mass extinction comparable to those observed in the fossil record,[3] the biosphere nonetheless will bounce back. If the geological past is any guide, then the recovery probably would take several million years; however, life on Earth is in no particular danger of being wiped out. It simply will continue to "create new versions of itself," as Adam Frank describes.[4]

Species extinction is a symptom of a broader misalignment between social processes and natural ones. Why do the two have to be out of alignment in the first place? What is to stop the human social system from becoming realigned with nature? The philosopher Arne Naess, writing about the basis of environmental norms, wrote: "the mature human individual ... seeks a social order, or rather a biospheric order, that maximizes the potential for self-realization of all kinds of beings."[5] If a state of biospheric order is how the world *should* be, then the question becomes how to get there. Naess suggested that the environmental crisis could yet turn out to be of great value to "the expansion of human consciousness" or, more concretely, to a re-emergence of social norms and attitudes acknowledging the interconnectedness of the world. As this book goes to press, the world has become consumed with the coronavirus pandemic: a suggestion from nature to humanity to change its ways.

In this, an enormous opportunity lies ahead. By imagining a prosperous economy in alignment with nature, it becomes possible to articulate pathways toward it. Not only can the economy attain long-term economic stability but also it can generate significant wealth in the process—such as

[1] Polak (1961), quoted in Boulding (1966).
[2] Smith et al. (1993).
[3] Kolbert (2014).
[4] Frank (2018).
[5] Naess (2008, p. 132).

through technological innovation—this wealth becoming the basis for widespread prosperity. The two go hand in hand; in fact, widespread prosperity is a prerequisite for economic alignment with nature. An economy that accomplishes both in the service of a large population is what I call a *planetary* economy.[6]

This book is the prescriptive companion to *Economics of a Crowded Planet*,[7] which describes how the field of economics needs to change to support a planetary economy, and the ways its economics might differ from today's. Just as the economy and nature today are locked in coevolution, so a coevolution between the methodology of economics and the prescriptions it recommends also needs to occur. One cannot describe an economics of a crowded planet without having an economy to refer to.

The specific kind of planetary economy described here is a 'market-planetarian' one. In principle, a variety of systems could be imagined to enable an economy to persist in material balance with nature, offering a minimum living standard to all. One example is a command-and-control approach. Another is a community-based system. This book argues that the most likely, and most desirable, is a market-based system. Markets emerge naturally from civilization. They can be tremendously powerful engines of social change, when harnessed properly. The difference between a planetarian society and today's is that, in a planetarian one, markets will serve planetarian goals. They will constitute a means to an end rather than the end in itself, as they are considered under current market liberalism.

The book is divided into three parts. Part I examines how and why the economy became locked into material coevolution with nature. Part II of the book is prescriptive. It asks where the economy needs to be in order to arrive in stable alignment with natural processes. In this, it addresses norms, institutions, policies and economic instruments. Part III weaves the threads from the previous chapters into a pathway from the present growth economy toward a planetary economy a century or two hence. It imagines how life in this future planetarian society might look.

Both *Economics of a Crowded Planet* and this volume employ a number of terms and conventions, summarized below.

'Nature' is the collection of all natural systems on Earth, both living and non-living. 'Planet Earth', for the present discussion, consists of nature plus all human social systems, including the economy.

[6] The term derives from Peter Berg's 'planetarianism', discussed in Chap. 2.
[7] Murison Smith (2019).

A 'crowded' planet is a densely populated one. A crowded planet may or may not also be materially or energetically intense. 'Material intensity' is taken to mean the total resource loading of the economy, or of any enterprise within it, upon nature. This loading is the sum of inputs from nature and outputs to it.

An economy that is 'large' relative to planetary processes is one that produces measurable effects upon natural processes at the global level, and is in turn affected by them. If a change in natural processes is labeled 'anthropogenic', then this means that there exists a broad scientific consensus that the primary cause, with a high degree of probability, is human activity.

'Natural capacity' is the capacity of nature to support the human economy, both in terms of the resources it provides and in terms of its ability to absorb and process economic effluents. This term is distinct from the familiar 'natural capital', being dynamic in nature, explicitly encompassing both the provisioning of inputs and the absorption of outputs.

The term 'alignment' refers primarily to the state of the interface between human and natural systems, the human system being principally the economy. When inputs to and outputs from the two domains fall within the bounds of natural capacity, we have alignment. To fall within the bounds of natural capacity, a material flow must be small enough to be supported by nature's regenerative capabilities, or to be absorbed and processed into naturally occurring substances, without any measurable anthropogenic change to natural processes. A policy of 'material discipline' is one that minimizes material intensity. Alignment of human processes with nature in this sense would be expected to result in long-term material stability of the economy.

'Alignment' at the individual level relates to norms, habits, attitudes and practices in the course of daily living. Ancient Eastern and indigenous worldviews speak of 'harmony' or 'balance' in one's dealings with the natural world. One of the challenges for a modern economy on a crowded planet is to reclaim a semblance of this balance in daily life, as part of a broader economic alignment with nature.

The 'prosperity' of the economy is manifested by an absence of poverty, a minimum living standard for all, productive work rewarded in a variety of ways—not merely through wages—and ample time for family, community and creativity. It is the kind of world in which most of us probably would like to live. As Chap. 18 describes, it is a stable, co-operative, tranquil society, accommodating also innovation, competition and change.

The terms 'environment' and 'sustainability' are avoided wherever possible, as their meanings never were very precise, even among the specialists who coined them, and they have in any case been politically co-opted over the years. This text uses the terms 'regeneration' and 'generative' because they refer specifically to practices that improve system health, whether of natural systems or social ones. They are the antithesis of 'exploitation' or 'exploitative'. As *Economics of a Crowded Planet* describes, the ability of nature to 'sustain' the human economy is inversely related to the economy's probability of collapse over any defined period. This ability is a function of natural capacity. The ability of nature to sustain itself, on the other hand, is taken as given.

References

Boulding, K. E. (1966). The Economics of the Coming Spaceship Earth. In H. Jarrett (Ed.), *Environmental Quality in a Growing Economy* (pp. 3–14). Baltimore, MD: John Hopkins Press.

Frank, A. (2018, June 12). Earth Will Survive. We May Not. *New York Times.*

Kolbert, E. (2014). *The Sixth Extinction: An Unnatural History.* New York: Picador, Henry Holt & Company.

Murison Smith, F. D. (2019). *Economics of a Crowded Planet.* New York: Palgrave Macmillan.

Naess, A. (2008). *The Ecology of Wisdom: Writings by Arne Naess* (A. Drengson & B. Devall, Eds.). Berkeley, CA: Counterpoint Press.

Polak, F. L. (1961). *The Image of the Future, Vols. I & II.* New York: Sythoff, Leyden & Oceana.

Smith, F. D. M., et al. (1993). Estimating Extinction Rates. *Nature, 364*: 494–496.

PART I

An Economy on a Crowded Planet

Chapter 1 describes the economy's coevolution with nature as a form of structural coupling at multiple scales of organization, a concept covered in technical detail in *Economics of a Crowded Planet.* It provides an overview of the human enterprise, seen over ecological timescales of hundreds to thousands of years, placing the economy's exponential growth into its biophysical context. It examines the role played by the economy's internal flows of money in magnifying the exchanges of material between the economy and nature. The perspective is explicitly from outside the economy looking in, as distinct from the conventional economic one, inside the economy, occasionally looking out.

Chapter 2 addresses the origins of this growth, focusing particularly upon changes in people's attitudes toward nature over the centuries and, in turn, toward their roles as economic agents. Even though the present economy is a product of Western modernity, alternate worldviews have persisted within Western society, and alternate futures imagined. Chapter 2 evaluates some of these as templates for a future worldview in which individual action would be generative rather than exploitative, and the economy managed as a stable planetary subsystem.

CHAPTER 1

The Economy's Coevolution with Nature

Much has been written over the past 50 years about the growing effects of people upon natural systems. A consensus about these effects exists to a much greater degree early in the twenty-first century than in the 1960s when the debate about 'the environment' first permeated into the modern public consciousness. The measurable effects of the economy on the planet's natural systems feed back into influences of nature upon the economy, so that the two have become embroiled in a global coevolution. On the timescales of natural systems, the global economy appeared suddenly, as if out of nowhere, even though on human timescales its emergence was brought about by internal forces operating over many hundreds of years. Prior to this state-change in Earth's systems, the economy was materially small relative to natural processes. There was no 'global' economy in the sense we know it today, but rather a collection of local and regional economies, many of which had local or regional effects upon nature, and were affected by those changes in turn, all of which, however, being characterized by relatively low material intensity.

A wide range of laws, policies and, more recently, markets have been created to try to reduce the effects of economic activity upon nature. Yet the effects continue to accelerate. Few during the twentieth century asked whether the whole economic and financial system—indeed, the Western modernist paradigm that underpins them—should be re-examined and reconfigured for a new state of the world in which the human enterprise is large relative to its planetary container. Those who did were largely

F. Murison Smith, *A Planetary Economy*,
https://doi.org/10.1007/978-3-030-49296-0_1

marginalized by the economic mainstream.[1] Yet in the first years of the twenty-first century a small but increasing number of writers has begun to tackle this question head-on.[2]

Our historical reluctance to tackle the question might have as much to do with its daunting size as with its perceived relevance. One way to make a start is to construct a hypothetical future state, based on the current understanding of natural systems, in which an economy would be relatively stable, both in its exchanges with nature and internally. Then, if the assumptions underlying such a model were sufficiently robust, the question could be asked, how to go from here to there.

It's an Economy, Jim, But Not as We Know It

So might Mr. Spock say to Captain Kirk as the starship *Free Enterprise* glides into orbit around a planet crowded with inhabitants.

On this planet, the economy is stable, the inhabitants enjoying high living standards and a good quality of life. The reason for this state, in spite of the large population, is because the economy exists in a form of dynamic balance with the planet's natural systems. For these economic conditions to prevail, the economy's material efficiency must be very high.

Such a state, writes the captain in his log, suggests that this society, at some time in its past, confronted its growing material size and concluded that, in the interest of long-term stability, its economic activity had to be reconfigured to minimize its effects on nature. The society must also have been forced to accept the thermodynamic reality of their planet being a materially closed system. Its only external input was free energy from the star around which it orbited. There was only so much material to go around.

Such a conclusion may have represented a fundamental alteration in perspectives on economic activity. The society on this planet would have recognized, perhaps for the first time, the prime importance of a long-term view, a shift that came about because of a growing recognition that there was nowhere else to go. Consequently, the inhabitants of the planet re-examined the way they thought about their economic activity, how they measured it, studied it and managed it. Their economics evolved as a

[1] Soddy (1926), Boulding (1966), Daly (1991), Georgescu-Roegen (1971) and Odum (1971) are examples.

[2] For example, Douthwaite (2000), Boyle and Simms (2009), and Panayotakis (2011).

result, from a field concerned with individual 'choices' to one oriented toward alignment with nature.

On the planet in question, writes the captain, evolution in the society's study of their economy stimulated a variety of fundamental changes in their management of it. A prime directive—to borrow another *Star Trek* phrase—became established to steer economic policy toward long-term stability. The economy underwent significant structural and dynamic changes away from a materially intense mode toward the materially efficient mode in which the *Free Enterprise* finds it. The inhabitants tell of certain difficult adjustments made along the way, even hardships; yet the society that the ship's crew encounters is prosperous, vibrant and largely peaceful.

So much is written in this day and age—either of the gloomy future that awaits society here on Earth if it keeps 'messing up the environment' or, in equal measure, of the complete denial of an 'environmental' problem in the first place—that there has been missing a narrative of the *opportunity* in attaining a life like the one our fictional captain described in his log.

Nature and the Economy as Complex, Hierarchical systems

The human economy is a complex system of trade among people, representing themselves as individuals or the institutions to which they belong. It is also a complex system of trade with nature, the exchange of material with natural systems. Internally, the economy circulates two kinds of flow: material and currency. Money is a convenient proxy for valuing material goods against one another; also for services rendered by people to one another.

The economy exists within a biophysical context because it extracts material from nature and expels effluents back to it. On a crowded planet such as ours, the economy must work in concert with its natural support systems if it is to persist. This necessitates understanding those systems in depth and operating the global economy according to their requirements. Ever since human economies emerged, they have engaged in local coevolutions with nature. The difference today is that the coevolution is global and there is nowhere to escape to if it goes wrong.

The natural world does not lend itself to description using the mechanical tools of conventional economics. Mechanical analysis assumes reversibility: it is ahistorical. The concept of elapsed time does not exist. Natural systems are inherently nonlinear, evolutionary and unpredictable, lending themselves to historical descriptions incorporating elapsed time. Social systems are analogues of natural ones in this respect. One of the main reasons why economics adopted a mechanical analogue during the nineteenth century is because that was the dominant form of mathematics at the time. The mathematics of nonlinear systems did not emerge until the first half of the twentieth century, and computer technologies to run simulation models employing it did not mature until the 1980s. By that time, the mechanical analogue had become embedded into economic orthodoxy and the economy had grown exponentially. Only since the 1990s has the mechanical analogue's grip on economics begun to weaken.

Mechanics implies the management of a system as a form of control, much the same way as you might turn a thermostat up or down to heat or cool your house. Management as control in this way does not apply to complex, nonlinear systems. Our inability to control the weather the way we would control the temperature of our dwelling is matched by our inability to control the world's ecosystems or our own economy in the same way. Although complex systems defy management in the form of control, they respond, however, to management as selection. Management as selection is a form of trial-and-error. Nudge a system a little bit this way or that, see how it responds, and adjust accordingly. This mechanism is analogous to artificial selection—the breeding of animals—and to natural selection operating in the wild.

Luckily for our large economy, nature is autopoietic: it is self-generating, subject to free energy from the sun. Left to its own devices, it will regenerate toward some stable, mature state. Nature's takeover of the abandoned city of Chernobyl provides a living testament to what would happen to the edifices of the human economy were people suddenly to vanish. The challenge for economic policy, then, is quite simple: it is to nudge the economy this way or that so as to provide the best possible conditions for natural regeneration. Over time, this management will steer the economy's coevolution with nature toward some form of dynamic alignment. Because the economy runs internally on a distinct flow of money, which does not circulate through nature, the challenge then becomes representing the economy's material interface with nature through its flows of money.

The reality of this challenge is complex, insofar as the economy interacts with nature at multiple levels of organization. *Economics of a Crowded Planet* draws upon hierarchy theory to create a framework for the field of economics to address this challenge. That same framework applies to the practicalities of governance, policies and economic controls. It states essentially that phenomena at any particular scale of observation will result from interactions among factors at lower levels of observation, these being *emergent properties* of the system under study. The behavior of any system under observation also will be constrained by *boundary conditions* imposed from higher levels of organization.[3] Think, for instance, of the interactions among the organs and tissues in your body giving rise to the emergent properties of you as a whole person. These properties—sentience, emotion, reason, homeostasis—are not properties possessed by any of your organs or tissues individually. Within the economy, entities at a variety of levels of organization can be identified, from individuals and households to private enterprises; also public entities at local, regional and national levels; and transnational entities, whether corporate, governmental or non-governmental. Any economic interaction, at any level, takes place within a context of both social and natural components. Economic entities at all levels interact with, and innervate with, natural processes.

By placing the economy within a context of natural processes, it becomes possible to construct a picture of economic activity as a hierarchy of complex systems. Hierarchy theory, as employed in systems science, is not the rigid, static hierarchy one might associate with a military institution but a dynamic hierarchy in which organization at one level emerges from dynamic interactions at another. It dovetails neatly with the evolutionary description of complex systems.

What Is the Economy For?

Human economies have arisen as a result of an increasing division of labor. Hunter-gatherer societies were characterized by minimal division of labor. As civilization took root in tandem with the spread of agriculture, the division of labor increased. People specialized in certain tasks or skills, such as spear-making, pottery or construction. Early forms of money, which were used to measure agricultural production, also served as a means of exchange among these various other labors. With this, we had an economy.

[3] Eldredge and Salthe (1984); Vrba and Eldredge (1984).

Division of labor makes life easier for all. A fletcher can just as easily make arrows for the whole community as for himself. For the extra time he spends making arrows, he gains in turn from the fruits of others' labors, whether in making pottery, growing vegetables or milking cows. In this way, as the economist John Kenneth Galbraith put it, an economy exists "to provide the goods and render the services that people want".

Galbraith, like many twentieth-century economists, also asserted that "the best economic system is the one that supplies the *most* of what people most want" (emphasis added).[4] This normative statement about the economy was symptomatic of the naïvety about the bounty of nature pervading the economic profession at the time. Economists were hardly alone in this. Human history is peppered with examples of societies that collapsed due to an overestimation or misjudgment of the capacity of nature to support them.[5] Galbraith's statement about the 'best' economy would only apply to a large economy on a crowded planet if people's wants encompassed long-term economic stability. Marjorie Kelly asks an alternate question: "what kind of economy will be suited for ongoing life inside a living Earth?"[6] That is an appropriate question for a crowded planet.

In the textbooks, the organizing purpose of a market economy is 'consumer sovereignty'.[7] The global economy is predominantly a market economy. Therefore, following the textbook orthodoxy, consumer sovereignty is the organizing principle of most of the economy. Although people have certain direct, non-market economic interactions with nature—such as felling a tree for wood or dropping litter by the roadside—most of the economy's material intensity is mediated through a market in some way. Yet the consumer is not as sovereign as the textbooks suggest. The presumption of consumer sovereignty is a convenience for a mechanical model of the market. In reality, large market players, such as transnational corporations, take actions that cannot be reduced simply to the benefit of individual people.[8] Dominant market players act to support their own persistence as much as to benefit their customers.

As a complex system, the economy evolves. In and of itself, an economy has no purpose as such, it simply exists as a complex system of exchange

[4] Galbraith (1973), p. 3.
[5] Diamond (2005); Motesharrei et al. (2014).
[6] Kelly (2012), p. 124.
[7] See, for example, Goodwin et al. (2009), p. 251.
[8] Galbraith (1973, pp. 3–4).

among people. However, to the extent that we identify a preferred state for the economy, given its already large and growing scale, it should be long-term stability. A large, unstable economy could lose its organization, and collapse. Although instabilities could—and do—emerge from within the economy, any deleterious effects from nature would represent exogenous sources of instability, rendered all the more probable, frequent or severe according to any number of prior effects of the economy upon nature. This situation would arise, and is already beginning to be seen, where the economy's material intensity upon natural systems becomes significant.

If long-term stability is a desirable state, then the primary purpose of economic activity becomes to adapt and evolve toward it. This is economy as *telos*: a goal-directed system. For a materially intense economy like ours, the purpose is to reduce its material intensity. The economy's 'prime directive'—to borrow the term from *Star Trek*—would become ***material discipline***, the material analogue of fiscal discipline. The 'best' economy now is the most persistent one. Within this normative framework, there remains ample scope for the economy to provide for people's wants, however frivolous they may be. Yet, by becoming materially stable over the long term, the economy would become resilient to perturbations, and adaptable. It would be a triumph of the adequate over the perfect.

How the Economy Became So Large

The present global economy's relationship with nature is parasitic. Parasitic relationships run the gamut from being lethal to the host to being lethal to the parasite. Some result in a kind of stalemate, such as a long-term, low-level infection. It is possible that certain commensal relationships, in which species coexist without harming one another, may have evolved in this way. Humanity has been a parasite upon natural systems for quite some time, although for many hundreds of years the infection incubated at a low level. Only in the past few generations, during which the world industrialized, has the infection flared up.

All species alter their environment to some degree. Certain groups of organisms had profound effects upon planetary systems over long periods. Cyanobacteria were responsible for releasing oxygen into Earth's atmosphere beginning around two billion years ago. This oxygen was a by-product of photosynthesis, a happy accident that laid the foundations for the emergence of oxygen-breathing animal life, including people.

Humanity's current release of carbon dioxide and other greenhouse gases into the atmosphere shares some parallels with similar releases recorded around major extinction events in geological history. The greatest extinction event of all, at the Permian-Triassic boundary about 220 million years ago, coincided with the release of massive amounts of carbon dioxide from the intrusion of hot magma from deep within Earth into sub-surface sediments rich in organic matter. Recent geological work has confirmed that the rate of that eruption was around ten times slower than current anthropogenic greenhouse-gas emissions. Nonetheless, on a geological timescale it occurred very rapidly. It was a giant burp of CO_2 that acidified the oceans and took out much of the marine biota. Global mean surface temperature rose approximately 6 °C to 7 °C in the process.[9]

Current greenhouse-gas releases so far are smaller in volume but more rapid than the Permian-Triassic one, leading to a likely average warming of 3 °C to 4 °C by 2100 at current rates. Some model scenarios forecast much higher anthropogenic warming, on the order of 7 °C to 12 °C over the next 300 years. Under that temperature rise, much of the land area of the tropics and subtropics, presently home to billions of people, would become so hot as to be physiologically lethal.[10]

Notwithstanding geologically short-term fluctuations, the overall mix of gases in the atmosphere has been relatively constant for hundreds of millions of years. At its peak around the Permian-Triassic boundary, the atmospheric concentration of carbon dioxide was about one hundred times higher than today. Even then, it accounted for only around 4 percent of Earth's atmosphere. The concentrations of the atmosphere's two major components, nitrogen and oxygen, have fluctuated little over the past billion years. While Earth has warmed and cooled over the eons, and biota have come and gone, the basic organization of natural systems on Earth has persisted. Against this backdrop, the hypothesis of Earth as a self-regulating organism is not all that far-fetched. It is consistent with many indigenous worldviews, as well as with those of some scientists.[11]

[9] Burgess *et al.* (2017).

[10] Sherwood and Huber (2010). A detailed discussion is in Chap. 5.

[11] James Lovelock was probably the first professional scientist to propose this kind of idea in his *Gaia Hypothesis* in 1979. See Lovelock (2000) for an updated account.

Population

The world's human population remained small, at around a few million, for many millennia before growing, following the advent of agriculture, to a few hundred million around a thousand years ago. As of 1800, early in the industrial revolution, the global population was estimated to be just under one billion.[12] Over the past 200 years, it has grown exponentially to around 7½ billion at the time of writing.[13] To put the rate of increase into perspective, the population has multiplied by seven times over a period about one-thousandth the lifetime of the human species. Over thousands of years, the increase looks like an almost vertical spike (Fig. 1.1). Even on a timescale of hundreds of years, the increase is very sudden (Fig. 1.2). It is characteristic of a chemical state change.

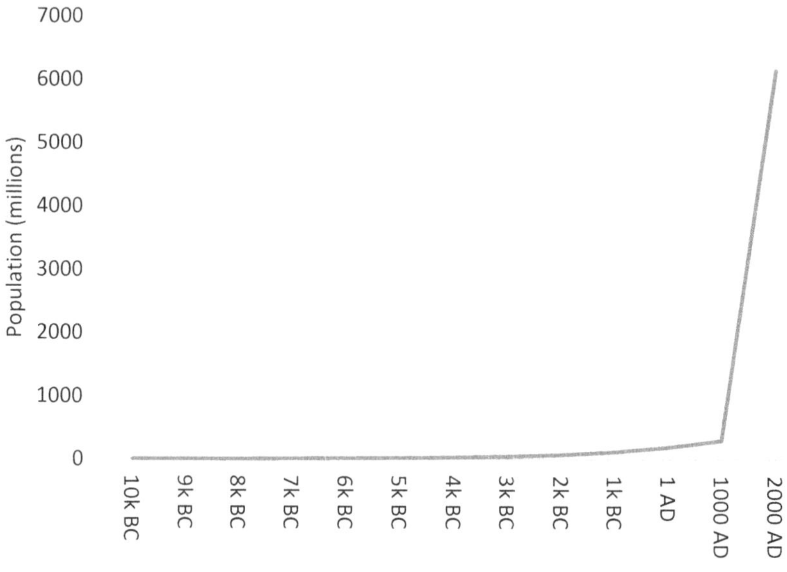

Fig. 1.1 Estimates of human population, in millions, from deep history to 2000: Population every 1000 years from 10,000 BC. Source: Author's creation, using data from Klein Goldewijk and van Drecht (2007), History Database of the Global Environment

[12] Klein Goldewijk and van Drecht (2007).

[13] United Nations World Population Prospects, https://population.un.org/wpp/.

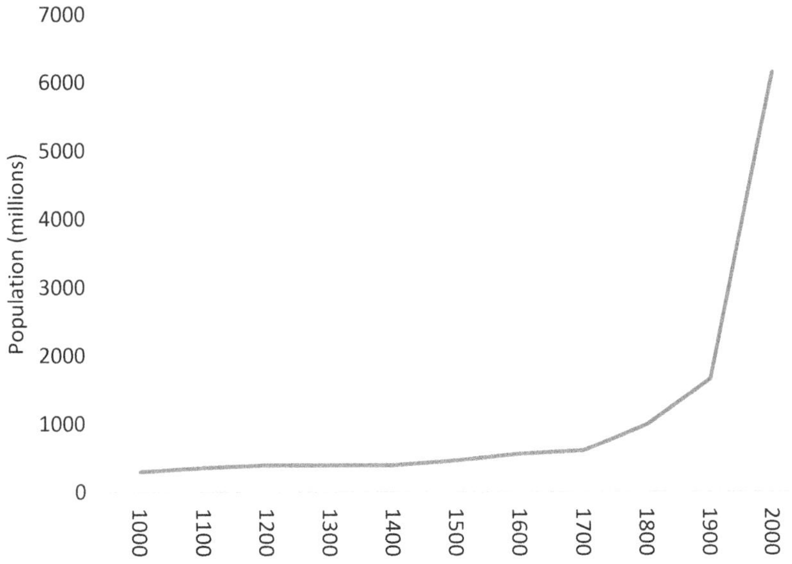

Fig. 1.2 Estimates of human population, in millions, from deep history to 2000: Population every 100 years from 1000 AD. Source: Author's creation, using data from Klein Goldewijk and van Drecht (2007), History Database of the Global Environment

Forward projections of population vary, due to differing assumptions about fertility rates. The United Nations' medium fertility projection puts the global population approaching a plateau of 11–12 billion by 2100 (Fig. 1.3).

The rate of population growth early in the twenty-first century already is decreasing. The peak growth year in absolute terms was 1988, when just under 93 million people were added. By 2012, that rate had dropped to 82 million. Percentage-wise, the growth rate peaked in the late 1960s, coinciding, perhaps unsurprisingly, with Paul Ehrlich's book, *The Population Bomb.*[14] Prior to 1988, it was unclear how large the population would grow. Now, early in the twenty-first century, we have a better idea what we are dealing with.

[14] Ehrlich (1968).

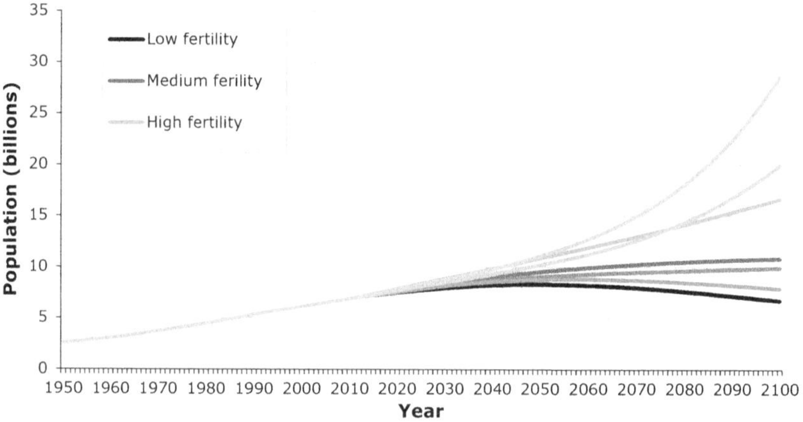

Fig. 1.3 Historical and projected global human population under various demographic scenarios. Source: Author's creation, using data from United Nations, Department of Economic and Social Affairs, Population Division (2013). World Population Prospects: The 2012 Revision

The 40 years from the end of World War II to the late 1980s saw the global population roughly double. This large increase caught the attention of scientists, policy analysts and development professionals. A rash of publications emerged during the 1960s and 1970s about possible ways to deal with the population problem. Some proposals, such as Ehrlich's to actively reduce population, were controversial. The economic development community favored solutions to lift people out of absolute poverty. Forty years later, as rates of absolute poverty have declined worldwide, so has the rate of population growth.

Writing in 1982, the philosopher Fritjof Capra identified his time as 'the great turning point', when geometric growth came up against natural limits, "the place where historical progress as growth is transformed into something else."[15] What that something else will be remains to be seen. It may take a while to perceive the late twentieth and early twenty-first centuries as a 'great turning point' in terms broader than merely demographic. If nothing else, Capra's observation was impeccably timed.

Population has been perceived as a vexing problem by some for the simple reason that having so many mouths to feed places a measurable

[15] Drengson (1995, p. 193), quoting Capra (1982).

burden on natural systems. This is true as long as per-capita material intensity remains constant or increases over time. If, on the other hand, it were to decline, even while population continued to increase, then overall material intensity would increase more slowly or even go down over time. This scenario would represent a decoupling of material intensity from population.

Material Intensity of the Economy

Up to the present, increasing population has been accompanied by increasing material intensity, as the following graphs show. As the sum of inputs to and outputs from the economy, material intensity correlates closely, about 96 percent, with population.

On the input side, Fig. 1.4 plots population against global primary energy consumption over years for which data are available.

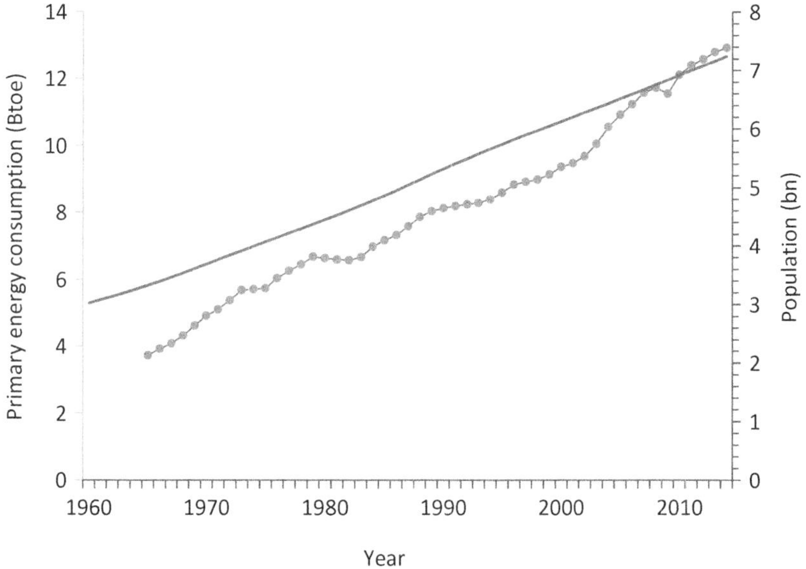

Fig. 1.4 Human population (solid line) plotted against total primary energy consumption in billion tonnes of oil equivalent (dotted line). Source: Author's creation, using population data from the United Nations and energy data from British Petroleum, Statistical Review of World Energy, 2015

Figure 1.5 shows a similar plot of population against total material resource extraction.

Combining the energy and material extraction data for years in which they overlap allows a plot of total material inputs to the economy (Fig. 1.6).

On the output side, atmospheric emissions and solid wastes are well quantified. Global data on liquid and semi-solid wastes are not as complete. Figure 1.7 shows population against the wastes for which good data are available, almost certainly an underestimate of actual waste flows.

Over the period from 1980 to 2008, average per-capita material intensity increased from about 18 tonnes per person to about 20 tonnes per person. Twenty tonnes is roughly equivalent to a dump truck of stuff per person per year. To put it into perspective, a blue whale eats about 200 to 300 tonnes of krill each year and voids about one-tenth of that in waste. On average, then, every person in the modern world processes

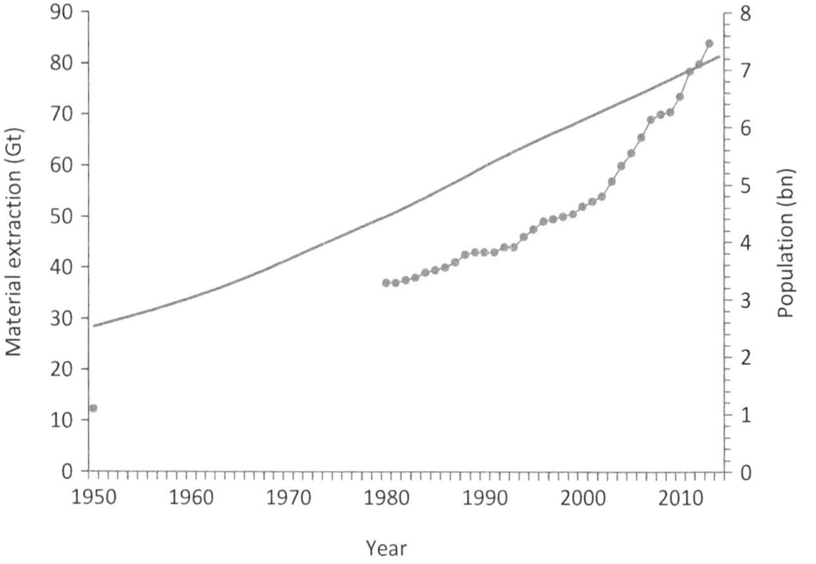

Fig. 1.5 Human population (solid line) plotted against total material extraction in gigatonnes (10^9 tonnes, dotted line). Source: Author's creation, using population data from the United Nations; 1950 material extraction value from Schaffartzik et al. (2013); and all other material extraction values from the Sustainable Europe Resource Institute (SERI), www.materialflows.net

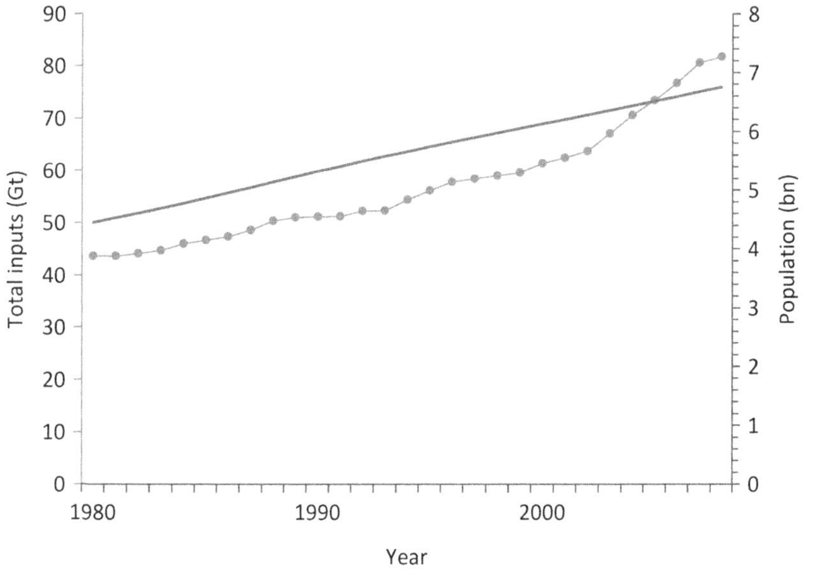

Fig. 1.6 Human population (solid line) plotted against the total material inputs to the economy in gigatonnes (dotted line), estimated as a sum of energy consumption and material extraction. Source: Author's creation

about one-tenth of a blue whale's worth of material. Yet a blue whale is 2000 times heavier than the average person. So, pound-for-pound, we are much more materially intense; and not all our effluents are biodegradable, unlike whale poo. People living in wealthy countries are responsible for much more material intensity than the global average, while people living in poverty or subsistence are responsible for much less.

Size of the Economy Relative to Nature

How big is the economy compared to nature? *Economics of a Crowded Planet* uses an estimate of this value as the basis for a model of the economy and nature as coupled systems. From the sources above and other data in the literature, combined with a handful of simple assumptions, the total mass of all artifacts in the economy is estimated at around 1650 billion tonnes or about half of the world's total biomass (Table 1.1). The economy's total material intensity—the amount of material flowing between it and nature—is estimated at around 106 billion tonnes per year,

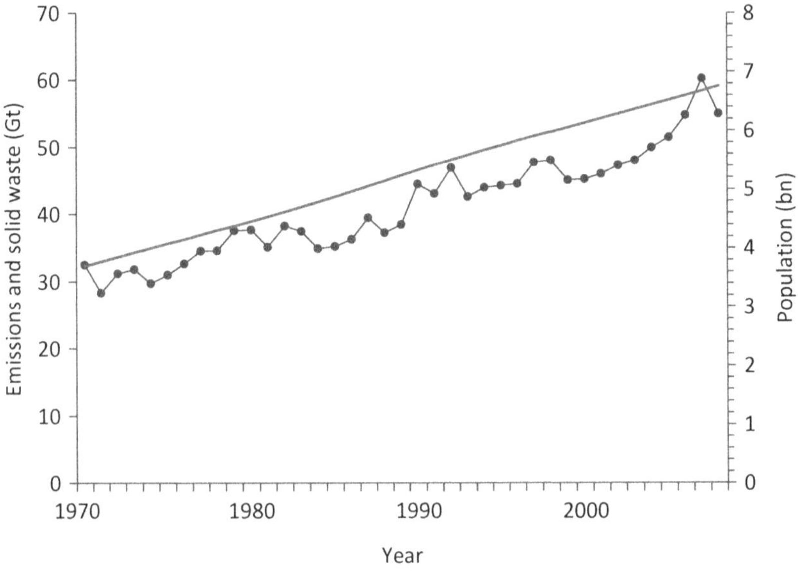

Fig. 1.7 Human population (solid line) plotted against the total emissions and solid waste in gigatonnes (dotted line). Source: Author's creation, using emissions data from Emissions Database for Global Atmospheric Research (EDGAR), http://edgar.jrc.ec.europa.eu, and solid waste data from Hoornweg et al. (2013)

Table 1.1 Comparison of the scale of Earth's biota against the scale of the economy

Measure	*Biota*	*Wet mass (Gt)*	*Economy*	*Mass (Gt)*
Stored mass	Biomass	3068	All artifacts	1650
Flow	Net primary production/ reabsorption	537	Material intensity	106

Source: Author's creation using data in billions of tonnes (Gt) from 2000 (Murison Smith, 2019)

which is about one-fifth of Earth's total net primary ecological production and reabsorption.

Not all of the economy's material intensity falls upon living systems, of course. Much of its resource extraction comes from the lithosphere in the form of fossil fuels or raw materials for the built environment. All the same, the numbers in Table 1.1 clearly show that we already have a crowded planet.

These masses, by the way, are tiny compared to Earth's abiotic environments: not much more than one-millionth by mass. Nonetheless, they have an effect. Total anthropogenic emissions in 2000 were about 44 billion metric tonnes,[16] which equals only 0.000003 percent of the total mass of the atmosphere.[17] Even at that tiny ratio, they have produced measurable effects upon climate. Climate clearly is very sensitive to changes in atmospheric chemistry.

Material Efficiency of the Economy

Although the economy continues to become ever-more materially intense, data allow us to ask whether it is also becoming more materially efficient. Figure 1.8 suggests the answer is yes, although the period for which data are available is short. Material efficiency is calculated as the inputs for one year minus the outputs for the same year, divided by the inputs. This estimate necessitates a simplifying assumption that all outputs in any given year originate from the inputs for the same year, which almost certainly is not true. The scatter in the data may arise partly as an artifact of this assumption. Nonetheless, an upward trend is clear, from between 15 percent and 20 percent efficiency during the period from 1980 to 1990, to around 25–30 percent efficiency from 2000 to 2010.

While material efficiency has been increasing, the economy, measured in money terms, has also been growing. Figure 1.9 shows global population since 1960 against the total value of merchandise exports. In this particular case, the correlation is weaker (about 89 percent). There are likely other, probably endogenous, forces at work driving the exponential increase in exports between 1980 and 2010, where the trend appears to level off.

Global gross domestic product—or, more succinctly, gross global product (GGP)—follows a similar path. Although the explosive growth in human GDP *per capita* since the industrial revolution mirrors the

[16] Emissions Database for Global Atmospheric Research (EDGAR), http://edgar.jrc.ec.europa.eu.

[17] US National Aeronautics and Space Administration (http://nssdc.gsfc.nasa.gov/planetary/factsheet/earthfact.html).

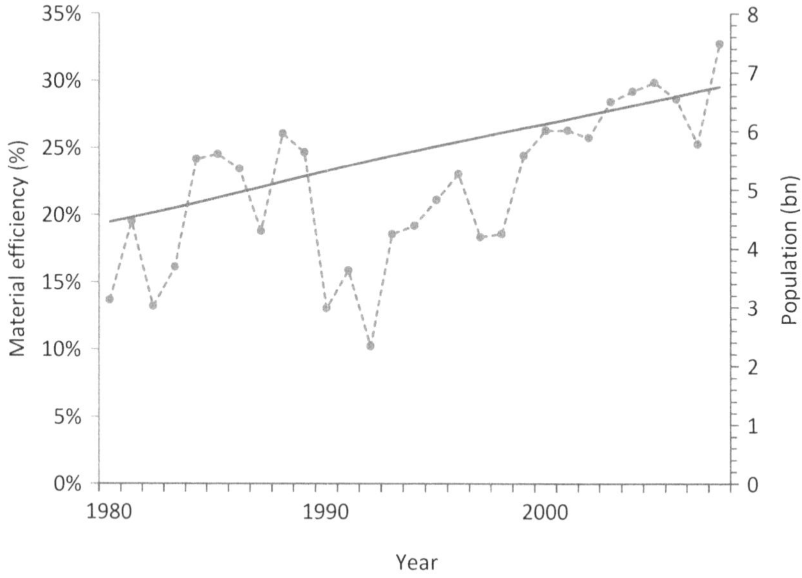

Fig. 1.8 Human population (solid line) plotted against material efficiency (dotted line). Source: Note: Material efficiency is estimated as the percentage of material not wasted by the economy (inputs minus waste outputs, divided by inputs)

hockey-stick growth in population,[18] it is not an exact mirror, as Fig. 1.10 shows. Although the correlation is strong (about 94 percent), the pattern suggests other forces may have driven the exponential growth in GGP. Notice again a slight leveling-off from around 2010, which may be due to the economic recession that took hold following the 2008–09 financial crisis.

One of the major forces driving the roughly 70-fold increase in gross global product has been the globalization of trade, indicated by the graph of exports in Fig. 1.9. Population growth also has likely been a driver, but probably a less important one.

In pre-industrial times, communication among geographical regions was pretty slow. It went about as fast as someone could walk, ride a horse or take a boat from one place to another. Consequently, the pace and volume of capital accumulation from economic production—as distinct from

[18] Beinhocker (2006, p. 10).

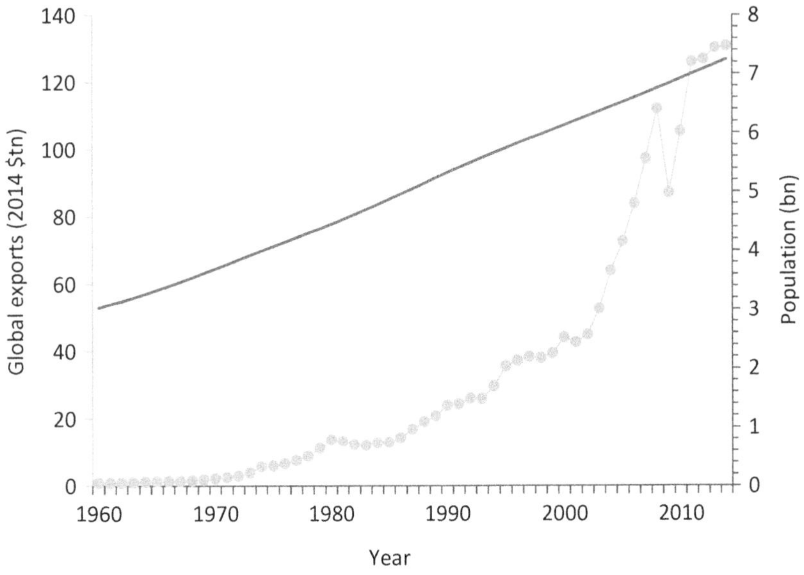

Fig. 1.9 Human population (solid line) plotted against the total value of merchandise exports in trillions of 2014 US dollars. Source: Author's creation, using population data from the United Nations and global exports from the World Bank, World Development Indicators (https://datacatalog.worldbank.org/dataset/world-development-indicators)

colonial looting, for instance—was correspondingly small. In the nineteenth century, new transportation and communication technologies, such as the railroad and the telegraph, allowed significant geographical expansion of capitalism and associated imperialism. However, compared to today's technologies, these were cumbersome and slow, and they limited how much could be accomplished in distant places. As the economic historian Douglas Dowd writes, "the great speed and reliability of contemporary transportation and communications combine with production technologies that are easy to transport. Thus, the lines between domestic and foreign production now tend toward obliteration."[19]

[19] Dowd (2004, p. 182).

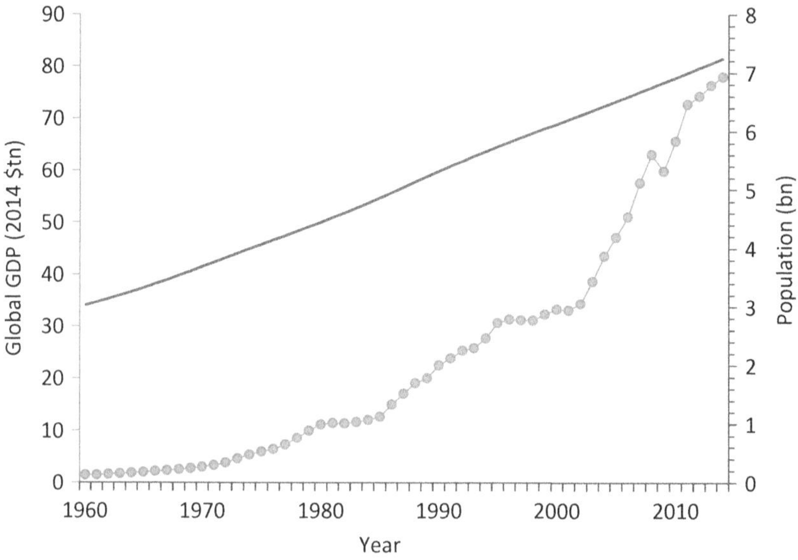

Fig. 1.10 Human population (solid line) plotted against total gross domestic product (or gross global product) in trillions of 2014 US dollars at market prices. Source: Author's creation, using population data from the United Nations and GDP data from the World Bank, World Development Indicators (https://datacatalog.worldbank.org/dataset/world-development-indicators). World Bank note: GDP at purchasers' prices is the sum of gross value added by all resident producers in the economy, plus any product taxes, minus any subsidies not included in the value of the products. It is calculated without making deductions for depreciation of fabricated assets or for depletion and degradation of natural resources

While globalization is a tendency toward transnational market integration and borderless economic crises, it is also an ideology.[20] Transnational organizations, such as the World Bank or the International Monetary Fund, tend to support it, as do transnational corporations, whereas nationalists and local or indigenous cultures often resist it. In this sense, globalization can be perceived as a kind of power play of the global against the local, the large against the small. As Chap. 4 argues, a degree of institutional globalization probably will be essential to the emergence of a

[20] Cypher (1999), cited in Dowd (2004, p. 170).

planetary economy, yet at the same time it would have to correct this imbalance of power.

That gross global product has grown so much during a period when overall material efficiency also seems to have increased suggests not only that society is beginning to adapt to its new resource-intense state but also that increases in material efficiency could be associated with economic growth. A more fine-grained analysis by the Sustainable Europe Research Institute (SERI) shows how economic growth, in terms of GDP, has diverged from material intensity over a 20-year period from 1990 to 2009.[21] SERI's work suggests that an incipient decoupling of material intensity and economic growth could be under way. If it persists, it will be important. It hints at a possible future of economic growth driven, at least in part, by improvements in material efficiency.

Nonetheless, material productivity has yet to really take off. The twenty-first century will reveal whether the global economy can become materially efficient. If so, the potential economic upside could be gigantic.

Role of Money and Credit in Economic Expansion

As Figs. 1.9 and 1.10 suggest, economic growth is driven not only by population increase and improvements in technology but also by other factors. The exponential population increase suggests that child mortality rates dropped precipitously, which in turn suggests higher standards of living. What drove these higher standards of living might be a convergence of factors: technology, for one, but also credit.

The history of money since the Renaissance has been one of ever-expanding forms of credit. Whereas for most of history prior to the Renaissance, credit was extended predominantly in coin, over the past 400 years it evolved into credit in the form of notes, which later became bank notes, then into credit in the form of checks and, most recently, in the form of electronic deposits. Throughout, banks typically lent more than the total value of coin in their reserves, a practice known as fractional reserve banking. In the aggregate, there was more credit in circulation than the supply of money.

Money, being in part a store of value, is relatively scarce, by definition. If it were not, it would not have value. Most money in the form of coin was issued by sovereign states. A sovereign, or their government, would

[21] SERI (2014).

control the amount issued. The limited amount of cash in circulation placed limits on production, because production costs money. With the advent of fractional reserve banking, new credit served the latent demand for money, which allowed production to increase. The industrial revolution took hold in eighteenth-century Britain as much due to the ready availability of credit as due to the convergence of enabling technologies: indeed, the evolution of new technologies fed off the availability of credit.

If more credit enables more production, then consumption also will increase. Whereas commercial debt may fuel production, household debt, in turn, can fuel household consumption. Over time, relatively more people were able to improve their material quality of life. By being able to better care for themselves, families found that more of their children were living long enough to have children of their own. Population increased. Population growth raised demand for economic production: more people wanting more of the things that made for a better life. More demand for production meant more demand for credit. The providers of credit—mainly banks—found more and more creative ways to leverage their reserves to meet the need. It was, after all a great way to make money, provided sufficient trust remained within the system. Economic growth thus was a form of social auto-catalysis—or autopoiesis—illustrated in Fig. 1.11.

For a long time, economic production was assumed to be enabled by unlimited natural capacity. This became no longer the case once the global economy entered into a material coevolution with nature. When natural capacity is limited, demand for economic production must be met through increasing material efficiency. This does not mean that the economy cannot grow in units of money, only that it cannot grow in physical units. This would represent a decoupling of monetary growth from material intensity. Wealth would be created through increasing material recirculation within the economy. The social consequences of this change are profound. As this book examines, they could be life-changing for people in beneficial ways.

In order for this to happen, though, the role of money within the economy would have to change. Just as nature does not accept cash, so money also does not—yet—account for nature.

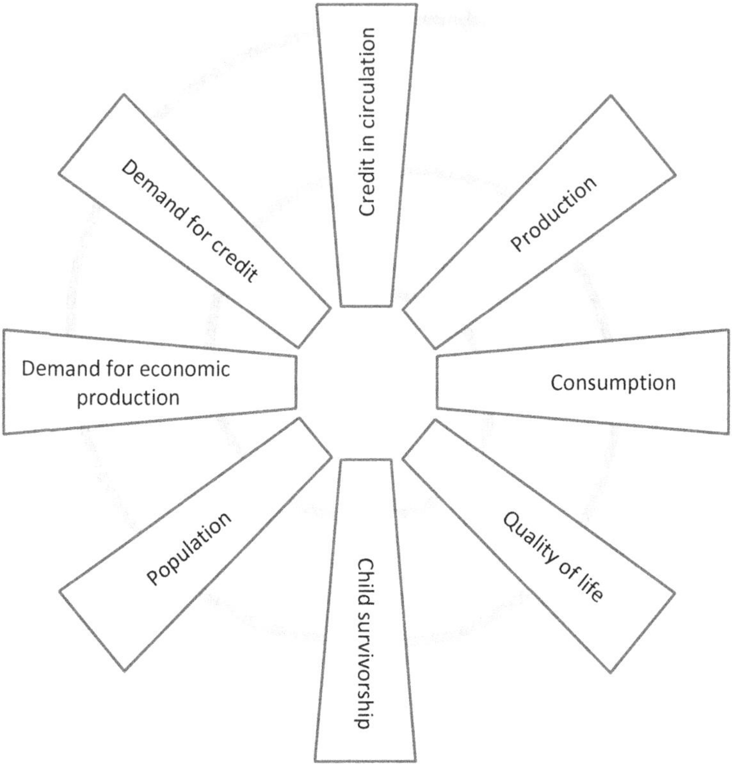

Fig. 1.11 Illustration of social auto-catalysis of economic growth. Source: Author's creation

Role of Belief Systems

The rise of the global industrial economy represents a convergence of factors occurring in Europe during the past 500 years or so: technologies for exploration and for the subjugation of other peoples, combined with increases in trade and the circulation of money through private hands. Industrialization catalyzed an exponential increase in material intensity, enabled in part by resources expropriated from European colonies. One other factor, though was pivotal. It was an abandonment of the long-held sense of place within the world—what Morris Berman called a 'participating consciousness'—in favor of the psychological detachment of the

Cartesian dualism, the separation of subject from object. It was this detachment that gave people permission to exploit the world's bounty.

"I've never met a child", notes the naturalist and broadcaster Sir David Attenborough, "who is not interested in natural history. Just the simplest thing: a five-year-old turning over a stone and seeing a slug, and saying, 'what a treasure'. Kids understand the natural world and are fascinated by it. So the question is, how did you lose it? How did anyone lose their interest in nature?"[22] The answer, for our modern times, is that there is no money in nature. Conversely, what we do with our time is determined too much by money. The vast majority of jobs in today's economy do not reward an interest in nature, nor do they reward a broader maintenance of natural capacity. In order for a large economy on a crowded planet such as ours to become stable in the long term, the maintenance of natural capacity would have to be rewarded, and money would have to become uncoupled from time, to a much greater extent. The rewarding of natural capacity within the economy would bring about alignment with nature and the decoupling of time from money would signify widespread prosperity. These characteristics would be emblematic of a planetary economy; normatively, people would expect them. A reconnection with the generative would not mean the abandonment of the profit motive but rather the recognition of its place within a broader panoply of self-enrichment.

Prospects

We Have the Physical Technology But Not the Social Technology

Human population growth in the historically resource-unconstrained world was enabled by two significant technological advances: agriculture and industrial manufacture. The development of these technologies is a distinguishing feature of the human species. Where the vast majority of animals employ what the ecologist Alfred Lotka called *endosomatic instruments*—such as eyes, ears, teeth or claws— to survive, a handful of species have an ability to fashion and use *exosomatic instruments*, or tools, to manipulate their environment to their advantage.[23] Some birds, for

[22] Sir David Attenborough in conversation with the US President Barack Obama, *The Guardian*, June 25, 2015: http://www.theguardian.com/tv-and-radio/2015/jun/25/sir-david-attenborough-and-barack-obama-face-to-face-in-tv-interview.

[23] Lotka (1945).

example, use twigs to get at insects inside small holes in trees. Chimpanzees do the same to feed on termites. *Homo sapiens* is distinguished by its sheer number and variety of exosomatic instruments. Consequently, humanity has become able to manipulate its environment in ways and to a degree that no other species on Earth did before. Given that the number of species on Earth, extant and extinct, is estimated in the tens of millions, it is reasonable to conclude that evolutionary events of this kind are rare indeed.

The evolution of an ability to develop a diverse suite of exosomatic instruments is a precondition for the evolution of a material economy. While individuals of other species engage in acts of reciprocity, such as mutual grooming, they do not trade objects. Objects, to animals who employ only endosomatic instruments, are not perceived as useful. To a species with an ability to create exosomatic instruments, objects have a potential use, such as to complete a task or to solve a problem. All of the social and technological evolution of human history was possible because early humans acquired an ability to create diverse exosomatic instruments.

Yet as Nicholas Georgescu-Roegen pointed out, there came a point at which exosomatic instruments grew in both size and complexity. A fishing boat or a flour mill is an exosomatic instrument whose construction requires the use of many other smaller tools, along with natural resources and the labor of several or many people. Once completed, it may remain the property of only one person or one family. It is not, as Georgescu-Roegen said, "a natural, indissoluble property of the individual person" like a spear made and used by a single hunter. Larger assets formed a basis of inequality among people or communities. "Distribution of communal income … thus became a social problem", he wrote, "the importance of which has never ceased to grow." And the distribution of income will continue to be a focus of social conflict, he contended, as long as human society exists.[24]

Technology has a central role in the dynamic alignment between the economy and nature. An economy poorly aligned with natural systems might be technologically primitive or advanced, but its technology is what would be responsible for the poor alignment. 'Technology' in this context is taken to mean not only physical technology, in the sense of tools,

[24] Georgescu-Roegen (1971, p. 308).

machines or communications networks, but also social technology, in the sense of how human society organizes itself to live on Earth.[25]

The physical technologies of the early twenty-first century probably are already sufficient to enable the global economy to reduce its resource intensity to become dynamically aligned with natural processes. The challenge becomes one of social technology: the norms, attitudes, habits, institutions, policies, laws, regulations and instruments that organize human economic activity. While the necessary physical technologies already exist, the corresponding social technologies are in their infancy.

This is an observation that flies in the face of mainstream economic orthodoxy, which says that 'technology' (meaning physical technology) will come to our rescue from whatever jam we get ourselves into with those so-called externalities. To the contrary: we do not need more physical technology, however welcome it may be. Better social technology is what is required. Better social technology means critically evaluating economics, rejecting the parts that are not relevant to the problem at hand and retaining the parts that are. It means also looking deeply at the attitudes and norms embedded within modern society, not necessarily with a view to rejecting them but with a view to orienting them toward a long-term goal. It means re-evaluating the 'market' not as an end in itself but as an engine for change. It means addressing the way people think about the economy and about economic activity.

As Donella Meadows et al. observed in 1972, "political and social institutions ... almost never change in *anticipation* of a social need, but only in response to one."[26] In the hope of breaking that pattern, this book proposes that the single most important need shaping social evolution for decades or possibly centuries to come will be the need for a social technology to allow society to prosper on a crowded planet. Although other social needs might appear more pressing over short periods, they will pale in comparison to this need over long ones. A stable, prosperous economy,

[25] Nelson (2003). Beinhocker (2006, p. 262) defines social technologies more generally as "methods and designs for organizing people in pursuit of a goal or goals".

[26] Meadows et al. (1972, pp. 154–55).

encompassing institutions, policies, accounting standards and instruments, will be grounded fundamentally in a system of norms.[27]

Some writers on humanity's relation to nature distinguish between a so-called reform philosophy and a 'transcendent' philosophy. A 'reform' philosophy acknowledges a disjunct between the individual interest and the collective interest, as exists today, proposing remedial action giving primacy to the collective interest. Institutions, policies and instruments would be established, within which the individual interest could express itself. To use Herman Daly's analogy, it is rather like putting a toddler in a playpen rather than allowing it to run amok around the living room.[28] Over time, a process of conditioning within this instrumental environment might lead to an emergence of new, 'transcendent' norms, which, in the words of Arne Naess, represent a way of living in harmony with the natural world, freely and unquestioningly. Yet such transcendence would not be required to approach a planetary economy; instead, a planetary awareness would be a feature of a planetarian existence.

That said, the 'reform'/'transcendence' distinction remains a uniquely Western one. Herman Daly and John Cobb proposed a 'precautionary principle' for economic development, which states that, in the face of uncertainty about the timing and scale of negative externalities from nature upon the human economy, policy should err on the side of caution.[29] This principle has recently started to gain traction in circles traditionally reticent to giving 'the environment' priority over economic growth, such as among political conservatives, because it represents sound risk management.[30] Yet the broader reason Daly and Cobb saw fit to advocate a precautionary principle was because people in modern society had abandoned whatever 'transcendent' relationship they may have had with nature in the first place, probably during childhood. Among many indigenous peoples, it remains very much part of the core identity. Indigenous peoples' principal negative externality is the market economy.

This book makes a case for the market as the main engine of a 'reform' program. A market, as a collective intelligence, is more efficient than a central planning committee because many brains rather than a few will

[27] Some organizations, such as the Wellbeing Economy Alliance (www.wellbeingeconomy.org), have made a head start in this direction.

[28] Daly (1991).

[29] Daly and Cobb (1989, 1994).

[30] Taylor (2019).

decide on a price that clears the greatest total value. However, unlike a central planning committee, a market lacks forward thinking. Markets act as if they can see far into the future only if the signals they receive from outside the market—such as from natural processes—bring that desired future into play.

Markets originally emerged as organizing institutions for exchange within increasingly crowded civilizations. Throughout history, societies with relatively high population densities generally possessed markets. Scattered populations, such as hunter-gatherer societies, did not organize into markets, except occasionally or transiently when groups of people met to exchange resources. As the global population becomes ever denser, markets are likely to become only more prevalent.

A huge economic opportunity awaits in reforming the economy for long-term stability in alignment with natural processes. Economic studies of the early twenty-first century are only beginning to recognize this potential.[31] The importance of the market is not only that much of the economy today and for the foreseeable future will be a market economy but also that markets are rather fundamental extensions of human evolution as social beings.

The Far Future

The transition from the state-change illuminated by the earlier graphs toward a materially stable, prosperous economy might unfold over a timescale similar to the one from the industrial revolution to the present, about 200 years. However long it takes, it likely will be on the order of many decades to a few centuries. Although this timescale might seem inordinately long from a standard economic viewpoint, it is but a wink from a natural-science one. Two hundred years is one-thousandth of the time *Homo sapiens* has existed as a distinct species. It illuminates the great pace of human social evolution compared to biological evolution. Even though the major social evolutionary events of human history, such as the emergence of agriculture and industry, have much shorter periodicities than major events in biological evolution, many of the same kinds of dynamics are likely at work: periods of consolidation and stability punctuated by episodes of breakdown, recycling and renewal. These same dynamics are

[31] See, for example, Neslen (2015).

at work in ecological processes, which unfold over middling periods of hundreds to thousands of years.[32]

If the human economy failed or refused to adapt to emerging signals from nature, then it likely would collapse long before any catastrophic ecological collapse. Although many ecosystems around the world are fragile in the face of human-induced stresses, the fact remains that, on a global level, the vulnerable system is the modern economy. Yet economic collapse does not mean human extinction. More likely it would mean a collapse of civilization and an associated contraction in population.

Economics of a Crowded Planet explores possible scenarios under which a model economy would adjust its flows of material in response to signals from nature about the state of natural capacity. That capacity would affect the economy's probability of avoiding collapse within a defined period, such as 200 years. The scenario analysis suggests that a relatively secure path toward long-term stability can be taken through a combination of supply and demand measures. If the supply of resources extracted from nature is constrained and the growth in demand for resources for production is held at zero or at a much slower rate than population growth, then the economy will respond over time by increasing its material efficiency and by recirculating a much greater proportion of material back into production. Flows of effluents to nature will decline accordingly. Adjustments made relatively earlier will reduce the risk of economic collapse over the medium term. Any adjustment should be the most aggressive that the economy can accommodate, for the same reason. Under this scenario, as Fig. 1.12 shows, the economy's chance of avoiding collapse does not fall very far in the early going and then, once the effects of the measures take hold, it increases quickly to a high, stable level.

The analysis also suggests that this 'material discipline' can be undertaken without diminishing the prospect of economic growth, as measured by gross global product per capita. In fact, the converse may be the case. Increases in material efficiency appear to stimulate economic growth, and they appear to do so almost immediately. Any concerns raised by advocates of economic growth about the purported negative effects of imposing constraints on the economy's appropriation of natural capacity may be spurious. To the contrary, by *not* imposing such constraints, an opportunity for significant economic growth might be missed.

[32] Holling and Sanderson (1996), ch. 4.

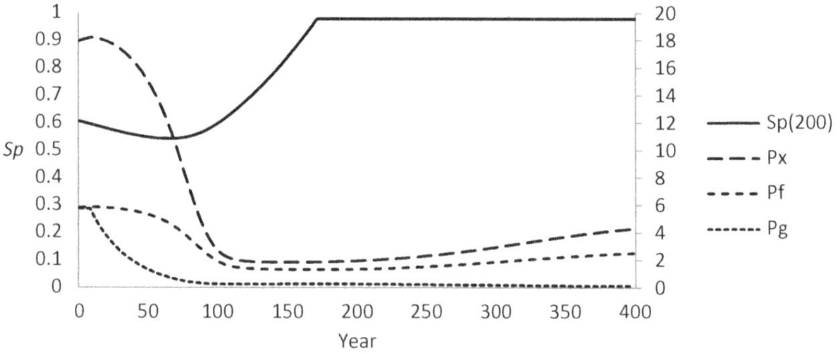

Fig. 1.12 Resource flows, gross planetary productivity per capita and a probability of avoiding economic collapse over 200 years in a nonparametric simulation model of an economy in coevolution with nature. Source: Murison Smith (2019). Note: Early reductions in economic demand and improvements in the economy's material efficiency pave the way for self-sustaining economic growth in the long term. The model tracks three principal resource flows: those into the economy from nature (R_x), out to nature from the economy (R_f) and regenerative circulation with the economy (R_g)

There are essentially two ways to constrain the economy's resource-loading upon nature: one is by adopting physical limits and the other is by applying financial instruments to the prices of material flows. Where the price of resources extracted from nature or the price of effluent flows is higher than the price of resources recirculated through the economy, an economic incentive emerges to recirculate material rather than for material intensity. The absence of such an incentive within the present economy is due to internal distortions to these prices as they are presented to the market, such as through perverse subsidies and taxes.

The suggestion that prolonged economic growth awaits an economy prepared to engage in material discipline has implications for the debate about growth and income distribution. Conventional wisdom in political circles influenced by neoclassical economics holds that economic growth will eventually close the income gap. While it is probably the case that economic growth will eventually lift an increasing number of people out of absolute poverty, it also will likely preserve or increase the size of the income gap. A wide income gap, in turn, may stimulate economic growth.

Because people's perception of affluence is relative, a wide income gap would increase demand.[33] Policies maintaining a high degree of distributional inequity likely will favor those at the top of the distribution, as they stand to benefit most from economic growth. A self-reinforcing cycle then takes hold.

While the political debate continues about the extent to which this trend is a social problem, there is another problem with it. The effect of instituting material discipline upon the economy may very likely raise prices of resources extracted from nature and effluents to it. Those in lower income brackets will be less able than higher earners to support these increased costs. In an economy with high income inequality, the proportion of households placed under financial strain by such rising prices would be relatively higher. The problem could be alleviated by saving households money elsewhere, for example, by reducing income tax. However, it also could be alleviated by narrowing the income gap. The conclusion is that reducing income inequality could be an important enabling factor in achieving material discipline which, in turn, would unlock long-term economic prosperity.

The scenarios also showed that improvements in material efficiency allowed for reductions in demand. This observation suggests that as material intensity decreases over time, so will income inequality. The relationships among material discipline, income inequality and growth merit further examination. The hypothesis that they are closely linked is consistent with other recent literature establishing a link between income inequality and natural resource intensity.[34]

References

Beinhocker, E. D. (2006). *The Origin of Wealth: Evolution, Complexity, and the Radical Remaking of Economics*. Boston, MA: Harvard Business School Press.

Boulding, K. E. (1966). The economics of the coming Spaceship Earth. In H. Jarrett (Ed.), *Environmental Quality in a Growing Economy* (pp. 3–14). Baltimore, MD: John Hopkins Press.

Boyle, D., & Simms, A. (2009). *The New Economics: A Bigger Picture*. London: Earthscan.

[33] See, for example, in Frank (1999) and Wilkinson and Pickett (2009).

[34] For example, Marglin (2008), Wilkinson and Pickett (2009), and Costanza et al. (2015).

Burgess, S., Muirhead, J. D., & Bowring, S. A. (2017). Initial Pulse of Siberian Traps Sills as the Trigger of the End-Permian Mass Extinction. *Nature Communications, 8*(164), 1–6.

Capra, F. (1982). *The Turning Point: Science, Society and the Rising Culture*. New York: Simon & Schuster.

Costanza, R., et al. (2015). *An Introduction to Ecological Economics* (2nd ed.). Boca Raton, FL: CRC Press.

Cypher, J. (1999). *Crisis Tendencies of the 1990s: Constraints on the Ideology of Globalization*. Unpublished.

Daly, H. E. (1991). *Steady-State Economics* (2nd ed.). New York: John Wiley & Sons.

Daly, H. E., & Cobb, J. B. (1989, 1994). *For the Common Good*. Beacon Press, Boston, MA.

Diamond, J. (2005). *Collapse: How Societies Choose to Fail or Survive*. Allen Lane, London: Penguin.

Douthwaite, R. (2000). *The Ecology of Money*. Schumacher Briefings, Book 4, UIT Cambridge Ltd., UK.

Dowd, D. (2004). *Capitalism and Its Economics: A Critical History* (2nd ed.). London: Pluto Press.

Drengson, A. (1995). *The Practice of Technology: Exploring Technology, Ecophilosophy and Spiritual Disciplines for Vital Links*. Albany, NY: State University of New York Press.

Ehrlich, P. R. (1968). *The Population Bomb*. Rivercity, MA: Rivercity Press.

Eldredge, N., & Salthe, S. N. (1984). Hierarchy and Evolution. In R. Dawkins & M. Ridley (Eds.), *Oxford Surveys in Evolutionary Biology* (Vol. Vol. 1, pp. 184–208). Oxford: Oxford University Press.

Frank, R. H. (1999). *Luxury Fever: Why Money Fails to Satisfy in an Era of Excess*. Simon & Schuster, New York: The Free Press.

Galbraith, J. K. (1973). *Economics and the Public Purpose*. Boston, MA: Houghton Mifflin.

Georgescu-Roegen, N. (1971). *The Entropy Law and the Economic Process*. Harvard University Press.

Goodwin, N., et al. (2009). *Microeconomics in Context* (2nd ed.). Armonk, NY: M.E. Sharpe.

Holling, C. S., & Sanderson, S. (1996). Dynamics of (dis)harmony in Ecological Systems. In S. S. Hanna, C. Folke, & K.-G. Mäler (Eds.), *Rights to Nature*. Washington DC: Island Press.

Hoornweg, D., Bhada-Tata, P., & Kennedy, C. (2013). Waste Production Must Peak This Century. *Nature, 502*, 615–617.

Kelly, M. (2012). *Owning Our Future: The Emerging Ownership Revolution*. San Francisco, CA: Berrett-Koehler.

Klein Goldewijk, K., & van Drecht, G. (2007). HYDE 3.1: Current and Historical Population and Land Cover. In A. F. Bouwman, T. Kram, & K. K. Goldewijk (Eds.), *Integrated Modelling of Global Environmental Change. An Overview of IMAGE 2.4*. Bilthoven, The Netherlands: Netherlands Environmental Assessment Agency (MNP).
Lotka, A. J. (1945). The Law of Evolution as a Maximal Principle. *Human Biology, 17*, 167.
Lovelock, J. (2000). *Gaia: A New Look at Life on Earth*. Oxford: Oxford University Press.
Marglin, S. (2008). *The Dismal Science: How Thinking Like an Economist Undermines Community*. Cambridge, MA: Harvard University Press.
Meadows, D. H., et al. (1972). *The Limits to Growth: A Report for the Club of Rome's Project on the Predicament of Mankind*. New York: Universe Books.
Motesharrei, S., Rivas, J., & Kalnay, E. (2014). Human and Nature Dynamics (HANDY): Modeling Inequality and Use of Resources in the Collapse or Sustainability of Societies. *Ecological Economics, 101*, 90–102.
Nelson, R. R. (2003). Physical and Social Technologies and Their Evolution. LEM Working Paper Series, No. 2003/09, Laboratory of Economics and Management, Sant'Anna School of Advanced Studies, Pisa, Italy.
Neslen, A. (2015). Limiting Climate Change Could Have Huge Economic Benefits, Study Finds. *The Guardian*, 30 March.
Odum, H. T. (1971). *Environment, Power & Society*. New York: Wiley-Interscience.
Panayotakis, C. (2011). *Remaking Scarcity: From Capitalist Inefficiency to Economic Democracy*. London: Pluto Press.
Schaffartzik, A. et al. (2013). Consumption-based Material Flow Accounting: Austrian Trade and Consumption in Raw Material Equivalents, 1995–2007. *Journal of Industrial Ecology, 18*, 1–11.
SERI. (2014). *Material Productivity and Decoupling*. Fact Sheet No. 4, Sustainable Europe Research Institute and Vienna University of Economics and Business Administration. www.materialflows.net.
Sherwood, S. C., & Huber, M. (2010). An Adaptability Limit to Climate Change Due to Heat Stress. *Proceedings of the National Academy of Sciences, 107*(21), 9552–9555.
Soddy, F. (1926). *Wealth, Virtual Wealth and Debt: The Solution of the Economic Paradox*. London: Britons Publishing Company.
Taylor, J. (2019). Conservatives Should Change How They Think About Global Warming. I did. *The Guardian*, 10 June.
Vrba, E. S., & Eldredge, N. (1984). Individuals, Hierarchies and Processes: Towards a More Complete Evolutionary Theory. *Paleobiology, 10*(2), 146–171.
Wilkinson, R. G., & Pickett, K. (2009). *The Spirit Level: Why More Equal Societies Almost Always Do Better*. London: Allen Lane.

CHAPTER 2

Economic Worldviews: Modernity and Its Alternatives

That the economy grew so large because of a self-reinforcing cycle of credit, technology and population growth does not answer the question of *why* it grew so large now, in this particular phase of human history, and why that growth happened to originate in Europe. To answer these questions, it becomes necessary to ask if or how Europeans were thinking differently from people in other regions about the world around them over the past few hundred years. The question is important for a future economy because it may reveal norms and attitudes incongruent with long-term stability and prosperity, which should be abandoned in favor of others. The discussion in this chapter therefore helps to frame the normative requirements to follow in the next. Along the way, a variety of alternate economic worldviews to the dominant Western one are encountered, which could prove useful in establishing a normative system in support of a future economy.

During the roughly 200 years since Adam Smith wrote *The Wealth of Nations*, the prevailing narrative within economics revolved around the maximization of individual utility. It was hoped that this would lead to the maximization of social welfare through allocative 'efficiency' within markets. Utility maximization became the most important criterion by which the economic process was evaluated, both within professional economic circles and in the advice economists gave to policymakers. Yet, rather than being "universal truths about human nature", individual maximization and the other foundations of mainstream economics actually reflected

F. Murison Smith, *A Planetary Economy*,
https://doi.org/10.1007/978-3-030-49296-0_2

"the implicit assumptions of modernity", as Marglin and others have chronicled.[1] These assumptions include the primacy of the individual and his or her associated freedoms, a position championed by such Enlightenment writers as Locke and Bentham.

In 1972, the Club of Rome called for "a Copernican revolution of the mind"[2] in which the individual ceases to be the primary unit of concern in economic activity. Instead, it should be the system. This worldview, it turns out, has dominated for most of human history. The individualist worldview, a characteristic of Western modernity, is an aberration within this context, one maintained for the convenience of free-market capitalism and its positivist economics. These are recent developments, which will have to develop soon into something else, in order for the economy to become aligned with nature and establish a broad basis of prosperity,

A Brief History of Human Attitudes Toward Nature

Although a survey of human attitudes toward nature from around the globe would make a fascinating volume in its own right, here it is necessary to limit the perspective principally to modern, Western attitudes, because these attitudes have come to dominate the world economy, and are in fact the basis for it. In establishing normative requirements for a planetary economy, it is important to understand the influence of the scientific revolution on modern attitudes, as they fundamentally influenced economics from its very beginning.

Detachment and Reattachment

Early people lived close to nature and were at its mercy. Hence, they treated nature carefully and with reverence, understanding that their survival depended upon it. Susan Hanna and Svein Jentoft define a pre-modern view of nature as the 'ecosystem portfolio' view, in which ecosystems provide a portfolio of goods and services.[3] Those pre-modern societies who speak of 'ownership' of nature mean belonging to nature, in a partnership with it, or a stewardship of it.

[1] Marglin (2008, p. 56); also McCloskey (1998).

[2] King et al. in Meadows et al. (1972, p. 199). One might quibble with their use of the term 'equilibrium', but the authors' intent was clear.

[3] Hanna and Jentoft (1996, pp. 36–38).

Over time, notions of a god or gods of certain aspects of nature emerged, along with a culture of placating such gods. Gods of nature were thought to have human qualities, so people engaged with them employing social tools, such as through conversation in the form of storytelling or prayer; or through altruism in the form of offerings. These practices persist today in many parts of the world.

People lived almost exclusively in highly egalitarian societies for most of the past two million years, the period during which humans have been anatomically modern. The hunter-gatherer existence lends itself to social structures dominated by sharing and reciprocity.[4] It would have rarely been in any member's interest within a hunter-gatherer group to act selfishly. However, with the emergence of agriculture and civilization—that is, within the last 1 percent of hominid history—societies have become more hierarchical, more stratified and less equal. They have also become very much larger.

One of the consequences of civilization—the practice of living in permanent settlements—was a sense of distance or removal from nature. People's environment was now as much social and artificial as natural. Nature was seen increasingly as a commodity, which could be owned, in the sense of holding title to land. As Hanna and Jentoft describe, it is through the commodity view of nature that property rights became established and land traded. The commodity view was already well established by Roman times, as the philosopher Seneca lamented: "How long shall we go on making demands upon the gods, as if we were still unable to support ourselves? [...] Did nature give us bellies so insatiable ... that we should outdo the hugest and most voracious animals in greed?"[5] The commodity view eventually enabled industrialization, since nature could be taken as no more than an input to production.

Nowhere has the sense of detachment from nature been more fully expressed than in 'modern' industrial society. The term 'modern' is used here to mean a philosophy of modernity, which has progressed hand in glove with industrialization. Modernity is manifested in the life most of us live, whether out of necessity or choice. Such detachment would have few serious consequences if the scale of industrialization were small. But now

[4] For example, Boehm (1999), Erdal and Whiten (1996), Wilkinson (2005) and Woodburn (1982), cited in Wilkinson and Pickett (2009).

[5] Seneca (c. 65 AD).

that the scale of human activity is detectably large, a philosophy of detachment from nature carries long-term risks.

People in early and pre-industrial societies had little ability to manipulate nature, so they were absolutely dependent upon it. It hardly needs emphasizing that, psychologically, they were highly attached to it, an attachment recorded many times among indigenous peoples during the colonial and modern eras. Viewed within the full sweep of human history, the past few hundred years represent a somewhat unusual period, during which attitudes and norms increasingly ignored natural phenomena. Human technological evolution provided the impression that society could transcend natural constraints. This mindset probably achieved its zenith in twentieth-century neoclassical economics, asserting as it did that human capital is a perfect substitute for natural capital. Yet the development of social technologies failed to keep pace with the ability of physical technologies to expropriate natural capacity. While society's dependence upon natural constraints has never actually gone away, this societal lag created a perceptual gap between apparent and actual progress, which lasted for several hundred years.

Today, early in the twenty-first century, with an increasing awareness of natural constraints, and of the interconnection of human and natural systems, that zenith is revealed for what it actually was: one of anthropocentric hubris. In recent decades, a need has become recognized to develop social technologies to manage the use of available natural capacity into the future. The future may yet hold a reattachment to nature, instrumental at first, with both physical and social technologies being deployed to enter into a kind of coevolution with natural processes toward greater symbiosis. If this were to happen, it could stimulate an evolution in philosophical attitudes toward nature, away from the adversarial or exploitative, back toward reverence and maintenance.

Western Epistemology and Ontology

The history of Western thought is a huge subject in its own right. Only a few highlights are possible here, enough to provide an impression of the evolution of the Western mindset since medieval times, in both science and society at large.

Epistemology is the study of knowledge: specifically, how and why we value knowledge as useful or not useful, and how we come to think we know about things. How we acquire and perceive knowledge determines how we relate to the world and describe it.[6] Ontology is the study of what exists or, more precisely, what is perceived to exist.[7] What we perceive to exist influences what we think we know and how we acquire knowledge. This in turn influences what we take to exist: the two concepts clearly are intertwined.

From the Medieval to the Mechanistic

In medieval Europe, the prevailing view of the world was based on resemblance. "The world was seen as a vast assemblage of correspondences", according to Morris Berman. "All things have relationships with all other things, and these relations are ones of sympathy and antipathy."[8] All entities, following the Aristotelian view, had natural tendencies, imbued as they were with spirit. An object falling to Earth accelerated because it was excited to be coming home. Lodestones attracted iron; oil repelled water. The cycle of life was seen as a cycle of breaking down and reforming: *solve et coagula*, to the medieval alchemist.[9]

The medieval ontology of resemblance existed across scales of observation. The alchemists distinguished between the microcosm and the macrocosm, projecting onto the natural world the image of the human body. "The rocks of the earth are its bones, the rivers its veins, the forests its hairs, and the cicadas its dandruff. The world duplicates and reflects itself in an endless network of similarity and dissimilarity."[10] An approximate modern analogue is the concept of self-similarity in fractal geometry. Branching patterns in nature are similar across multiple scales, from the veins in leaves or in blood vessels to the branching of trees, to the branching of river systems. And where patterns are self-similar, so often are processes. The difference between medieval and modern thinking, however, is that today we would interpret a sentence about Earth's rocks being its bones as metaphorical, whereas to the medieval mind, the rocks *were* Earth's bones.

Equipped as we are with our modernist way of thinking, the way we arrive at a description of the world is through measurement and analysis, ideally with no preconceptions about what we might find. To the medieval

[6] Honderich (1995, pp. 242–245).
[7] Honderich (1995, pp. 634–635).
[8] Berman (1981, p. 74).
[9] Berman (1981, pp. 89–90).
[10] Berman (1981, p. 74).

person, the relation of the microcosm to the macrocosm, indeed the whole "assemblage of correspondences", was never questioned in the first place. It was always thus and was taken as given.

The medieval reality was just as valid in its time as the modern reality is today: it produced verifiable results. One was so much a part of the world, wrote Berman, that it was a part of us. "The view of nature which predominated in the West down to the eve of the Scientific Revolution was that of an enchanted world." The animism held by many indigenous peoples would have been very familiar to the medieval person: all things were alive and wondrous. The concept of a separate observer did not yet exist; instead, a sense of belonging prevailed. Berman called this mindset a 'participating consciousness': one's destiny was inexorably bound up with that of the cosmos. The relationship gave life meaning and bespoke "a psychic wholeness that has long since passed from the scene".[11]

To question this worldview would be tantamount to questioning oneself. Yet that is exactly what happened during the historical period in Europe known as the Renaissance. The church's hold over the popular consciousness already had been slipping for decades, and in its place emerged a secular society. "[T]wo of the most powerful activities in modern life, technology and science", wrote Alan Drengson, "developed within a spiritual and moral vacuum", free to serve individual self-interest.[12]

At the Venice Arsenal in the 1590s, Tartaglia, Galileo and other investigators had to solve a ballistics problem, which was how to calculate the range for firing a cannon in battle? The Aristotelian paradigm, which had held sway all through the Middle Ages, stated that an object, when propelled into the air, will follow a straight trajectory until its desire to return to Earth takes over, whereupon it will drop straight down. Never mind that actual objects did not quite conform to this picture: the ideal was a straight line up at an angle, followed by a straight line vertically down. This was the ideal trajectory because that is what projectiles *wanted* to do. (The similarity between the Aristotelian paradigm and the neoclassical economic orthodoxy is not lost.) Galileo made a conscious decision to step outside this self-fulfilling way of thinking. He did not initially seek to reject it, only to find out what projectiles *actually* did, unencumbered by any preconceptions. He found out that projectiles follow parabolic

[11] Berman (1981, p. 16).
[12] Drengson (1995, p. 189).

trajectories—typically asymmetrical ones, due to wind resistance—a finding which prompted him to refute the Aristotelian paradigm.

That Aristotle had been wrong was the lesser revelation. Having been dead for nearly 2000 years by that point, the old sage could be cut some slack. The greater revelation was how Galileo went about his project. Conceptually—psychologically, in fact—Galileo separated himself, as the observer, from nature, the observed. In this way, he catalyzed a fundamental shift in European thinking. Others who followed, principally Francis Bacon and René Descartes, saw the philosophical importance of Galileo's experimentation, and molded it into a new epistemology. Descartes' epistemological and methodological innovation was to explicitly separate subject from object, the observer from the observed, so that the object could be perceived as if from without. This separation later became known as the Cartesian dualism. The days of the magical, numinous medieval worldview now were numbered. During the first few decades of the seventeenth century, Berman wrote, it was "irrevocably dismantled".[13]

By separating subject from object, the Cartesian dualism removed from objects any form of anthropomorphism. No longer did an apple fall from a tree because it 'wanted' to; instead, another, objective, explanation had to be found. Apples were inanimate objects: they fell from trees because something was causing them to. It may seem obvious to us today that that something is gravity; we have Isaac Newton to thank for the concept. But in his day, it was not at all that obvious. The concept of gravity troubled Newton. It was mystical, in the sense that it could not be seen, heard or touched, but it was real because it could be measured. Newton, observed Gregory Bateson, did not so much 'discover' gravity as invent it.[14] The constant for acceleration due to gravity, g, was a mathematical necessity for his laws of motion. Its existence was verified by dropping objects and watching them accelerate. Yet gravity's precise nature eluded Newton, and it remains to some extent a mystery today. Newton, perhaps more than anyone, solidified the Cartesian epistemology; yet he remained himself a mystic, fascinated with alchemy, believing himself to be one of a long line of master magicians chosen to help solve the riddles of nature set by God.[15] Although the underlying nature of gravity may not be fully explained,

[13] Berman (1981, p. 40).
[14] Berman (1981, ch. 5).
[15] Berman (1981, ch. 5).

today we never question the *concept* of gravity—a force arising from differences in mass—as an explanation for why apples fall to the ground.

Soon after the publication of Descartes' *Discourse on Method* in 1637, a tension emerged between the new rationalism and the older medieval worldview. Rationalism found new allies in Protestantism and in capitalism, which had been growing an as economic force since the Renaissance. Here was a natural synergy: a scientific epistemology that separated the human observer from nature; allied to a religious denomination that separated God from nature and advocated personal salvation through earthly achievements, including the accumulation of wealth; allied, in turn, to a new economic paradigm promoting ownership of the means of production and capital accumulation therewith: essentially a 'winner takes all' system. This nexus of beliefs formed a powerful social force that, during the seventeenth century, overcame centuries of dialectical holism in Europe. Unlike the medieval worldview, this worldview could accomplish new things, such as exploit nature for profit. "Cultivation [of nature] became rape", wrote Berman, "and that ... is most clearly what industrial society—not just capitalism—represents."[16] Modernity was born.

The industrial revolution may have changed how humanity uses Earth's resources but the modern paradigm itself "is ultimately the child of the Scientific Revolution".[17] In the 1770s, when industrialization in England was beginning to gather momentum, Adam Smith gave birth to economics by applying a rational and empirical analysis to such topics as labor, capital, the distribution of wealth and government policy. His was an era of technological advancement accompanied by colonial expansion from Europe to many other parts of the world, bringing with it European attitudes and norms inspired by the Cartesian dualism. René Descartes himself, writing 140 years before Smith, concluded that we can employ our understanding of natural phenomena for practical purposes, "[so making] ourselves *masters and possessors* of nature"[18] (emphasis added).

Descartes and others writing early in the seventeenth century saw that "the attempted domination of nature depended on cognitive clarity"[19], necessitating a conceptual separation of subject from object. Berman called this separation a 'nonparticipating consciousness', contrasting it

[16] Berman (1981, p. 126).
[17] Berman (1981, p. 24).
[18] Descartes (1637), quoted in Berman (1981, p. 25).
[19] Berman (1981, p. 106).

with the 'participating consciousness' of medieval times. In a Cartesian world, we are conscious of nature but we do not participate within it; instead, we observe it dispassionately from outside. Fact becomes separated from value.

European exploration and colonial expansion from the late sixteenth century onward was accompanied by the emergence of a worldview that saw nature—and native peoples whose lands were colonized—as an asset to be possessed and controlled. This worldview represented a significant departure from the earlier, medieval European worldview. It was also directly counter to the worldviews of the native peoples whom European explorers encountered. These peoples for the most part lived within a participating consciousness, placing the person within nature, belonging to it, rather than it belonging to them. Indeed, the Cartesian dualism has been perpetuated within modern, industrialized economies by the perception that Earth's resources are not particularly limiting, a perception rooted in these societies' economic colonialism. A perceptional shift away from dominance over nature toward coexistence with it consequently has been delayed.

It is illuminating, and indicative of the degree to which the Cartesian mindset is so deeply rooted within the Western consciousness, that the very terms used to describe the periods of European history following the Middle Ages are the 'Renaissance' (beginning around the middle of the fifteenth century), the 'Enlightenment' (beginning around 1600) and the 'Age of Reason' (beginning around 1700). They charted a progressive rejection of the spiritualism and superstition of earlier times. Yet in throwing out medieval superstition, the new thinkers also jettisoned any notion of belonging to the world in which they lived. Instead, the world was theirs for the taking. People steeped in this mode of thinking since the seventeenth century have taken more and more. Within the context of a contrasting viewpoint recognizing nature's ultimate control over humanity—a viewpoint appropriate for a crowded planet—the philosophical developments of seventeenth-century Europe must seem less an enlightenment and more of a 'disenchantment', to use Berman's term.

Yet human nature was not fundamentally altered by the scientific revolution. People did not suddenly begin running around like little emotionless automata, devoid of any human instinct. The 'participating consciousness' of medieval times was rejected but not refuted.[20] We did

[20] Berman (1981, p. 135).

not, and still have not, become *Homo economicus*, the imaginary caricature who populates the world of the standard economic model. There remains to this day a tension within modern society—first in the West and increasingly around the world, as the economy has globalized—between modernity's rational actor and the nonrational (intuitive, holistic) actor who has always been there since we were digging up roots and hunting mammoths. It manifests within modern society as differences between people's professions and their proclivities: a currency trader by day, for example, and a musician by night.

The modern state of being is a state in which more than one worldview or mindset coexist, sometimes harmoniously, sometimes uneasily, within the same human life. It is as much a part of us as any other way of interacting with the world, yet it becomes particularly expedient and necessary for survival in a modern world, in the same sense that a dialectical, holistic mindset would have been essential to survival in the Middle Ages.

Epistemologies—ways of thinking and knowing—can be thought of as analogous to genes in the genome. Their expression over time will depend in part on environmental factors. The social environment became conducive to the emergence (or 'expression') of a Cartesian epistemology during the 1500s. Cartesianism spread through the 'meme pool',[21] while the prevailing medieval-holistic meme declined in frequency. By this line of thinking, there may be many other human epistemologies still waiting to be expressed. Which ones actually become expressed will depend on the social and natural contexts impinging upon people's cognitive and psychological states. Michael Polanyi argued that in committing to a certain scientific methodology, we make a nonrational commitment, an act of faith.[22] A system of thought is coherent not because it is true in some objective sense but because it is socially stable. Systems of thought, including science, are social processes.

From the Mechanistic to the Integrative and Beyond

As a guiding epistemology, the Cartesian dualism was remarkably useful to science, as an exercise in discovering mechanistic explanations of natural phenomena, provided that any object of study was not significantly affected by observer interaction. A mechanistic approach underpins the scientific method as it applies to experimentation. To understand the

[21] Dawkins' (1976) term.
[22] Polanyi (1946).

mechanism of a process, one must change a single factor of interest at a time, holding all other known factors constant. Controlled experimentation assumes an absence of observer interaction. It necessitates the creation of a self-contained system that allows factors of interest to be manipulated so that their dependence upon one another can be understood. This understanding has practical use, in that it enables the development of technologies relating to the mechanisms thus elucidated. René Descartes probably would never have dreamed of the dazzling variety of innovation spawned by his methodological advancement, from dish soap to jetliners.

All well and good for a mechanistic world. But the real world is not mechanistic, at least not in a Euclidean sense. Euclidean mechanics describe an idealized, reversible state. In the real world, an apple's fall to the ground is not reversible. Newton's laws of motion were ahistorical. Other, non-Euclidean tools emerged during the twentieth century appropriate to the systems they studied, such as quantum thermodynamics, as a statistical theory of matter and energy, and chaos theory for the study of all manner of deterministic nonlinear systems.

Then there is the issue of observer interaction. In quantum physics, the measurement of either the spin on an electron or its momentum precludes the measurement of the other quantity. The measurement process thus interacts with the object being measured, violating the Cartesian duality. At a much larger scale of observation, ecological field research regularly contends with observer interaction. In the field, unlike in the laboratory, variables cannot be manipulated in a completely controlled manner, and the best an observer can do is to minimize their effects on the system of study. Results are statistical, expressed in degrees of confidence. Attempts to recreate ecological communities in the lab so as to manipulate them in a controlled setting, while valuable in their own right, are beset by the problem of representation. Is the microcosm created in the lab representative of an actual system in the field? To what extent can the results it produces be confirmed through fieldwork?

As science has embraced an increasing methodological pluralism, the dominant view of scientific progress has evolved away from the Popperian ideal of systematic falsifiability toward a social view of science. Thomas Kuhn's 'scientific revolutions' and Michael Polanyi's description of science as a social process were early examples.[23] The writings of Bruno Latour

[23] Polanyi (1958).

and others have built upon this foundation.[24] Where in the nineteenth century, observer interaction would have been frowned upon as a kind of methodological error, today it may be acceptable, provided one is explicit about its possible implications for the findings in question.

The growth of planetary sciences—climatology, oceanography, soil sciences, macroecology—has further stretched the credulity of the Cartesian dualism. In these areas of investigation, the observer is located not outside the experiment, as in the laboratory, but is inescapably in the middle of it. Observer interaction is logically unavoidable. The experimental boundary in these areas of study is the whole planet, and the experiment is, of course, not reproducible. Berman called this kind of study 'participant observation': the observation of a phenomenon or system by a participant within it, rather than by a nonparticipant outside it.[25] Yet the quality of the science in these fields is accepted as robust and defensible, not only by the community of scientists but also by an increasing number of policymakers.

Broadly speaking, the trend in natural sciences over the past 200 years has been from the mechanistic to the stochastic toward the integrative. In Isaac Newton's and Adam Smith's time, the popular conception was of a clockwork universe. The mechanistic paradigm became increasingly strained during the nineteenth century, as first Darwin in biology and later Planck, Maxwell, Rutherford and others in physics and chemistry made advancements that threw into question the universality of linear determinism. Einstein showed that the linear, mechanistic universe was really only a special instance of a wider, relative universe. Then Schrödinger, Pauli and others pulled the rug out altogether from under mechanism by showing that what appears to be mechanistic at the level of visible matter is actually stochastic at the subatomic level.

In parallel, scientific epistemology became less monolithically Cartesian and more plural. Scientists take great pride in analytical rigor; however, if science followed the Popperian model, little if any actual progress would be made. Science is not the objectivist endeavor many might imagine it to be, or that economics attempted to emulate. As Polanyi, Latour and others have described, it is a social process, albeit a very systematic one. Scientific investigation mirrors Beethoven's description of musical composition as '10 percent inspiration and 90 percent perspiration', the

[24] See, for example, Latour and Woolgar (1986) and Latour (1987).
[25] Berman (1981, p. 147).

'perspiration' in question being the application of rigorous methods to test, refute, refine or validate whatever came out of the inspiration. But the inspiration—the leap of intuition—is essential. Scientists also hold a great deal of tacit knowledge, acquired simply through immersion and experience, like everyone else. In effect, science is a coexistence of three epistemologies: an intuitive one, a participatory or immersive one, and a rational, Cartesian one. Each has its particular role to play: the three together form a kind of epistemic toolkit for advancing scientific understanding.

Sciences differ from one another in their use of this toolkit. Theoretical physics relies more heavily on the intuitive one than on the immersive one. Empirical sciences, on the other hand, are heavily immersive. Polanyi's description of tacit knowledge used an example of the way medical students learned to interpret x-ray images of the body through repeated exposure and pattern recognition, as if by osmosis.[26] Field ecologists who spend a long time in certain natural habitats become steeped in them: they acquire a depth of knowledge and understanding comparable only to that of local people. However, all sciences share a reliance upon the rational, Cartesian episteme to transform the inspirational or the experiential into the publishable.

Scientific thinking is important socially because it is a kind of indicator of the way we all think about the world. We accept that planetary motion is predictable but that the weather a month from now is not. The epistemic coexistence found in science prevails within society at large. Returning to the gene analogy, the subject-object distinction was not 'born' with the ballistic experiments of Galileo and the philosophizing of Descartes; rather, the distinction, which was there all along, simply became overtly expressed where previously it had been suppressed. Jared Diamond recounted an ornithological expedition to a rainforest in New Guinea, in which the visiting scientists observed and identified more than one hundred species of bird. The local people along with them identified all the same species, resulting in a nearly one-to-one relationship with the Linnean classification system.[27] The Linnean system, used by biologists around the world is a rational-empirical system, "a logical order imposed on nature", according to Berman.[28] Yet the congruence between the Linnean

[26] Polanyi (1958, p. 101).
[27] Diamond (1966).
[28] Berman (1981, p. 132).

classification and a New Guinean one suggests that rational distinction must in some way be innate.

If we employ more than one different way of understanding our world, then it should hardly come as a surprise that we may hold more than one worldview. Epistemic pluralism is paralleled by ontological pluralism. It is well known to biologists working in New Guinea—this author included—that while local people may worship a Christian god introduced by missionaries by day, many revert to their older animism by night. They cannot be convinced to assist field biologists on nocturnal expeditions into the forest, even when offered twice the going rate in payment, because the forest at night is full of bad spirits. Within Western society, similar ontological dualities can be found. A sizable proportion of scientists, for instance, are devoutly religious.

In short, we wear different ontological hats for different occasions. Some philosophers and psychologists have claimed that the subject-object dichotomy of the Cartesian epistemology created a 'divided self', leading to higher rates of psychosis within modern society compared to past ones.[29] Leaving aside the problem of measurement error, our ability to adopt more than one worldview or way of learning probably has always been there. Although an overemphasis on mechanistic, atomistic or materialistic views of the world may be to blame for many of the problems of the modern economy and society, these views cannot be wholly replaced by holistic, integrative, participatory ones, nor should they be. In making a clean break from medieval mysticism, the Cartesians neglected the importance of holism, even while they rejected such crackpot practices as alchemy. A planetarian society needs both a participatory consciousness as a normative guide and a Cartesian epistemology for technological development.

Individualistic, exploitative behavior may appear to dominate modern society, but only because its reward structures are themselves based on tenets and norms established during the early modernist program by the nexus of scientific rationalists, Protestants and capitalists of the seventeenth century. Society is set up to put a large percentage of the population into positions where they have to compete for limited available resources.[30] If the reward structures within society were different—for

[29] Berman (1981, pp. 130–132).

[30] Robert Frank (1999) documented the Darwinian nature of the modern economy in his book *Luxury Fever*.

example, discouraging exploitation and waste, while encouraging stewardship and coexistence—the modernist epistemology would seem less fundamental, more like a veneer. Our Cartesian selves would not melt away and disappear altogether, because they have a role to play, but they would cease to dominate our lives.

Reward structures create selection pressures: certain behaviors are rewarded at the expense of others. Both biological and social selection pressures led to certain distinctly modern human traits, such as longevity, menopause and language, all of which seem to have emerged within the last 50,000 years. These were physical and physiological adaptations for survival within small hunter-gatherer groups whose success depended upon knowledge and long memories of their environments among elders, and upon communication of this knowledge to younger members of a group.[31]

If nature continues to exert sufficiently strong selection pressures upon the human economy, whether directly or mediated through economic instruments, then individuals better able to adapt to those pressures would be conferred a social advantage. Over time, instrumental changes in behavior would bring about cultural changes—such as for material efficiency and widespread prosperity—which, in turn, would render the new behaviors 'acceptable'. At that point, which could arrive some time during the twenty-second century, most of humanity would have become socially adapted to a world in which the economy and nature are coevolving. Yet humanity still would not be *biologically* adapted to a crowded planet. Our adaptation to our new reality would be conscious rather than innate. On an evolutionary timescale, the world has become crowded with human beings so rapidly that a time-lag has appeared in the emergence of biological adaptations within the human species. Over the course of centuries, it is possible that certain innate behaviors, being advantageous within a society on a crowded planet, would spread through the human gene pool, thereby conferring upon the species a new biological trait, a participatory *sub*conscious, if you like, or a way of being in the world which comes naturally, and to which the extractive individualism of 'modern' society would seem completely alien.

Today, a great deal of human knowledge and experience about how to coexist with nature remains in the world, some of which has been

[31] Diamond (1991).

documented by anthropologists.[32] As a cultural resource, much of this knowledge has been lost to economic development, and continues to be. For a long time, the perception within the major development agencies was that these 'old ways' represented an obstacle to economic development.[33] It is easy to see, however, that a conservationist ethic extends naturally from conserving nature to conserving non-Western knowledge and experience. The preservation and dissemination of this knowledge, in addition to supporting its intrinsic value, would stimulate the early normative development of a planetary economy. "It is not, as Keynes suggested", writes Stephen Marglin, "that we can rethink the premises of society once we have 'enough.' Rather, we shall have enough when we re-think the premises of modernity."[34]

Deep Ecology

The distinction between a conscious, instrumental adaptation to humanity's expanding role on planet Earth and an intrinsic one, in which participatory or generative behaviors would 'come naturally', was first developed by the deep ecologists. Deep ecology is intrinsically holistic. Although it identifies 'society' and 'nature' as distinct entities, it does not consider them separate in the way Western modernism does. Similar to the hierarchical framework outlined in the previous chapter, it considers them homologous and overlapping: distinct enough to warrant separate terms yet also deeply innervated with one another.[35]

Deep ecology can trace its roots back through a line of Western thinkers dissatisfied with the Cartesian worldview, including John Muir, Walt

[32] For a few examples, see LaChapelle (1995, pp. 57–58).

[33] Sachs (1995, p. 430).

[34] Marglin (2008, p. 222).

[35] Neil Evernden (1978) made a key distinction between causality and intermingling in his definition of the 'interrelatedness' of social and natural systems. A Cartesian or Newtonian mindset might equate interrelatedness with causality, one system causing changes in another, whereas ecology, wrote Evernden, tends to look at the interrelations of natural and human systems as a commingling. It is in this latter sense that the term 'innervated' is used to describe the connections between natural and human systems. Think of the innervation of the nervous system with muscle tissue, for instance. Latour (2004) takes the line of thinking a stage further, dismantling the dichotomy between nature and society, erecting in its place a collective of the human and non-human. This ontology is consistent with the framework for an economics of a crowded planet outlined in the companion volume (Murison Smith 2019).

Whitman, Henry David Thoreau and Thomas Jefferson; also through such Eastern traditions as Taoism, Sufism and Zen Buddhism.[36] The field is most commonly identified with the writings of the Norwegian philosopher Arne Naess (1912–2009) who coined the term. To Naess,

> [t]he emergence of human ecological consciousness is a philosophically important idea: a life form has developed on Earth which is capable of understanding and appreciating its relations with all other life forms and to the Earth as a whole.[37]

Deep ecology is not the same as environmental radicalism. It is a philosophy rather than a form of direct action. While some radical environmental movements have co-opted aspects of the deep ecology philosophy for political purposes, and perhaps given it a bad name by association, David Rothenberg pointed out that deep ecology cannot be a doctrine because, by definition, it is a personal way of engaging with the world.[38] The deep ecologists seek to give their detractors an opportunity to experience their worldview for themselves. It is as much a literary or rhetorical exercise as a scientific one. The overlap between science and liberal arts is intentional. Deep ecology uses the term 'ecology' in a philosophical sense rather than a scientific one. Although wary of philosophical attempts to co-opt the term, scientific ecologists would not dispute many of the premises of deep ecology, even though they might question their practicality.

In a 1973 article, Naess characterized the awakening of an 'environmental' sensibility in the 1960s as 'shallow ecology'.[39] It was then, and remains today, an ethical or instrumental movement. Environmental protection is to the benefit of people: therefore, we should be good stewards of the natural environment to safeguard our future. Also known as the 'reform' ecological movement, it seeks to modify people's practices and behavior through laws and regulations.[40]

'Deep ecology', in contrast, was defined by Naess as intrinsic and ontological. It is a way of being. It is an existence characterized by practices and habits which come naturally and are not questioned. Whereas shallow ecology urges action through a sense of obligation, deep ecology urges a

[36] Devall and Sessions (1985, ch. 1).
[37] Naess (1989, p. 166).
[38] Rothenberg (1989).
[39] Naess (1973).
[40] Naess (1989, p. 162).

more fundamental enlightenment, in which actions mostly harmonious with nature come naturally, while exploitive actions are unnatural. John Rodman elaborated on this distinction by referring to 'shallow ecology' as a form of 'moral extensionism' ascribing rights to non-human entities and obligations to humans to honor those rights. 'Deep ecology', to Rodman, is an 'ecological sensibility', "a theory of value that recognizes intrinsic value in nature without (hopefully) engaging in moral extensionism".[41] An evolution from moral extensionism to ecological sensibility, Rodman noted, is an evolution in perception. "When perception is sufficiently changed", he wrote, "respectful types of conduct seem 'natural', and one does not have to belabor them in the language of rights and duties."[42] The key to a 'sustainable' culture, added Dolores LaChapelle, is not a matter of "*opposing* nature or *trying* to be in communion with nature; but of *finding* ourselves within nature..." (italics hers).[43]

'Deep ecology' or 'ecological sensibility' refers to a sense of being deeply embedded within Earth's natural systems, as distinct from considering oneself somehow separate from them. In the language of the social sciences, deep ecology represents the 'internalization' of a norm acknowledging the intrinsic value of nature. The 'social' and the 'environmental', distinctly identified within modern society, would become one and the same in people's minds. Such a paradigm change, Naess wrote, represents not "a slight reform of our present society but a substantial reorientation of our whole civilization",[44] not to mention also our consciousness, as Berman argued.

Ever somewhat on the fringes of modern social discourse, deep ecology was ignored by the mainstream more often than it was critiqued. In a rare exception, a poll on environmental attitudes within the US in the late 1970s concluded that, because deep ecology "rejected economic growth and the assumptions underlying Western science, [advocating instead] the subordination of human society to natural processes, and the doctrine that humans share a profound identity with nonhuman nature", it was "a far more radical position than that taken by most supporters of alternative technology".[45] This was a fairly accurate assessment.

[41] Rodman (1995, pp. 248–250).
[42] Rodman (1995, pp. 252–253).
[43] LaChapelle (1995, p. 62).
[44] Naess (1989, p. 45).
[45] Rothenberg, p. 17, in Naess (1989).

Yet deep ecology appeared radical only relative to Western modernism. Its interest within the present discussion is its perception of something beyond Western modernism, perhaps many decades into the future, inspired in part by pre-modern ways of living. Deep ecology saw the modern age as a phase within a longer social evolution. Alan Drengson, for instance, conceived of eight metaphorical 'cities' representing various past and possible future stages of society, beginning from early Babylonian times. His 'city of technocracy' represents modern times, which he proposed would be followed by a 'city of planetary consciousness' as a kind of reformist society, falling within a coming 'Age of Ecology'. This age eventually would give way to a final 'Age of Ecosophy', occupied by a 'city of being', in which harmonious existence with nature would come as a matter of habit.[46] The distinction between these two ages is that the first is characterized by ecological living as a conscious or reasoned decision (the *logos* in 'ecology') whereas the second represents ecological living as a philosophy or as a way of life (the *sophos* in 'ecosophy').

Ecosophy: Personal Norms

Naess further distinguished ecosophy, as a system of personal norms, from 'ecophilosophy', which is the study of humanity's place within nature using ecological concepts.[47] Deep ecology's ecophilosophy posited that injuring nature wantonly "would mean injuring an integral part of ourselves", according to David Rothenberg. Humanity and nature were inseparable.[48]

Ecosophy imagined an ideal lifestyle in harmony with nature. Leaving aside the degree to which its norms could be acted upon practically within current society, the value of ecosophy is its ability to highlight certain existing structural and institutional impediments to living "as if nature mattered".[49] Yet ecosophy is not a set of personal prescriptions. It is a

[46] Drengson (1995).

[47] From the Greek *sophia*, philosophy is an affinity with wisdom or insight. Naess (1989, p. 37) points out that sophical insight should be relevant for action. Through actions people demonstrate wisdom (or not). *Sophia* is to be distinguished from *logos*, whose Greek root is the verb *legein*, to count or tell, and which refers to structure and order. Sciences are 'ologies': as such, they aim for impersonal or abstract results, independent of personal value to the greatest extent possible.

[48] Rothenberg, p. 2, in Naess (1989).

[49] Devall and Sessions (1985).

system of thinking to help people develop their own systems in their own ways.[50] Naess understood that change in humanity's engagement with the natural world can begin only with changes in each person's perception of it and interaction with it. There is no point in trying to impose a dogma upon people. Although we tend to act and decide as if we have a clear view of the world, he argued, a complete verbal articulation of such a view is impossible. At some point, we all fall back upon intuition. Ecosophy, to Naess, was not the reified pontificating of philosophers but an actual practice.

In ecosophy, a state of living is sought where altruism toward others or toward nature becomes equivalent to actions in the interest of the self. Ecosophy, Naess argued, should use humanity's special capacities among all other living things not for domination and mistreatment but "as a premise for a universal care that other species can neither understand nor afford".[51]

The ecosophical critique of modernism was that its emphasis on individual self-interest, competition and dominance, are "but partial impulses of the human being which, in modern times", according to Paul Shepard, "have become organizers of the whole of human life". Civilization, he argued, keeps people in a stage of petulant early adolescence, always demanding more for themselves.[52] Modernity's emphasis on 'self-realization', 'self-expression' and 'self-interest', wrote Naess, "assumes the ultimate and extensive incompatibility of interests of different individuals".[53] This being the case, it follows that modernity is fundamentally incompatible with a culture of harmony between the individual and their surroundings.

Naess outlined a 'deep ecology platform' as a system of norms for ecosophical living. He maintained that these norms are intuitive concepts that are taken as given.[54] In economic parlance, they are foundational and axiomatic. The platform states that "the flourishing of human and nonhuman life on earth has inherent value. [This] value is independent of the

[50] Rothenberg, p. 5, in Naess (1989).
[51] Naess (1989, p. 171).
[52] Berman (1981, p. 189).
[53] Naess (1989, p. 85).
[54] Martin Heidegger suggested a reversal of analytical thinking to rely instead upon intuitive power as a way to reconnect with the world as an interconnected whole, thereby escaping what he saw as a modernist trap (Devall and Sessions 1985, pp. 98–99).

usefulness of nonhuman life for human purposes".[55] In other words, the value of non-human life is invariant to the existence of people. By 'life' Naess meant the whole biophysical environment, not only biotic life. "One cannot quantify adequately what is important for the quality of life", he wrote, "and there is no need to do so."[56] David Rothenberg built on this proposition to state a new concept of value, "not *of* the Earth but *for* the Earth, originating in human choice" (emphasis added).[57] "The anthropocentrists, who proclaim many things to which people have rights", Naess wrote, "may find it sobering to be faced with a norm about something to which people have no right."[58] Subscription to the norms of the deep ecology platform, Naess proposed, obligates the individual to implement the necessary changes.

Ecophilosophy

As a response to modernity, deep ecology advocates not only a system of personal norms but also certain structural modifications to society. Specifically, it advocates decentralization of communities, allowing people once again to place themselves within groups where all are known to one another. This structure, the deep ecologists argue, would allow decision-making to become much more democratic and nonhierarchical. It would foster self-responsibility and local autonomy. Whereas 'primal' societies, to use Devall and Sessions' term, already live a deep-ecological lifestyle—perceiving only a connected whole rather than 'society' and 'nature'—modernity has become characterized by a culture of dominance:. "dominance of humans over non-human nature, masculine over feminine, wealthy and powerful over the poor, [and] the West over non-Western cultures."[59] Western society exists in a state of high population density where anonymity is commonplace. Because not everybody is known to one another, it has become customary for social order to be maintained in part through dedicated institutions, such as the police and the legal system. Such institutions existed in the earliest cities. The decentralization advocated by the deep ecologists would undercut the dominance hierarchy of modern

[55] Naess (1989, p. 29).

[56] Naess (1989, p. 114).

[57] Rothenberg, p. 19, in Naess (1989).

[58] Naess (1989, p. 30).

[59] Devall and Sessions (1985, p. 66). See also Robert Aitken Roshi, Appendix C, p. 233, in the same volume.

society and promote a more egalitarian one. In its idealized state, the ecophilosophical society they imagine would, in conventional economic parlance, be a kind of Pareto-optimal state of altruism.

Deep ecology also advocates biocentric equality. This norm emerges directly from ecosophy. Biocentric equality is essentially an even-handedness in humanity's treatment of nature. Whereas equal-rights movements in the social sphere have sought equality on the basis of ethnicity, gender and other human attributes, the deep ecologists advocate equality on the basis of being alive, pure and simple. "Any ethic which restricts the right to live to humans alone", argued Naess, is "an anthropocentrism with detrimental effects upon the life quality of humans themselves."[60] A biocentric philosophy, argued Devall and Sessions, thus would "place the best of the reformist [environmental] response into a coherent philosophical perspective".[61]

Commentary

The personal norms of ecosophy contain much that may become prevalent within a future, stable society having achieved alignment with natural processes. Yet ecosophy also supports a decrease in the human population so that non-human life can flourish. Further, it says nothing about how to reach the imagined harmonious state. In reality, the twenty-second century likely will witness a human population of 11–12 billion people trying to figure out how to persist and stabilize itself in the face of surprises from nature. Although many people would gladly adopt a more or less ecosophical lifestyle in principle, they cannot, in practice, within today's predominantly modern society. The value of ecosophy is normative rather than practical, in that it could help to guide reformist policies over the coming decades.

Deep ecology could arise only in Western culture, being as it is a response to modernity. Many other cultures, not only indigenous but also Eastern ones, have maintained philosophies similar to deep ecology all along. Primal societies adopted ways to self-manage: they had no other choice, being highly dispersed and self-reliant. Everyone within a group was known to everyone else. Where a group had a leader or a chief, that

[60] Naess (1989, p. 26).
[61] Devall and Sessions (1985, p. 61).

individual was not dominant by force, as Devall and Sessions explained, but typically had the noncoercive qualities of a mentor.

In a world having a large, growing and increasingly urban population, it is unrealistic to imagine human demography and social organization becoming more dispersed any time soon. However, as Chaps. 4 and 5 describe, there may be ways to establish governance structures for a crowded planet that emulate some of the decentralized community decision-making imagined by the deep ecologists. A key criterion of success for any planetarian governance will be inclusion. A global challenge necessitates substantially full social participation, a requirement that has significant implications for governance.

Deep ecology is somewhat contradictory about the meaning of 'biocentrism'. Scientific ecologists, Naess noted, do not ask 'what kind of society would be best for maintaining a particular ecosystem?' The reason he gave is that it is a question of ethics rather than science.[62] That is as may be; yet the scientific ecologists probably would argue that the best human society for maintaining any particular ecosystem is no society. Devall and Sessions made the same mistake as Naess. They criticized the 'reformist' environmental movement for being 'anthropocentric', without realizing that deep ecology itself is anthropocentric.[63] To be wholly biocentric would be to advocate for human extinction.

The alternate concept of 'biocentric equality' is more practically useful because it is an ethic that treats natural and human systems as equals. It is anthropocentric only insofar as it is also biocentric, appropriate for a world where the anthroposphere and the biosphere are locked in a coevolution. It is the penetration of a deep-ecological sensibility into societal norms—a sensibility that modern, Western society has not thus far been ready or willing to accept. But it might.

'Environmental reformism' has come in for criticism from the deep ecologists because it does not address underlying attitudes.[64] They do not defend reformism as an essential stage toward a more fundamentally biocentric existence. Yet any professional biologist would acknowledge that all organisms are to a degree self-centered. What distinguishes the human species from others is our power, our ability to affect natural systems on a global scale. Consequently, our anthropocentrism will likely compel us

[62] Devall and Sessions (1985, p. 65).

[63] Devall and Sessions (1985, p. 52).

[64] Devall and Sessions (1985, p. 196).

over time to become more biocentric, thereby establishing a degree of balance between the two. The path from today's excessively anthropocentric modern society to a more balanced one can only be reformist because only the incentives created through reform will begin to reinforce behaviors that eventually will become deep-seated norms and values.

The challenge of the twenty-first century will be one of management: that is, to help push society out of Drengson's 'city of technocracy' toward the 'city of ecology'. The subsequent challenge, and potential reward, of the twenty-second century and beyond would then become a new way of living.

Future Societies: A Brief Survey

The philosophical and normative context for a possible future economy is not developed in isolation. It is part of a long history of similar social and economic narratives. Although that history would make an interesting volume in its own right, two particular past narratives share certain ideas with the present one, deserving a brief review. One is Thomas More's *Utopia* and the other is a collection of writings on 'conserver' societies.

'Visioning' exercises such as these serve the useful purpose of imagining a *desideratum*: a desired state or goal. They ask the question, 'what if the world were this way?' The answer to that question gives us perspective upon our own world. It helps to frame policy options and outline future pathways. These form requirements, statements proposing what would have to happen for such a society to come into existence.

With this in mind, where better to start than the original vision of a future economy and society, Thomas More's *Utopia*?

Utopia *by Thomas More*

The English lawyer, philosopher and statesman Thomas More (1478–1535) lived during a period of English history when dissent against the crown, particularly from an individual of More's standing, could be a death sentence. In More's case, it was, although not for writing *Utopia*. Twenty years before he was beheaded by King Henry VIII for refusing, as Lord Chancellor, to endorse the king's founding of the Church of England, More penned an allegory about an imaginary country discussed among a group of friends in a garden in Antwerp. It was during a trip to Flanders in 1515 as Undersheriff of the City of London that More had the idea for the book, which was published the following year.

More relates being introduced by his friend Peter Giles to a traveler, Raphael Hythloday, recently returned from the New World. Hythloday had, fictitiously, attached himself to one of Amerigo Vespucci's voyages there. In real life, on the last of these voyages in 1503 to 1504, twenty-four Portuguese sailors stayed behind at a fort in Cape Frio in Brazil. In More's account, Hythloday was one of these men. From there, he visited the city-state of Utopia, whose name in Greek means 'no-place'. The title of the book may have been inspired by Plato's *Republic*, in which Socrates asks whether the wise man will participate in real-world public life, with all its messy imperfection, or remain in his idealized community, "a city which is placed in words and can be found nowhere on earth".[65] More's 'city of words' was a veiled way to engage in the 'city of actions': Hythloday was a foil for More's opinions. In this way, he could reasonably claim, if questioned, that the opinions expressed were not his. Hythloday gave More free rein to imagine a society untroubled by the problems of his own day, without fear of censure.

Certain features of this imaginary Utopian society stand out. First, Utopia has no money. Resources are distributed more or less evenly from large warehouses managed and staffed by the people on behalf of the state. There is no accumulation of wealth, nor is property privately owned. Houses are large and may accommodate several families. Adults dress plainly whereas children are allowed to decorate themselves. Adornments, such as gold or jewels, are seen as childish. Like toys, they are eventually left behind as individuals grow toward adulthood. Gold acquired by Utopians through trading with other nations is the property of the state and is used by the state only in an emergency. Indeed, relates Hythloday, "Nature, ... like an indulgent mother, has placed all wholesome things like air, water and earth itself, within our reach; those that are vain and unprofitable she hides away in inaccessible places."[66]

Consistent with a Platonic ideal, Utopia was described as a society established by a philosopher-king, Utopus, who originally seized the island that later bore his name, shaping it to his own conception. Utopus was able to begin with a blank slate and to proceed purposefully with a goal in mind. Understanding perhaps that increasing crowding could lead to a

[65] Baker-Smith (2012, p. xviii).
[66] More (1516, p. 75).

competitive response resulting in the definition of property rights—as Cicero acknowledged in ancient Rome, and which Renaissance Europe inherited from the Romans—Utopus ensured that practices and institutions were established to ensure that his society evolved from a primitive state of nature without necessitating a culture based on property rights. The Utopians thus evaded the whole process by which natural law becomes subordinated to civil law and to private interests as civilization grows.[67] Yet, as Dominic Baker-Smith describes, "they live in splendid cities and they do [not] toil like European peasants." Utopian society, he notes, "mark[s] an alternative line of development from the one which produced Renaissance Europe".[68]

Utopia succeeded through a form of collectivism. Hythloday described the Utopians' communal meals, their communal living, and frequent pre-dawn lectures. Privacy and individualism were suppressed in the interests of inclusion and social participation.[69] The Utopians acted as if imbued with a wisdom from past mistakes, rather than the way actual societies develop, which is mostly organically, without much foresight. Their normative system was one of living in accordance with nature's precepts. To the Utopians, nature prompts the course to be followed as a voice of reason, supplemented by moral intuitions that act as social constraints.[70] Through Hythloday, More conveyed an understanding of how the unequal society in which he actually lived had arisen through a lengthy divorce from the natural world.

Consequently, the Utopians understood the risks of overcrowding. They agreed upon a maximum population above which they would establish colonies off the island. They even forced mainland people off the lands they colonized if they believed they could steward those lands better. For all this, More depicted the Utopians as cultivators rather than exploiters. More's description of the Utopians' response to population growth was far-ahead thinking for the early sixteenth century. It came nearly 300 years before Thomas Malthus' discourse on population. More's description of the Utopians' takeover of other lands for stewardship might be the first written example of a concept of environmental vigilantism.

[67] Baker-Smith (2012, p. xxvii).
[68] Baker-Smith (2012, p. xxvi).
[69] Baker-Smith (2012, p. xxviii).
[70] Baker-Smith (2012, pp. xxviii–xxix).

Utopia stood as much in stark contrast to the society of Renaissance Europe as it stands to the capitalist economy of the present day. Many ancient societies also could not compare to it. Writing 2000 years ago, Cicero complained how society was "a conspiracy of the rich … impervious to high-minded interventions" that could help to make life better for all.[71] *Utopia* itself ends with Hythloday's critique of the 'ungrateful society' of More's time: a society which divorced social convention from natural ethics, distributed wealth and power unevenly through private ownership of property and financial capital, and celebrated an aristocracy of conspicuous consumption. In Appendix 1 to *Utopia*, this society was contrasted to the first of Erasmus' *Adages* (1515), in which he reviewed the work of all previous known authors on the merits of commonly held property. That adage is titled, 'Between friends, all is common', implying that a norm of treating property as common could arise within an environment either free of conflict or having effective mechanisms to resolve it.

Appendix 2 to *Utopia* presented an interesting account of the first European contact with the Taíno people during the voyages of Columbus, written by Peter Martyr d'Anghiera in his 1511 book *On the New World*.

"It has been discovered with [the Taíno] that the earth, like the sun and water, is common, nor do 'mine and yours', the seeds of all evil, fall among them. For they are content with so little that in that vast earth there is an excess of land to farm rather than a lack of anything. Theirs is a Golden Age: they do not hedge their estates with ditches, walls or hedges; they live with open gardens; without laws, without books, without judges, of their own nature, they cultivate what is right."[72]

That this account was included as an appendix to *Utopia* is both curious and significant. It is curious because it did not really add anything to Raphael Hythloday's fictional account of Utopia. It is significant because it creates a reference point for Utopian society. Utopia was resource-constrained and crowded. The Taíno, by contrast, lived in a subsistence agricultural economy. Like the Utopians, they had no need for money although, unlike them, they had no need for the 'laws, books and judges' maintaining structure and order in a resource-limited condition. The Taíno's 'Golden Age' was equivalent to Kenneth Boulding's 'cowboy economy'.[73] Theirs, like Utopia, was a 'grateful society', although an

[71] Cicero's *De Oficiis*, cited by Baker-Smith (2012, p. xxv).
[72] Martyr d'Anghiera (1511), quoted in More (1516, pp. 128–129).
[73] Boulding (1966).

original, primal one rather than a rediscovered one. The genius of *Utopia* was to connect a primal state, in which gratefulness predominates, through a contemporary, resource-limited state, beset by individualism, competition and inequality, to an imaginary, possibly future 'grateful society', in which people would have found a way to reappreciate the gifts bestowed upon them by nature.[74] More never articulated an actual pathway from the society of his day toward such an ideal, perhaps because it would have been too prescriptive and politically risky. Instead, he left the reader to connect the dots.

'Conserver' Societies, Ecotopias and Planetarianism

Without perhaps knowing it, Thomas More started a meme. In the 500 years since *Utopia* was published, all kinds of utopian and dystopian imaginings have followed. One class of utopias particularly relevant to the present narrative is a loose agglomeration of ideas variously described as 'ecotopias', 'green utopias' or 'conserver societies'. These were influenced equally by deep ecology.

Conserver Societies and Ecotopias

'Conserver' societies, as the name implies, are societies geared toward conserving as many resources as possible, both material and energetic. "The economy", wrote Morris Berman, "will be a steady state, a mixture of small-scale socialism, capitalism and direct barter." In a conserver society, he argued, nothing would be wasted. A great emphasis would be placed on regional self-sufficiency. "There will be little interest in profit as an end in itself. The posture toward others, and toward natural resources, will be one of harmony rather than of exploitation or acquisition."[75]

It is not hard to find elements of conserver societies everywhere. People living close to nature, particularly in some of the world's less hospitable climates, such as in the Arctic or the Kalahari, have a history of extracting maximum value from the resources they harvest from nature. Even within modern society, which is wasteful in so many ways, conserver behavior is alive and well. In addition to such obvious examples as municipal recycling, there exists a thriving secondary economy in used goods, whether sold through internet exchanges, donated to charities for resale, or simply

[74] This was the same trajectory that Berman (1981) drew 460 years later.
[75] Berman (1981, p. 277).

left on the street for others to take. Of course, this behavior, in and of itself, does not a conserver society make. The society imagined by the deep ecologists and ecotopians is one in which the dominant social norms are for conservation and 'harmony' with others and with nature, rather than these attitudes being the exception.

In 1972, Edward Goldsmith and other editors of the journal *The Ecologist* published a *Blueprint for Survival*, which drew upon Meadows et al.'s *The Limits to Growth* and other contemporary studies of population and resource use, to propose a 'strategy for change toward the stable society'. This society they defined as one that "can be sustained indefinitely while giving optimum satisfaction to its members". The authors identified four principal conditions for this society:

(1) Minimum disruption of ecological processes; (2) maximum conservation of materials and energy—or an economy of stock rather than flow; (3) a population in which recruitment equals loss; and (4) a social system in which the individual can enjoy, rather than feel restricted by, the first three conditions.[76]

Conditions 1 and 2 are roughly equivalent to material discipline practiced by a society having measurable effects upon natural systems. Condition 3 may be met by around 2100, if the United Nations' population projection bears out. Condition 4 is the subject of chapters to follow.

Goldsmith et al. proposed a conversion from an economy of flow to one of stock, citing Boulding's observation that gross national product actually is a measure of gross national cost when one considers the natural stocks from which flow the resources for GNP.[77] The authors proposed a handful of fiscal measures to effect this conversion. First, a raw materials tax would be levied in proportion to the availability of any given raw material so as to allow such materials to last for an arbitrary length of time. Second, an amortization tax would be levied on products inversely proportional to their expected life. The shorter the life, the higher the tax. This measure would discourage throwaway products and encourage design for longevity. It would also encourage craftsmanship and increase employment, the authors claimed. A third measure would be a 'power tax' to penalize power-intensive processes that cause relatively more pollution.[78] The authors also made various proposals to protect areas of high

[76] Goldsmith et al. (1972, p. 18).

[77] Boulding (1966).

[78] Goldsmith et al. (1972, pp. 25–28).

biodiversity and the land rights of indigenous peoples within them, as well as proposals to revise economic accounting systems.[79]

Like Boulding and Howard Odum, Goldsmith et al. took the normative position that the fields of ecology and economics should not be in conflict.[80] "ecology should provide the approach, the framework, for an understanding of the interrelationships of social and environmental systems, and economics should provide the means of quantifying those interrelationships in the light of such an understanding, so that decisions on courses of alternative action can be made without undue difficulty."[81]

In the same year as Goldsmith et al.'s *Blueprint*, the Norwegian engineer Erik Dammann published a different argument for a conserver society. In *The Future in Our Hands*, Dammann framed a conserver society as a necessary precondition for all people on Earth to have a quality of life enshrined in the United Nations Universal Declaration of Human Rights, which included as fundamental rights access to water, nutrition, shelter, healthcare and education. Dammann argued that a persistence of large differences in material wealth between rich and poor nations hindered poor nations' ability to rise out of absolute poverty. He proposed a normative shift in rich countries away from self-reinforcing consumerism and economic growth toward a conservationist ethic in which excess production capacity within the wealthy, industrialized regions would be deployed to assist those in poor regions directly. Rather than compete with one another for material status, people in wealthy countries would cooperate to narrow the gap between rich and poor.[82] To Dammann, the challenge for modern society was to transcend the human tendency to compete, at least with regard to financial wealth and material status. Only then would the excesses of modernity be left behind for a more equal, stable society.

Gary Snyder argued for economics to be "a small sub-branch of ecology". We have the power, he claimed, to change not only ourselves but also our culture. "If man is to remain on Earth he must transform the five-millennia-long urbanizing civilization tradition into a new ecologically sensitive, harmony-oriented, wild-minded, scientific/spiritual culture … nothing short of total transformation will do much good. What we envision is a planet on which the human population lives harmoniously

[79] Goldsmith et al. (1972, pp. 28–32).
[80] See Odum (1971, 1994).
[81] Goldsmith et al. (1972, p. 32).
[82] Dammann (1979).

and dynamically by employing a sophisticated and unobtrusive technology in a world environment which is 'left natural'. [...] Master the archaic and the primitive as models of basic, nature-related cultures—as well as the most imaginative extensions of science—and build a community where these two vectors cross."[83]

There is much to admire in this proposal. First, Snyder's recognition of the need for both the scientific and the spiritual in a cultural transformation. Second, the notion of dynamically harmonious living: that is, in concert with the dynamics of natural systems. Third, a retention rather than a rejection of technology, which is unobtrusive: surely a more sophisticated technology than one that calls attention to itself. Fourth, a recognition of science as a pursuit of the imagination, not as the enemy of an ecologically conscious society, as some writers suggested. And fifth, an active deployment of the wisdom of indigenous societies. Those that have avoided elimination by the spread of Western modernity have, to-date, been largely ignored by the mainstream in this regard.

Paul Shepard offered a more radical ecotopia, in keeping with his unorthodox idea about civilization stunting our psychosocial growth. Shepard proposed a world in which eight billion people would live in roughly 160,000 cities of about 50,000 each, all strung around the perimeters of continents, thereby leaving the interiors of the continents wild and tractless. People would venture into the interior to connect with nature and to hunt; otherwise most food would be grown in urban environments using microbial food technology. "Perhaps three-quarters of Earth could be freed from its present destructive use", he wrote.[84] Devall and Sessions questioned whether the wild parts of Earth would not be recolonized by people who wanted to live away from the urban environment. There is also the practical question of how all these people would be relocated, not to mention completely restructuring the food supply.

In addition to the ruminations of these and other authors in the non-fiction arena, a wide literature of fiction can be found on utopian and dystopian themes. Much science fiction and fantasy writing falls into this area, as well as some more conventional literature, such as George Orwell's *1984*. Dystopian and post-apocalyptic themes have long been favorites of Hollywood movie TV producers. Yet relatively few fictional works focused on ecotopias specifically. Ernest Callenbach's *Ecotopia* and *Ecotopia*

[83] Snyder (1969), quoted in Devall and Sessions (1985, pp. 171–172).
[84] Shepard (1973), quoted in Devall and Sessions (1985, p. 174).

Emerging are two exceptions. Published in 1975, *Ecotopia* tells the story of the first outside visitor to the land of the same name, long after it had formed through the secession of northern California, Oregon and Washington from the US. That visitor, an initially skeptical journalist, is gradually brought around to the Ecotopian way of life, eventually falling in love with an Ecotopian woman.

The novel is unusual as a work of literature, being a fully utopian narrative rather than a dystopian one. Aside from the love story, the parallels with More's *Utopia* are evident. Ecotopia itself is described as being divided into numerous small, energy-efficient communities to eliminate urban sprawl. Pollution is not tolerated, trees are worshipped, farms and businesses are employee-owned, and the government is a matriarchy overseeing a twenty-hour work week. Ecotopians fundamentally value stability, both natural and social. Ecotopia, as imagined by Callenbach, is a conserver society, reusing and recycling as much as it possibly can.

Planetarianism

In his vision of Ecotopia, Callenbach may have been inspired by the work of his contemporary Peter Berg. In the 1970s, Berg promoted the idea of adopting bioregional political units rather than the sociopolitical units we have today. He proposed, for instance, for northern and central California to become a separate political unit from southern California, since it is a distinct bioregion.[85]

Impractical though this proposal may appear, it was grounded in a distinctly ecological perspective. As Berg himself explained, he sought "to encourage discussion of a regional/planetary identity as opposed to a national/global one" [86] The key concept here is one of identity, which is to any normative system what a stem cell is to a body tissue. Norms arise from identity, and if one's identity were bioregional or planetary—as opposed to national or global, in a social sense—then a distinct system of norms would follow.

In a 1978 interview with Michael Helm, Berg expounded his 'planetarian' vision:

> We are in the midst of a crucial struggle between globalist and planetarian interests regarding our ongoing mutual inhabitation of this planet. Globalism

[85] Berg (1977).
[86] Berg (1978, p. 42).

> perceives our species as possessing the primary intelligence and consciousness in the biosphere. Wants and needs of human beings are important to the degree that they should actually control the entire biosphere. The planet is a warehouse of resources in which agricultural and industrial production proceed to deliver commodities. [...] Centralization and monoculture are necessary to keep the warehouse-factory at peak efficiency. Globalists are rushing to complete an edifice-habitat.
>
> Planetarians, people who view themselves from within the biosphere rather than on top of it, extend importance beyond the human species to include other life and the processes by which all other life continues. The planet is organismic and expresses itself through diversity. [...] Planetarians are anxious about maintaining distinct regions, cultures and species, and look forward to experiencing full ranges of planetary diversity without destroying them.[87]

Berg contrasted his planetarianism with 'globalism' by identifying globalism as a technological ideology treating the whole planet as a single unit: a unitary problem to be 'solved' in the sense of the *Limits to Growth* endeavor. Like other ecotopians, Berg stressed the importance of developing a sustainability of communities, both social and natural, at a bioregional scale. The implication was that a planetary sustainability would emerge from this. One can see where the bumper sticker 'think globally, act locally' might have come from.

Berg's bioregionalism expressed itself programmatically through what he called 'ecologics', which he saw as a replacement of economics for a planetary civilization. In 1982, he stated, "we are entering a period when the planetary biosphere will become the major subject of our civilization's concerns."[88] Thirty-five years later, that period is much closer, as Bruno Latour argues.[89] Berg himself offered scant detail as to what ecologics actually entailed beyond any other existing deep-ecological or ecotopian ideas.

Discussion and Critique

Two recurring themes emerge from the various deep-ecological and ecotopian writings. One relates to the physical structure and organization of society and the other to human behavior.

[87] Berg (1978, p. 42).
[88] Berg (1982).
[89] Latour (2018).

First, most of the authors, with the exception of Shepard, made an assumption that a state of low per-capita material resource intensity is contingent upon an organization of society into small, localized communities exhibiting a degree of self-sufficiency. Little effort appears to have been made to examine this assumption critically. Second, the ecotopians appealed to people's cooperative nature to argue for the establishment of societal norms based on harmony with one's fellows and with nature, as opposed to the runaway competition so prevalent within modern society. At the risk of misinterpreting the message, it seems as if the ecotopians wished to deny or somehow assume away people's competitive urges. Those urges are real: they are part of our biology. Since time immemorial, people have competed with one another for status. Of course, many measures of status exist, only some of them material or financial. Erik Dammann recounted a story of living among Polynesians on remote Svaii Island in Western Samoa where people competed, if at all, to be sociable.[90] Yet for every such place, one can readily find a counterexample of a people who, in spite of having lived more or less in harmony with nature for millennia, compete with one another for material status on a regular basis. The Haida people on Haida Gwaii (formerly the Queen Charlotte Islands) are an example.

Ecotopias, like any utopias, are articulations of a state already attained in the imagination, without necessarily incorporating a vision for how to attain such a state. Although the material differences among people in a planetary economy would necessarily be significantly less than they are today, this does not automatically imply that competition among people, in material or any other terms, would be less intense. There is no clear basis for an assumption that it would.

Goldsmith et al. articulated not only a kind of ecotopia in their four conditions for a stable society, but also an outline of some programs for attaining such a society. Yet conditions three and four, namely that population must be stabilized and people must enjoy living under the first three conditions, do not appear to be prerequisites for the stability described by conditions one and two. A biophysically stable society can have a growing population, provided that per-capita resource intensity is declining at the same rate. This may present a significant challenge, because it necessitates sustaining increases in material efficiency over time, although it is theoretically possible. Also, in a pinch, we may not like the conditions we impose

[90] Dammann (1979, p. 140).

upon ourselves, but if the alternative is recognized as far worse than we would have to put up with them. Enjoying living a certain way is not a normative requirement for a materially stable economy. It may be sufficient but it is not necessary, and it is not included in the normative requirements described below.

Future Market Economies

Surprisingly little has been written about possible future economies from a market-centric perspective. Writers' dreams about possible future societies seem to have been dominated by themes of equality, social justice and harmony with nature, rather than by the cut-and-thrust of the competitive market environment. One of the few exceptions was written by Stephan Schmidheiny and Federico Zorraquín. Following the 1994 formation of the World Business Council for Sustainable Development (WBCSD), Schmidheiny, an industrialist, and Zorraquín, a business leader, asked the WCBSD's 123 national members whether the financial markets are "a force for sustainable human progress or ... an impediment against it".[91] They examined the workings of the financial markets and the roles of key sectors within those markets—investment, banking, accounting and so on—to assess where and how financial markets could be altered to meet the goal in question. Like the WCBSD, Schmidheiny and Zorraquín took as their starting-point the Brundtland definition of sustainable development which is anthropocentric rather than biophysical.[92]

Notwithstanding this caveat, the authors painted a persuasive picture of a future, 'eco-efficient' market economy:

> "Imagine for a moment that the majority of environmental costs are internalized so that they are borne by companies and passed along to consumers. Imagine that governments make greater use of economic instruments to reward continuously companies that are becoming increasingly more eco-efficient, while punishing those that are not. Imagine that growing numbers of governments revise national accounting systems to reflect environmental damage and resource depletion accurately. Finally, imagine tax shifts toward the discouragement of pollution and resource overuse:
>
> Then it is not hard to imagine that the balance sheets of companies would also change strikingly [...] The financial markets would change the

[91] Schmidheiny and Zorraquín (1996, p. xi).

[92] World Commission on Environment and Development (1987).

> basis on which they decide to invest in, lend to and insure companies. Financial markets would not have to care about 'the environment'; ... if a company were financially successful in a world of internalized environmental costs ... then it must also be eco-efficient.[93]

Admittedly, this vision statement takes an 'inside looking out' perspective, in that it sees the internalization of environmental costs as the key to eco-efficiency. In spite of this limitation, the authors made a number of credible proposals, which will be revisited in the following chapters. Few if any of these proposals have yet been adopted.

References

Baker-Smith, D. (2012). *Introduction to Thomas More's Utopia*. London: Penguin Classics.

Berg, P. (1977). *Reinhabiting California*. In C. Glotfelty & E. Quesnel (Eds.), *The Biosphere and the Bioregion: Essential writings of Peter Berg* (pp. 35–40). London and New York: Routledge.

Berg, P. (1978). Globalists Versus Planetarians: An Interview of Peter Berg by Michael Helm. In C. Glotfelty & E. Quesnel (Eds.), *The Biosphere and the Bioregion: Essential writings of Peter Berg* (pp. 41–51). London and New York: Routledge.

Berg, P. (1982). *Bioregion and Human Location*. Transcript of a talk given at the University of North Carolina, available at http://www.haven.net/deep/council/berg.htm.

Berman, M. (1981). *The Re-Enchantment of the World*. London: Cornell University Press.

Boehm, C. (1999). *Hierarchy in the Forest: The Evolution of Egalitarian Behavior*. Cambridge: Harvard University Press.

Boulding, K. E. (1966). The Economics of the Coming Spaceship Earth. In H. Jarrett (Ed.), *Environmental Quality in a Growing Economy* (pp. 3–14). Baltimore, MD: John Hopkins Press.

Dammann, E. (1979). *The Future in Our Hands*. Oxford: Pergamon Press.

Dawkins, R. (1976). *The Selfish Gene*. Oxford: Oxford University Press.

Descartes, R. (1637). *Discourse on Method* (Laurence J. Lafleur, Trans.). The Liberal Arts Press, Indianapolis, USA (1950).

Devall, B., & Sessions, G. (1985). *Deep Ecology: Living as If Nature Mattered*. Salt Lake City, UT: Gibbs M. Smith, Inc..

[93] Schmidheiny and Zorraquín (1996, p. 27).

Diamond, J. (1966). Zoological Classification System of a Primitive People. *Science, 151*(3714), 1102–1104.

Diamond, J. (1991). *The Rise and Fall of the Third Chimpanzee*. London: Hutchinson Radius.

Drengson, A. (1995). *The Practice of Technology: Exploring Technology, Ecophilosophy and Spiritual Disciplines for Vital Links*. Albany, NY: State University of New York Press.

Erdal, D., & Whiten, A. (1996). Egalitarianism and Machiavellian Intelligence in Human Evolution. In P. Mellars & K. Gibson (Eds.), *Modelling the Early Human Mind*. Cambridge: McDonald Institute Monographs.

Evernden, N. (1978). Beyond Ecology. *North American Review, 263*(4), 16–20.

Frank, R. H. (1999). *Luxury Fever: Why Money Fails to Satisfy in an Era of Excess*. Simon & Schuster, New York: The Free Press.

Goldsmith, E., et al. (1972). *Blueprint for Survival*. New York: Signet, New American Library, Inc..

Hanna, S. S., & Jentoft, S. (1996). Human Use of the Natural Environment. In S. S. Hanna, C. Folke, & K.-G. Mäler (Eds.), *Rights to Nature*. Washington DC: Island Press.

Honderich, T. (Ed.). (1995). *The Oxford Companion to Philosophy*. Oxford: Oxford University Press.

LaChapelle, D. (1995). Ritual—the Pattern That Connects. In G. Sessions (Ed.), *Deep Ecology for the 21st Century: Readings on the Philosophy and Practice of the New Environmentalism* (pp. 57–63). Boston and London: Shambhala.

Latour, B. (1987). *Science in Action: How to Follow Scientists and Engineers through Society*. Cambridge: Harvard University Press.

Latour, B. (2004). *Politics of Nature: How to Bring the Sciences into Democracy* (Catherine Porter, Trans.). Harvard University Press.

Latour, B. (2018). *Down to Earth: Politics in the New Climatic Regime*. Cambridge, UK: Polity Press.

Latour, B., & Woolgar, S. (1986). *Laboratory Life: The Construction of Scientific Facts*. Princeton: Princeton University Press.

Marglin, S. (2008). *The Dismal Science: How thinking Like an Economist Undermines Community*. Cambridge: Harvard University Press.

Martyr d'Anghiera, P. (1511). *The First Decade of the Ocean: On the New World*. In Eatough, G. (ed. & trans., 1998) *Selections from Peter Martyr*, Repertorium Columbianum, Vol. 5 (Brepols, Turnhout, Belgium).

McCloskey, D. N. (1998). *The Rhetoric of Economics* (2nd ed.). Madison, WI: University of Wisconsin Press.

Meadows, D. H., et al. (1972). *The Limits to Growth: A Report for the Club of Rome's Project on the Predicament of Mankind*. New York: Universe Books.

More, T. (1516). *Utopia*. Thierry Martens, Louvain (2012 Translation with Introduction by Dominic Baker-Smith, Penguin Classics, London).

Murison Smith, F. D. (2019). *Economics of a Crowded Planet.* New York: Palgrave Macmillan.
Naess, A. (1973). The Shallow and the Deep, Long-range Ecology Movement: A Summary. *Inquiry, 16*, 95–100.
Naess, A. (1989). *Ecology, Community and Lifestyle* (David Rothenberg, Trans. and Ed). Cambridge University Press.
Odum, H. T. (1971). *Environment, Power & Society.* New York: Wiley-Interscience.
Odum, H. T. (1994). *Ecological and General Systems: An Introduction to Systems Ecology*, Revised Ed. University Press of Colorado.
Polanyi, M. (1946). *Science, Faith and Society.* Oxford: Oxford University Press.
Polanyi, M. (1958). *Personal Knowledge.* Chicago: Chicago University Press.
Rodman, J. (1995). Four Forms of Ecological Consciousness Reconsidered. In A. Drengson & Y. Inoue (Eds.), *The Deep Ecology Movement: An Introductory Anthology* (pp. 242–256). Berkeley, CA: North Atlantic Books.
Rothenberg, D. (1989). Ecosophy T: From intuition to system. In Naess, A., *Ecology, Community and Lifestyle* (D. Rothenberg, Trans. and Ed.), pp. 1–22. Cambridge University Press.
Sachs, W. (1995). Global Ecology and the Shadow of 'Development'. In G. Sessions (Ed.), *Deep Ecology for the 21st Century: Readings on the Philosophy and Practice of the New Environmentalism* (pp. 428–444). Boston and London: Shambhala.
Schmidheiny, S., & Zorraquín, F. J. (1996). *Financing Change: The Financial Community, Eco-Efficiency and Sustainable Development. With World Business Council for Sustainable Development.* Cambridge, MA: MIT Press.
Seneca, L. A.. (c. 65 AD). *Ad Lucilium Epistulae Morales* (*Moral Letters to Lucilius*), no. 60. English translation by Richard M. Gummere, William Heinemann, London, 1925. Online at https://en.wikisource.org/wiki/Moral_letters_to_Lucilius.
Shepard, P. (1973). *The Tender Carnivore and the Sacred Game.* New York: Scribner's.
Snyder, G. (1969). Four 'Changes'. In G. Debell (Ed.), *Environmental Handbook* (pp. 323–333). New York: Ballantine.
Wilkinson, R. G. (2005). *The Impact of Inequality.* New York: New Press.
Wilkinson, R. G., & Pickett, K. (2009). *The Spirit Level: Why More Equal Societies Almost Always Do Better.* London: Allen Lane.
Woodburn, J. (1982). Egalitarian Societies. *Man, 17*, 431–451.
World Commission on Environment and Development. (1987). *Our Common Future.* Oxford: Oxford University Press.

PART II

Requirements for a Planetary Economy

1.1 Overview of Part II

Where Part I discussed how and why the economy grew so large as to put its long-term future at risk, Part II acts like a requirements document. Since the future coevolution of the economy and nature is uncertain, there is little point in trying to predict it, not least because the timescales involved are so long. However, it is possible to articulate certain requirements that may have to be met for the system to achieve long-term stability, a process similar to the familiar requirements-gathering process in technology and business. The requirements would bound the event-space within which any actual coevolution might proceed.[1]

Forming the bulk of the book, Part II is divided into 13 chapters, whose ordering reflects a cycle of economic development, mentioned in the introduction and depicted in Fig. 3.1. Social norms shape our institutions which, in turn, frame policies, these leading to the development of economic controls. Controls influence agents' behavior within the economy. As behaviors become conditioned, they become normal, and so social norms evolve.

Chapter 3 addresses normative requirements in relation to nature and the economy. Chapters 4 and 5 describe institutional and governance requirements, outlining a hierarchy of participatory governance. Chapter 6 describes the corporate and financial institutions of the future as

[1] The requirements overlap to some extent with John Fullerton's *Regenerative Economics* (Lovins et al. 2018, ch. 3) and Kate Raworth's *Doughnut Economics* (Raworth 2017).

primarily generative rather than exploitative enterprises. Chapter 7 articulates policymaking requirements, and Chap. 8 outlines a collection of policies. Chapter 9 discusses financial and accounting standards. Chapter 10 describes the roles of money and finance on a crowded planet, following which Chap. 11 outlines a set of monetary and financial reforms. Chapters 12, 13, 14, and 15 propose certain economic controls for a stable, prosperous economy, encompassing currency mechanisms, fees, taxation, subsidies and incentives. The behaviors entrained by these new instruments would give rise to new economic norms over time, which in turn would influence new institutions and policies. This is the cycle of economic evolution.

The chapters of Part II each cover subjects that are vast in their own right. Their reference to existing literature inevitably must be selective, their analysis cursory and their proposals preliminary. The requirements they generate should be taken as hypotheses, subject to testing and refinement, as would be routine within any development project.

References

Lovins, L. H. et al. (2018). *A Finer Future: Creating an Economy in Service to Life*. Canada, BC: New Society Publishers.

Raworth, K. (2017). *Doughnut Economics: Seven Ways to Think Like a 21st-Century Economist*. White River Junction, VT: Chelsea Green Publishing.

CHAPTER 3

Normative Requirements

The Cartesian dualism gave rise within Western society to a culture of modernity that has shaped economic policymaking, propelled the positivist program of economics, and pervaded the very structure and functioning of the economy itself. The first, and possibly most difficult, step in the transformation of the economy is to extricate society from modernity's grip. "The ultimate test of a set of economic ideas", wrote J.K. Galbraith, "is whether it illuminates the anxieties of the time."[1] People need to see a compelling reason to embrace something new—a reason compelling enough to change how they see themselves. "If you have been raised with an instrumental view of life", wrote Morris Berman, "you will relate to your social and natural environment in that way." In the event your premises are not validated, "you will probably not abandon your worldview, but classify the negative response … as an anomaly. In this way you remove the threat to your view of reality, which is also your character structure."[2]

This chapter lays out normative requirements relating to nature and to the economy. They are ways people would have to perceive nature and their relation to it, and perceive economic activity and their roles in it, in order for a stable, prosperous economy to emerge. These requirements are foundational, in the sense that everything else in such a future society would be built on them. However, as the cycle in Fig. 3.1 illustrates, their

[1] Galbraith (1973).
[2] Berman (1981, p. 217).

F. Murison Smith, *A Planetary Economy*,
https://doi.org/10.1007/978-3-030-49296-0_3

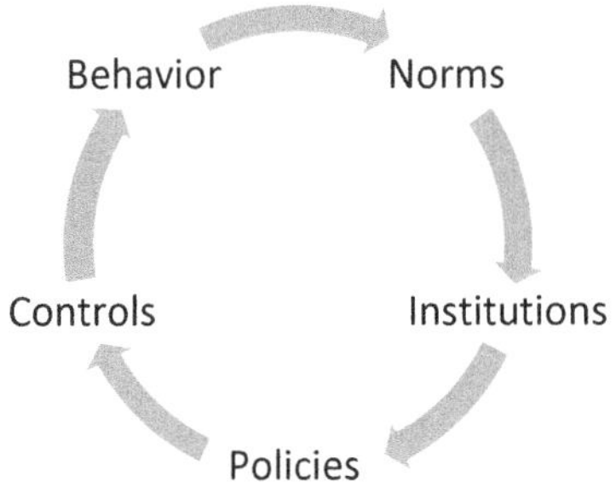

Fig. 3.1 Cycle of social norms to institutions, policies, controls and behavior, back to norms. Source: Author's creation

emergence will be iterative, nudged along by changes to economic controls, these in turn influenced by new policies and institutions. Since we must start somewhere, it will be with how people see the world.

Perceptions of Nature

What ways of thinking about nature would help to improve the economy's alignment with it? If, as Berman, Naess and others argued, the Cartesian worldview impairs our thinking about people's relationship with nature, then not only might a future epistemology prompt different kinds of questions, but also, asking different kinds of questions might represent the first steps toward a future epistemology.

In asking the question 'how would a materially stable, prosperous economy look if one started with a blank sheet of paper?', a value judgement is made that this question is worth asking. That value judgement—a normative statement—is motivated by an epistemology of observing the economy from the outside looking in: from nature's perspective, rather than from an anthropocentric one. The Cartesian dualism is purposely kept alive in this epistemology because the 'economy' and 'nature' are identified as distinct entities. The question could not be asked in this form if the Cartesian dualism were collapsed because then there would be no *conceptual* differentiation between the two systems. In the *real* world, the boundary between the two systems is fuzzy and indeterminate, even though, beyond the boundary, the economy and nature are indisputably distinct. The Cartesian distinction serves the purpose of *conceptually* demarcating a boundary for the purpose of asking the question. If or when the economy actually became better aligned with nature, such a crisp conceptual distinction might cease to be useful. Our norms and attitudes by that time would have become more integrative or holistic, less Cartesian. Our language might have evolved accordingly. "In a post-Cartesian mode of thinking", argued Berman, "'in here' and 'out there' will cease to be separate categories and thus ... will cease to make sense."[3]

Yet there will be no single 'post-Cartesian' epistemology, as argued above. We employ a variety of epistemologies and ontologies for different circumstances. The path toward understanding how to live within planetary constraints will be through an *integrative* epistemology. Normative requirements established for a planetarian economy must employ a kind of epistemic toolkit, along with a guide on when and how to use the tools.

In their book, *The Tree of Knowledge*, Maturana and Varela extend their discussion of structural coupling among autopoietic systems into epistemology and cognition. They develop a theory—founded in the science of biological systems—of knowledge as the structural coupling of an individual to their social environment, through language. To Maturana and Varela, we are both observers and describers of our world, balancing between objectivism on the one hand and subjectivism, or idealism, on the other. There is no external reference point, as Descartes would argue, nor is the world around us a figment of our imaginations. Our biological history, including our large brains and our tendency to form into social groups, traits shared with other apes, gave us the capacity for language. A circular iteration then unfolded in which language and society coevolved into ever more complex forms—an example of structural coupling.

Varela calls the theory an *enactive* epistemology, in the sense that "what is known is brought forth".[4] An enactive epistemology is circular but nonrepeating, like a spiral through time. As autonomous living systems, we humans evolve, in a cognitive sense, through cycles of effective action, arising from our immersion in networks of interactions.[5] The process is iterative and recursive. Humanity's social and political evolution in response to signals from nature is, and will continue to be, an evolution in enactive epistemologies, generating an evolution in social technologies.

Whereas Descartes' axiom *cogito ergo sum* (I think therefore I am) led him to conclude that the entity doing the thinking, the subject, is distinct and separable from that which is being pondered, the object, what he missed is that the observer is ineluctably part of the system they observe,

[3] Berman (1981, p. 149).
[4] Afterword to Maturana and Varela (1998).
[5] Maturana and Varela (1998, p. 241).

and is structurally coupled to it. To Maturana and Varela, the acknowledgement of an ability to think provides a basis from which to engage with others through language to develop systematic explanations of thinking. "The world everyone sees is not *the* world but *a* world which we bring forth with others", they wrote.[6] "Knowing how we know does not consist of a linear explanation that begins with a solid starting point and develops to completion as everything becomes explained. [...] On the contrary, [cognition] constitutes the starting point that enables us to explain it scientifically."[7] Nicholas Georgescu-Roegen made a similar point: "in being in the actual world, there is an order which ... cannot be represented by an analytical (strictly causal) formula."[8] The perception of nature perhaps best-suited to society's long-term persistence on Earth is as a participant in it rather than as a dispassionate, disconnected observer.

Normative Requirements Regarding Nature

Norms, then, are not plucked out of thin air: they have a basis in cognition, in how people perceive and understand the world. This understanding becomes reflected in behavior which, at a social level, becomes tradition. A society's norms—its principles, standards or rules—derive from an understanding of its traditions. In many Western societies, for instance, there exists a tradition of free speech. Norms and institutions exist within these societies defending the right to it.

The present analysis, looking as it does into the future, does not have any particular prior tradition to draw upon. The normative statements articulated here are simply requirements: principles, standards or rules that may need to exist for a stable, prosperous economy. The norms, of course, would imply certain social traditions, some of which have yet to arise. Numerous traditions in 'modern' society, carried over from the modernist mindset of the past two centuries, may be at odds with long-term economic and societal stability. By articulating normative requirements, the intent is to stimulate reflection upon some long-held modernist traditions, thereby potentially catalyzing a process of normative evolution. From that would naturally follow an identification of required institutional and political changes.

[6] Maturana and Varela (1998, p. 245).
[7] Maturana and Varela (1998, p. 244).
[8] Georgescu-Roegen (1971, p. 59).

If the mind plays a role in shaping what it perceives, as Emmanuel Kant postulated,[9] then a modernist may perceive a rainforest being chopped down as 'progress' because it results in economic production which generates financial wealth. To anyone steeped in a naturalistic or holistic existence, such as a member of an indigenous forest people, such a clear-cutting would be perceived as a loss of part of the self: a senseless personal tragedy. In framing normative requirements regarding nature, it therefore becomes necessary to outline ways the world would have to be perceived in order to support a planetary economy. This is the 'normative space' within which a planetary economy may take different forms, all of them in alignment with natural processes. A variety of cultures would exist within this space. They may evolve over time due to social-selective pressures; however, if the economy and society were in stable alignment with nature then there would exist limits beyond which that evolution would not go, limits imposed either consciously or unconsciously by individuals. "We must become the agent for something the Earth has not seen before", writes Robert Frank, "a biosphere that is awake to itself and can act for its future."[10] The culture then would exhibit a certain transgenerational stability presently absent.[11]

Table 3.1 summarizes the requirements. It describes the kinds of attitudes toward nature that might prevail within an economy and society well aligned with natural processes.

It is hardly surprising that these norms appear similar to pre-modern or indigenous norms. Yet they say nothing about how society would be structured, in a physical, political or cultural way, nor do they presuppose any particular scenario for reaching a stable, prosperous future. They simply lay the foundations for other requirements to follow. For a planetary economy to emerge, these norms would have to be recognized and reflected in conscious decision-making. Through repetition and use over a long time, they would become internalized. We in modern societies will have to unlearn many of the things about humanity's relationship with nature to which we became conditioned during and after the industrial

[9] Berman (1981, p. 317).

[10] Frank (2018).

[11] Maturana and Varela (1998, p. 201) define cultural behavior as "the transgenerational stability of behavioral patterns ontogenically acquired [i.e., not genetically inherited] in the communicative dynamics of a social environment". Through social interaction, common ideas are formed about the world we inhabit—including about society—and, if these ideas are more or less stable across generations, they give rise to cultural behavior.

Table 3.1 Normative requirements for a planetary economy: attitudes toward nature

Norm	*Description*
Natural capacity is finite	A society having a stable, planetary economy will exhibit an attitude toward natural capacity recognizing it as finite. The notion of natural capacity being unlimited will be dismissed as incomprehensible or as a relic of a distant, uncrowded past.
Natural capacity is society's life-support	Human technology is not a replacement for natural capacity but a means to maximize it. Nature provides services to support human life and human society, for which no substitutes need be created within the economy. By maximizing natural capacity, society maximizes its own persistence.
Society and all people within it are embedded within nature	If the artifacts of society were lifted away for a moment, the individual would see only nature beyond. The relationship between people and nature has always been present, yet over the centuries it has become mediated and somewhat obscured by the artificial environment. People within a planetarian society will recognize that all individuals remain embedded within Earth's natural systems and there is no escape, nor is there any need for one.
Natural stability is the foundation of economic stability	A state of high natural capacity globally more or less ensures the stability of natural systems over socially and economically meaningful timescales. These are timescales over which economic plans may be made and put into action: years to decades. Whether natural systems are stable over many thousands or millions of years is irrelevant to a planetary economy. Although economic instability could arise for social reasons, it would be trivial compared to economic instability arising from natural instability.
Preservation and maintenance of natural capacity is what people do	Because human beings have power over natural systems on Earth, it is normal for people to care for natural capacity. No other species on Earth has the ability to do this. Individuals of other species, with little power over nature, tend to exploit natural capacity unthinkingly. For a powerful, sentient species, such as human beings, it is in their best interests to maintain natural capacity.

Source: Author's creation

revolution, in the process relearning other conditioning that was with us for thousands of years before. Looking back as if from a future holistic society, wrote Berman, the precepts of 'modernism', such as mastery over nature, might "make school children giggle, and produce blank, uncomprehending stares in adults".[12] One lives in hope.

[12] Berman (1981, p. 255).

Attitudes Toward the Economy

In a planetary economy, what would people think the economy is for? As Chap. 1 noted, the conventional economic refrain over the past two centuries has been that it provides for individual wants. When the economy was materially inconsequential, an individualistic purpose for economic activity posed little risk to the economy, even if, ethically, it was still irresponsible. On a crowded planet, however, the purpose of economic activity must become the persistence of the economy, by whatever means necessary.

The scenarios under which a model economy persists or collapses explored in *Economics of a Crowded Planet* inform a conceptual framework and requirements for an economics of the future. That economics, operating in the service of the future economy outlined here, would reflect many of the norms and attitudes toward the economy described below.

Individualism vs. Collectivism and the Satisfaction of Wants

The individualist orientation of conventional economics is a product of the modern age. It traces its origins to the writings of such Enlightenment philosophers as Thomas Hobbes and Jeremy Bentham, who asserted individualism as a repudiation of the static feudalism of the Middle Ages and as a way toward a more meritocratic society. Although the Hobbesian conception of 'every man for himself' was hardly egalitarian, it formed a philosophical justification for the arguments of the early economists that the greatest social wealth is produced when markets are allowed free rein and competition is encouraged.

This view of the economy was at odds with the prevailing mercantilism of the eighteenth and early nineteenth centuries, which was protectionist. It dovetailed neatly with the notion that the purpose of economic activity, particularly when mediated through markets, is the satisfaction of individual wants. The more wants are satisfied, the more wealth is created, thereby improving standards of living. This economic worldview served the emerging industrial age perfectly, as figure 1.11, depicting the social auto-catalysis of growth, illustrated.

Improving the lives of individuals by serving their desires for consumption was an economic paradigm that worked for a global economy is its earliest stages of exponential growth. A by-product of all the naked individualism, though, was a dramatic increase in economic inequality. Where before the industrial revolution only a handful of individuals within society were fabulously wealthy—in Europe, mostly royal families—the new meritocracy produced an explosion by the mid-nineteenth century, in the number of millionaires. Most of those who left the farms for the factories, on the other hand, underwent a diminution in quality of life. This inequality prompted some writers, such as Marx, to question the underlying individualism, instead proposing *collectivism*: the idea that the production of economic wealth should be in the service of all, not only those who happen to rise to the top.

In many of today's market economies, the word 'collectivism' has negative connotations because of its association with communism. Yet the two are not synonymous. Collectivism is a social philosophy whereas communism was a political program. Other political programs that are not communist but are nonetheless collective in purpose can be conceived. On our crowded planet, whose economy is locked into a material coevolution with nature, the collective imperative becomes very much apparent. The long-term persistence of the economy now places bounding conditions upon individual wants. The interests of the individual become subsumed within these bounding conditions. The reformist view would be that this bounding of wants is in the interests of the individual. However, a reformist justification really is a precursor to a transcendent one. In a materially stable, widely prosperous economy, it would be normal to live and act conscious of the systems in which we participate. Individuals, households and institutions would be expected, as a matter of course, to satisfy wants in the most materially efficient way possible, giving it hardly a second thought. The economics of such a society would reflect this norm.

The norm would emerge if, as this book proposes, increasing material efficiency of the economy propelled it toward long-term stability. The satisfaction of wants in a materially efficient way has implications for economic controls at work within markets, in that they would influence both the wants that can be satisfied and the way economic actors may satisfy them.

Attitudes Toward Value

Modernity and its positivist economics ignored intrinsic value, focusing only on instrumental value. Instrumental value misses the whole system whereas intrinsic value acknowledges it. Value, wrote Georg Simmel early in the twentieth century, is not ready-made but instead accrues to a desired object through our coveting of it and through the sacrifice needed to acquire it in exchange.[13] In an economy mediated by money, quantity triumphs over quality: it is "a world in which something must be measurable against something else to be deemed to have value".[14] Money, according to David Orrell and Roman Chlupatý, makes value measurable. However, it can lead to an obsession with number that obscures other forms of value in the real world.[15]

Although the concept of value is a wholly human construct, it can nonetheless be applied to non-human systems. If or when a life-bearing planet were discovered in a nearby solar system, there would be little dispute that ecosystems on that planet would have some form of value by virtue of their existence, even if human beings did not have the technology to reach the planet and derive instrumental value from it.

Intrinsic value captures the relationships among entities, as distinct from the behaviors of entities considered in isolation, such as self-maximization. If I have intrinsic value to myself and you have intrinsic value to yourself, and we are all at least directly or indirectly connected within some kind of system, then it follows logically that we all have intrinsic value. Furthermore, at the level of organization of the whole system, the system itself has intrinsic value. This value can be recognized independent of any instrumental or utilitarian value any one entity may recognize in any other entity.[16] As Freya Mathews observed, individuality and interrelatedness "are not mutually exclusive; … they entail each other".[17] It is the relations among entities that bring a system into existence, and allow it to persist over time.

The implication for an economy on a crowded planet is that, once the relationship between society and natural systems has stabilized to a degree, the human system may itself have become more self-regulating through the establishment various norms giving due accordance to intrinsic values: individual, social and natural. Economic incentives directing behavior are

[13] Dodd (2014, p. 28), citing Simmel (1900, 1907).

[14] Dodd (2014, p. 30).

[15] Orrell and Chlupatý (2016, p. 176).

[16] As Drengson (1995, pp. 89–90) describes, the philosopher A.N. Whitehead earlier arrived at a similar conclusion (Whitehead 1938). The proposition is similar to the 'Great Spirit' of Native Americans or the transpersonal 'Self' of Eastern philosophy, yet it is grounded in Western reasoning. It is concrete and not mystical in any way, not necessitating any kind of ritualized transcendence to comprehend.

[17] Mathews (1995, p. 127).

only the instrumental part of the solution. The ethic at play here calls for a re-evaluation of personal attitudes, away from 'this is mine and I can do what I like with it' toward 'we all share this and we all decide what to do with it'. As many primal peoples demonstrate, it is possible to appreciate the value of other beings through respect for them.[18]

Attitudes Toward the Market

In conventional economics, the market is an end in itself. If the purpose of economic activity is the satisfaction of individual wants, then the consumer is sovereign. The real world, of course, is not as cut and dried as this. The consumer is not actually sovereign, because of the effects of corporate monopolies and oligopolies, and of government regulation. It may be truer to say that, within policy circles in market economies, the market itself is sovereign. The mantra is to not curtail the 'free' operation of markets, conventional economic theory dictating that market freedom delivers the greatest aggregate economic benefits.

Even if this statement were true, the aggregate measure of economic benefit ignores its distribution. If fair distribution were given primacy then the market would no longer be sovereign or an end in itself, but instead a means to an end, the end in question being fair distribution. Similarly, if material discipline of the economy were given primacy then the market would become a means to that end as well. To the extent economic policy gives primacy to the market, it is a product of twentieth-century economic dogma.

Economics of a Crowded Planet makes a case for a 'principle of conservation of value', which states that, in a materially closed system such as planet Earth, value cannot be created or destroyed, only moved around. A prevailing assumption within capitalist economics is that an opportunity for profit is available to everyone. If all economic agents were to profit then this would mean that some agent outside the market or the economy would incur a loss, implying further that the economy is not an isolated system, as conventional economics assumes. The cost of all this economic profit-making would become externalized.

The entity incurring a loss, of course, is nature. If the degradation of natural capacity were fully accounted for within market or economic activity, then what appears to be economic profit might not be. The

[18] Drengson (1995, pp. 92–93).

thermodynamic basis for economic activity states that economic wealth may increase only through the effective use of solar flux not otherwise captured by natural systems, since almost all of nature's gift which is actually free in a *physical* sense, and therefore also in an *economic* sense, is solar flux.[19] Where the economy is materially small, its expropriation of natural capacity would be small, and consequently the long-term risk of viewing the market as an end in itself would be minimal. This is no longer the case when an economy is materially large. The collective impetus places the market in service of one or more directives, thereby bounding it and directing it.

Attitudes Toward Growth and Stability

Herman Daly referred to the prevailing culture of economic growth as 'growthmania'.[20] Economic growth undoubtedly is population-driven, yet it is also autocatalytic. The internal structure and organization of the modern market economy depends upon continuous growth.[21] Politically, growth is 'good', not least because it helps politicians become re-elected.

From outside the economy looking in, growth is acceptable provided it does not increase material intensity. While the population continues to grow, economic growth, in money terms, probably should be encouraged. Yet by becoming more materially efficient, both in terms of inputs and in terms of outputs, an economy would be characterized by increasing *innovation*, which would stimulate gross economic product. Economic growth within a materially disciplined economy would be achieved through improvements in material efficiency. It follows that the transition from today's material indiscipline toward a state of material discipline would represent a major opportunity for economic growth in money terms. However, assuming that actual costs from nature were well reflected within economic processes, there would come a point where the marginal cost to nature of economic activity would equal the marginal benefit to nature of increasing material discipline. This would be a point at which the economy would become materially aligned with its planetary container.

A state of material discipline implies a culture of 'enough', in terms of material wealth. The prevailing norm within a planetary economy would

[19] See Georgescu-Roegen (1971).

[20] Daly (1991, 1996).

[21] Douthwaite (2000).

be an acceptance of relative material constancy. For much of human history, the accumulation of wealth beyond subsistence has been largely an impossibility. Various technological developments over the past few millennia, however, enabled this wealth accumulation. Now the technological challenge is to decouple its creation from material intensity so as not to grow too much too quickly, flaming out in an economic collapse. Recognizing this need is a first step toward a culture of enough. Within the market economy of the early twenty-first century, a few seedlings are sprouting.[22] Above all, growth must be seen as a means to achieve stability. Long-term economic stability is dependent not upon continuing material growth but in fact upon the absence of it.

The idea that economic development may be steered toward a goal of long-term economic stability creates a definable future, in contrast to today, where the economy appears to lurch aimlessly from one market crisis to another. In a volatile system, the future is uncertain, and therefore more highly discounted than in a relatively stable system. It follows that a relatively smoother path of economic growth toward a state of alignment with natural processes presents a trajectory having a lower discount, at any point along its path, than a volatile one. In a materially stable economy, discount rates should be expected to remain close to zero. People would trust that the material balance between economy and nature would not change significantly over time. Confidence in the macrosphere thus being assumed, the discount rate principally would reflect social trust.

Normative Requirements Regarding the Economy

The requirements below are written as if from the perspective of an individual within a materially stable, broadly prosperous society, articulating the prevailing norms toward the economy:

- Societal persistence is primary:
  - On a crowded planet, it is normal to view society as a global whole. With nowhere to go beyond planet Earth, people are acutely aware of the need for society to persist within the home it has. This outlook in no way implies a monolithic society, nor does it necessarily imply a single global government. It does, however,

[22] See, for example, Kho (2015).

imply that regional or cultural differences or disagreements be subsumed within the realization that it is in the interests of all to support the interests of all.

  - Self-interest is acceptable or desirable, provided it does not come at the expense of societal persistence. An example of an undesirable self-interest is free-ridership: an assumption that others, not oneself, will act for societal persistence. On a crowded planet, undesirable self-interests are heavily penalized.

- Social and economic stability are the most desirable states to approach:

  - In a planetary economy, the growthmania of the twentieth century will have become a meaningless anachronism. Economic growth is a function of population growth and per-capita resource intensity. In a planetary economy, population growth is approaching, or has arrived at, a plateau.
  - The pressures for physical growth would continue if a significant portion of the population remained in absolute poverty; however, if the vast majority are above this level around the time the population reaches its plateau, then the critical question becomes whether Earth's natural capacity could support the total resource intensity demanded of it. If so, the economy should enter a period of relative stability. If not, then per-capita resource intensity would have to be reduced to avoid economic instability.
  - The economic opportunity lies in creating social value, or prosperity, out of physical shrinkage. Social value can be approximated in money terms but also has many other non-monetary components relating to the quality of life. It accrues not only to individuals but also to groups: households, communities and institutions. It should not be expected that a stable régime would be approached smoothly, nor that the economy would settle there. However, if the economy oscillated within a generally stable domain then it would be stochastically stable. In a society whose population is no longer growing by any appreciable amount, this stability is viewed as desirable; indeed, normal.

- Recognition of the unity of Earth leads to a recognition of intrinsic value:

- A planetarian society acts according to a recognition of intrinsic value as an essential counterpoint to instrumental value. Intrinsic value arises through simple existence. An entity having the characteristics of an autopoietic unity has an intrinsic value unto itself as well as an instrumental value to others. Consequently, living systems as a whole, being interrelations of functional unities, also have intrinsic value. The concept extends to encompass also nonliving natural systems, since living ones interact with and mutually influence them. The whole system of nature on Earth therefore has intrinsic value.
- Economic value denying the existence of intrinsic value, capturing only those components of instrumental value that can be measured monetarily, is understood to lead in the long run to economic instability, due to a failure of alignment with natural systems. On the other hand, economic value capturing both instrumental and intrinsic value is a precondition for long-term economic stability in alignment with natural processes.

• Material discipline in relation to nature is a foundation of economic stability:

- In such a society, it is normal to exercise degrees of material discipline that in today's economy would be accomplished by only the most zealous conservers. In a materially disciplined economy, the social and economic incentives to do so are well established, as is the infrastructure. Physical technologies have evolved to support and maintain material discipline. If any kind of perturbation upon the economy or society were to destabilize it in any way, the social norm of material discipline would help to drive the system back toward a relatively more stable state—provided, of course, that such a perturbation was not so large as to cause economic collapse.
- In this way, material discipline as a norm becomes a foundation for economic stability, providing a degree of resilience to social or economic perturbations.

• Economic growth is desirable provided it does not come at the expense of material discipline:

- A state of physical stability maintained by norms of material discipline by no means precludes economic growth, in terms of social value created. Innovations viewed as socially desirable stimulate economic growth within existing industries, or the emergence of new ones, just as they always have. The difference is that it will be normal for such growth to occur without any increase in material resource intensity of the economy. When developing an idea, an entrepreneur asks themselves two key questions: how will the idea improve—or not diminish—natural capacity, and how will the idea add social value—including, but not necessarily limited to, money value?

- Social and economic stability are predicated upon a moderate distribution of well-being:

  - Economics fundamentally is about human well-being. An economics of a crowded planet will recognize the multivariate nature of well-being as a nexus of personal safety, social standing, physical and emotional health, and financial security, among other measures, discarding the twentieth-century analytical convenience of a narrow money measure.
  - Large differences in well-being across the social system create conditions for instability, analogous to the effects of large differences in temperature and pressure in the weather. Also, like the weather, the distribution of well-being will never settle to an equilibrium. Yet the prevailing view within a planetary economy will be that a certain range of distributions of well-being is best suited for long-term economic stability. It is not a particularly even distribution nor is it a highly skewed distribution. It is uneven enough to perpetuate social and economic innovation and even enough to maintain social and economic stability. Further, if any single measure of well-being became skewed then it is reasonable to expect that the others would, up to a point, become relatively more even to compensate. Beyond that point, they would not be able to. All measures would have to remain within certain bounds at all times in order for long-term stability to be maintained.
  - In an economy where well-being has multiple measures, it will be normal for economic instruments, or non-economic social ones, to be applied to the distributions of the measures of well-being, so as to maintain them within desired bounds.

- Prosperity, not wealth, confers economic stability:

  - The foregoing requirement implies that collective prosperity becomes a more socially desirable goal than individual wealth. A prosperous society is one where individual wealth is attainable, provided it does not come at the expense of the poverty of others. In a planetary economy, there would exist a prevailing norm of maintaining prosperity, which means supporting the wealth of all while leaving open the opportunity for individual enrichment.

- The market is the most powerful driver of material discipline, yet it cannot act alone:

  - This normative requirement is also a hypothesis to be tested over the coming decades. The hypothesis states that the only economic organization able to transform the normative landscape over a short time (years to decades) is the market. Markets reflect actual social dynamics, which are themselves inherited in part from our evolutionary past. Over periods in which evolutionary forces typically act—thousands of years and above—it is possible that social selective pressures may cause human instinct to evolve toward a more cooperative state. If society were to attain a long-term state of relative stability in which the needs of all were more or less adequately met then it is reasonable to expect that prevailing norms within society would become less competitive and more cooperative. This change to the normative landscape could represent a first step toward a biologically adaptive change. It is important to maintain distinct timescales in perspective. Norms may change over years and decades whereas genetic changes to the human population would take centuries or millennia.
  - For this reason, the market as an economic institution is likely to persist for a very long time. It is an economic arena within which people or firms compete with one another. The changes to society it produces are deep-rooted in a way that no imposition from a command-and-control structure could be. Within a planetary economy, it will be viewed as normal and preferable for the dynamics of the market to be guided toward material discipline, and for the market to be a principal engine for this change. As Kenneth Boulding concluded, "there seems to be no absolute reason why a free-enterprise society could not survive its own successes in accu-

mulation", and therefore that "the task of preserving a free economy is not economically impossible." The present volume goes further to propose that the only kind of economy that could survive at a large material scale over the long run is a market-based one.

- A companion norm here is one that views the market as an instrument for change toward a social goal, rather than an end in itself. This norm is quite distinct from the market-fundamentalist paradigm of late-twentieth-century capitalism. Viewing the market as one of a suite of economic or social platforms recognizes the mutual reinforcement of platforms—markets, regulation, social welfare and distribution—as necessary for effecting rapid change through both cooperation and competition.

References

Berman, M. (1981). *The Re-Enchantment of the World*. London: Cornell University Press.

Daly, H. E. (1991). *Steady-State Economics* (2nd ed.). New York: John Wiley & Sons.

Daly, H. E. (1996). *Beyond Growth: The Economics of Sustainable Development*. Boston, MA: Beacon Press.

Dodd, N. (2014). *The Social Life of Money*. Princeton: Princeton University Press.

Douthwaite, R. (2000). *The Ecology of Money*. Schumacher Briefings, Book 4, UIT Cambridge Ltd., UK.

Drengson, A. (1995). Shifting Paradigms: From Technocrat to Planetary Person. In A. Drengson & Y. Inoue (Eds.), *The Deep Ecology Movement: An Introductory Anthology* (pp. 74–100). Berkeley, CA: North Atlantic Books.

Frank, A. (2018). Earth Will Survive. We May Not. *New York Times*, 12 June.

Galbraith, J. K. (1973). *Economics and the Public Purpose*. Boston, MA: Houghton Mifflin.

Georgescu-Roegen, N. (1971). *The Entropy Law and the Economic Process*. London: Harvard University Press.

Kho, J. (2015). Karma Marketing: Can the Idea of 'Branding Like a Buddhist' Change Corporate Mindsets? *The Guardian*, 4 June.

Mathews, F. (1995). Conservation and Self-realization: A Deep-Ecology Perspective. In A. Drengson & Y. Inoue (Eds.), *The Deep Ecology Movement: An Introductory Anthology* (pp. 124–135). Berkeley, CA: North Atlantic Books.

Maturana, H. R., & Varela, F. J. (1998). *The Tree of Knowledge: The Biological Roots of Human Understanding*, Revised Edition. Shambhala, Boston & London.
Orrell, D., & Chlupatý, R. (2016). *The Evolution of Money*. New York: Columbia University Press.
Simmel, G. (1900, 1907). *The Philosophy of Money* (original and revised editions). 2004 Third Enlarged Edition, Routledge, London.

CHAPTER 4

Institutional Challenges and Legal Institutions

From norms follow institutions and governance structures. Institutions are the social phenotypes of dominant ideas and norms within society. They exist or perish based on the norms they reflect. It is logical, then, to proceed from normative requirements to institutional ones. Institutions in turn create and enact policy.

Institutions emerge because they serve an agreed-upon purpose. They often persist because they consolidate power. Yet they also weaken and are replaced by other institutions as social norms evolve. Institutions exist for groups of people to make decisions. As such, they follow certain rules. A major part of the current disconnect between the economy and nature arises because the institutions and organizations currently regulating economic activity are not set up to recognize, account for, or adjust to natural processes. In order to assess the viability of the current economy within its planetary context, and the applicability of current economics to the economy, it becomes necessary to first outline how the institutional and governance structures of the economy limit its ability to adjust to planetary dynamics of its own accord, such as through the market.

Institutional Challenges

There are some apparently intractable questions to consider. The relative absence of accountability among large institutions with global influence, such as transnational corporations, presents a risk to a growing economy

F. Murison Smith, *A Planetary Economy*,
https://doi.org/10.1007/978-3-030-49296-0_4

having significant effects upon its planetary container. If the normative requirements regarding nature and the economy are to be fulfilled then it would necessitate institutions and governance with global reach, delivering accountability to all. A common path would have to be agreed upon in order to avoid excessive freeloading. Yet in the farther future, as the deep ecologists have suggested, the practice of material discipline, or 'ecosophy', as part of normal life might permit the emergence of much smaller, more localized institutions, since the underlying norms of living within natural capacity would be more or less the same wherever one went. Whether institutions became more global or, eventually perhaps, more local, both institutional environments would be significantly different from the prevailing nation-state one that has characterized the modern world.

Who Protects the Right to Life-support?

Howard Odum asked this question in his 1971 book, *Environment, Power and Society*. The question is intentionally anthropocentric. To Odum, writing not long after the apogee of modernity, the question was valid because he saw the only way toward the kind of economic self-regulation for harmony with nature was through some kind of overarching authority. The philosopher John Rodman called this the 'moral extensionist' phase of societal evolution, leading, he hoped, later toward an 'ecological sensibility'.[1]

Odum observed that access to breathable air, drinkable water, edible food and safety in movement are the most basic human rights of all, yet they are not constitutionally protected.[2] Biological diversity in particular is an essential component of the life-support system because the functional specialization and redundancy of mature ecosystems provides an abundance of chemical pathways to process and mineralize wastes. Echoing the contemporary emerging analysis of the 'commons', Odum noted that Earth's water and mineral cycles, the chemical complexity of organisms, and the planet's diverse ecosystems, which "organize and manage the earth's surface" are not individual property but part of the basic right of life-support. Current legal systems, he observed, are flawed in allowing

[1] Rodman (1995).

[2] Odum (1971, pp. 299–301). They are not protected under the US Constitution, but they are partially enshrined in the UN Declaration of Human Rights.

landowners to assume special rights over what is actually a public domain. "The situation is no longer one of seemingly unlimited reserves of environment," he wrote. "Now any action to foul the life-support system threatens the life of individuals everywhere."[3]

Odum proposed that laws should be revised to allow conviction of anyone who willfully releases waste that disorders the life-support system, even though in practice this would be very hard to prove. Another, more workable proposal was that economic development should incorporate the cost of interfacing with Earth's life-support systems. New ventures, he argued, should pay the public sector for the energy values "heretofore regarded as free ... such as water, air mineral reserves and ecosystem area". Development, in and of itself, is not a "God-given right", Odum argued, and that any right to land ownership "does not include the right to take [land] out of the life-support system without due process". In fact, "once the basic facts of planetary existence are understood, laws and court actions can protect individual rights, and through this principle incorporate nature and man into the same system of right".[4]

This is a powerful concept because it confirms that in a crowded world, in which the realities of planetary limits are recognized, what is good for nature is by definition good for humankind: therefore, it is logical to protect both under a common constitutional umbrella. Odum noted that the US would have to pass a constitutional amendment to accomplish this. Any global agreement would require new legal mechanisms. They would have to be enforceable, which raises the institutional question, who does the enforcing? If the US were to pass a constitutional amendment affirming the right to life-support, it at least would send a clear message to the international community that the US administration were serious about planetary health and would be prepared to do its part to ensure it. Such an amendment could pave the way for other global agreements to come later.

A constitutional right to life-support would, by definition, encompass the rate and quality of both human outputs and inputs to the economy. As Odum noted, "the right to inject fuel into the overheated world economy must be governed. The safety of the citizens of the world requires that oils be ... released at a rate consistent with long-range public safety. [...] Oil depletion allowances [over and above such a rate] clearly will be unconstitutional..." He further argued that "advertisements to accelerate power

[3] Odum (1971, pp. 298–299).
[4] Odum (1971, p. 300).

consumption, and proclamations ... for expansion are menacing to public safety".[5]

Odum modeled these proposals using his energy circuit language. He drew attention to a missing reward loop from the human economy to the life-support system which would serve to regulate human activity within planetary constraints.[6] The goal of an economics for a crowded planet essentially is to design and implement such a reward loop.

From Tragedy of the Commons to Triumph of the Commons

The 'tragedy of the commons' identified in 1968 by Garret Hardin,[7] and since refined into a more comprehensive institutional theory by others, is a tragedy because there exists no mechanism for those who take from the commons for their own gain to be brought to account. Herein lies the main institutional deficiency of modern society with respect to nature. Modern society has well-established institutions to prosecute those who commit theft of property where legal ownership of that property is clear. It has no structures or mechanisms to prosecute those who commit theft of natural capacity. Yet natural capacity is an asset in which everyone alive today and yet to be born has a share.

Similarly, modern society has no overarching structure or mechanism to reward those who enhance natural capacity. In principle, the same structure and mechanism could accomplish both. If it did so effectively, what once would have been a tragedy of the commons could become a triumph of the commons.

The present institutional deficiency has been acknowledged for decades. Writing in 1966, Adrian Kneese admitted that "we know almost nothing about how to design institutions which could systematically search out and implement measures for ... action with respect to [environmental problems]".[8] By the 1990s, a growing consensus had emerged among environmental and ecological economists of the importance of institutions in the protection and maintenance of ecological resilience and carry capacity.[9] The prevailing norms and attitudes that could propel such institutional

[5] Odum (1971, p. 300).
[6] Odum (1971) fig. 10-9, p. 301.
[7] Hardin (1968).
[8] Kneese (1966, p. 86).
[9] For example, Arrow et al. (1995); Hanna et al. (1996).

evolution were, by the 1990s, beginning to shift, although they have not yet translated into significant institutional change.

C.S. Holling and Steven Sanderson highlighted temporal disconnects between institutions established to manage natural resources and the resources themselves. "Until modern human institutions are built on ecological dynamism, and [are] designed to flex with natural variability", they wrote, "their principal impact will be to impede nature, not to sustain it."[10] Holling and Sanderson identified how management strategies tend to focus on a narrower set of variables than are likely in operation within an ecosystem at any one time. Institutions exist in large part for the purpose of reducing uncertainty, so management strategies that limit the time horizon to the relatively short term and manage for a reduced set of variables likely will ignore other ecological dynamics that later could cause surprises.[11] Management myopia, the authors argued, arises because of the radical uncertainty about the future. It is hard to justify imposing costs on today's society based on presumptions about a future society that exists in the abstract.[12] Institutional design, the authors argued, should foster opportunism, innovation and entrepreneurship, all the hallmarks of what Joseph Schumpeter called 'creative destruction'.[13] In this way, they argued, institutional evolution would follow some of the same dynamics as the evolution of the natural systems whose useful products institutions are set up to manage: exploitation, growth, consolidation, perturbation, disintegration, release and reorganization.

As various authors have concluded, there exists an appropriateness of scale of institutions at different levels of organization. Yet several kinds of groupings can be identified, which cut across one another: demographic, political, geographic, geopolitical and ecological, to name a few. As Kenneth Boulding remarked in relation to trade, "There are ... aggregates of trade which have enough utility to serve as objects of discourse, but they seldom coincide with national boundaries."[14] Joseph Stiglitz expressed a similar sentiment 40 years later: "global public goods and externalities ... need to be dealt with at the global level [whereas] local public goods and externalities ... should be dealt with at the local level."[15] Both Stiglitz and

[10] Holling and Sanderson (1996, p. 79).
[11] Ostrom (1990); Sanderson (1994).
[12] Holling and Sanderson (1996, p. 77).
[13] Schumpeter (1950).
[14] Boulding (1962, p. 188).
[15] Stiglitz (2003, pp. 313–314).

Steven Marglin noted that forces operating at these different levels may work against one another. "If we reject the primacy of the [nation-state]", wrote Marglin, "the claims of local communities become more legitimate. Starting from a different set of assumptions, we may well discover a different set of priorities..."[16]

For a global economy having significant interactions with natural systems, global regulation of some kind becomes necessary to ensure its long-term persistence. This does not mean a 'controlling authority', in the sense of Adolph Lowe's concept of political economics.[17] Without agreed-upon jurisdiction and legal standing, such an institution would have neither control nor authority. It follows, then, that global governance will emerge, and can only emerge, from the ground up.

Elinor Ostrom, in critiquing and refining Hardin's work, proposed an eight-point framework for the governance of common-pool institutions, from local to national scales. The principles the framework articulates—paraphrased here—also apply more broadly to Earth system governance:

1. Clear definition of rights
2. Spatial and temporal congruence of rules to natural capacity
3. Broad stakeholder participation in decision-making
4. Monitoring, verification and accountability
5. Graduated sanctions for violations
6. Conflict resolution mechanisms
7. Rights to organize fitting to the specific institutional level, rather than organization being imposed from above
8. Organization and enforcement through nested institutional levels.[18]

Institutions enabling the economy's alignment with nature may take many forms, organizing themselves geographically, demographically or bioregionally, for instance. In today's economy, these groupings often create mismatches in scales. A collection of smaller socioregions, such as local communities, may interact with a larger bioregion partially or fully containing them; or, alternately, several or many bioregions could be contained within a large socioregion, such as a nation. Interactions among these regions can create common-access or common-use problems, which

[16] Marglin (2008, p. 263).
[17] Lowe (1965), cited by Georgescu-Roegen (1971, pp. 347–348).
[18] Ostrom (1990), summarized in Costanza et al. (2015, p. 64).

arise from multiple small social groups attempting to use a natural resource extending across their boundaries.[19] Institutions have been suggested to enable constituent socioregions to manage their interactions with a containing bioregion in which they are vested. Ostrom and Schlager suggested a federated structure, for example. Locally-devised governance systems, they argued, regulate local resources better than an external authority, although they cannot easily regulate spill-over effects to other localities. A federated structure would mitigate this weakness, in that larger-scale issues could be ceded to an umbrella organization of which they are all members.[20] The intent would be to create a kind of meta-socioregion whose geography overlaps approximately well with the bioregion in question. The same line of thinking, argued Margaret McKean, could be applied to very large bioregions or even global systems, those she termed “physically indivisible or unbounded”.[21]

Where several or many bioregions are partially or wholly contained within a large socioregion, an inverse problem can potentially arise: one of shared access. It arises where people or institutions from outside a bioregion seek to avail themselves of its capacity, resulting in competing forms of demand upon it. Tourism, for instance, supports the conservation of a region's biota: people come to view a bioregion's top predators and other 'charismatic' species. Yet those very same species can be perceived as a threat to livestock or crops by ranchers and farmers.[22]

An institution representing a single bioregion would have to not only regulate the activities of its constituent socioregions but also accommodate outside interests or, in the case of resource imperialism, protect it from them. The problem becomes more knotty when considering fugitive resources, such as pelagic fish and migratory animals. They are 'unbounded' in essentially the same sense as the interconnected, global systems of atmosphere, water, biota and soils. Whereas at the regional level there may exist a mismatch between natural boundaries and social ones, at the global level there is no social institution to begin with. As of the early twenty-first century, planet Earth has an institutional vacuum.

A planetary economy would have figured out how to fill this vacuum with an institutional superstructure representing global natural capacity.

[19] For example, Capistrano et al. (1997); Toledo (1997).
[20] Ostrom and Schlager (1996, pp. 146–148).
[21] McKean (1996, pp. 228–231).
[22] Ashley and Barnes (1997); Adger (1997).

This is perhaps the most significant and complex challenge facing a global society; yet, from a management standpoint, it is theoretically very simple: you only have to manage the economy in relation to one other entity. Such an institution would not replace the nation-state; it would necessitate the participation of nation-states. An increasing number of institutional scholars recognize the importance of 'vertical interplay' among institutions at different scales in governing planetary-level problems.[23] In a connected world, increasing transnational pressure may be brought to bear upon national governments failing to act with requisite urgency. Although pressure may come in part from transnational corporations or from transnational interest groups, ultimately most of it would come from people.

What should be the socioregional components of such an institution? We are so used to dividing the world up into the little colored shapes on a political map that we might overlook the possibility of other socioregional compositions potentially being more effective. As the world makes the long transition toward a biophysically aligned economy, this is one reason why the nation-state as we recognize it today might partially dissolve, becoming a more open, less rigidly defined institution. Institutions at a variety of scales—local, regional and national—all may be stakeholders and all could play a role in a global governance structure for the economy's use of natural capacity. While national governments participate in the establishment and implementation of top-down goals, such as for reducing emissions, local governments have an ability to employ different approaches to encourage residents and businesses in material efficiency, irrespective of whatever national targets may be in effect. Large cities are especially attractive for programs to encourage material efficiency because local authorities have jurisdiction over many people at once. When residents of a large urban area pull in the same direction, whole supply chains may be affected, having downstream and upstream effects in industrial and other commercial practices. These effects have the potential to be more significant than those imposed from above by regulation.

[23] Mitchell (2013) and references therein. A pioneer among these writers is Oran Young, whose body of work addresses the institutional challenges of the onset of the Anthropocene, emphasizing governance over government and institutional fit with governance problems. Young also was instrumental in the establishment of the Earth System Governance Project, www.earthsystemgovernance.org.

The two approaches are, of course, complementary. It is neither necessary nor desirable that a strict, nested hierarchy of institutions should be observed. To the contrary, in a networked, globally aware world, component or stakeholder institutions may represent more than one kind of grouping and may operate at more than one level of organization. The challenge—which, like all challenges, presents an opportunity—is how to bring these voices together.

Legal and Economic Agency for Nature

"I am the Lorax. I speak for the trees." In Dr. Seuss's parable, the Lorax claims agency on behalf of a forest. He does not, however, have jurisdiction over the forest. In fact, nobody does: the forest in question is an open-access resource. The Lorax, not actually owning the forest, is powerless to prevent its clearcutting by others seeking to meet demand for its resources. This children's tale contains an insight for adults: agency may be claimed but it cannot be enforced unless it has legal standing. Further, even if agency has legal standing, it is useless if it cannot be enforced in the field. The only legal system in place today giving any nongovernmental entity legal standing with respect to a natural resource is the system of property rights. Unfortunately, this system is limited to most terrestrial and some coastal resources. Even for these resources, it is not universal. It provides no protection to a natural resource against a property owner seeking to liquidate it, unless other governmental legislation is enforced to restrict it. In large areas of the Amazon Basin, for example, such government enforcement is absent. Further, the system of property rights remains largely helpless in the face of the common appropriation of global natural capacity, such as through atmospheric pollution.

The need Odum raised for protections to the planetary life-support system has been raised many times at the local and regional scales. In legal disputes over land rights and development, nature has been discriminated against by the legal system due to a lack of standing under the law. This is primarily a procedural issue, rather than an institutional or political issue.

In the early 1970s, a California wilderness area known as Mineral King became threatened by a proposed ski resort. The Sierra Club filed suit on behalf of the wilderness area. The case went all the way to the US Supreme Court in 1972, which denied standing to the Sierra Club because its members had no 'interest' in the proposed development and therefore would not be injured by it. The court, by convention, took a narrow view of the word 'interest' to mean specifically a financial interest or other direct effect upon people's lives, such as to their health. To the court, 'interest' meant an instrumental value, not intrinsic value. Further, as Alan Drengson explained, by denying standing to the Sierra Club as a representative of the wilderness, "the court in effect said that ... only humans or the fictional persons of ships and corporations can be given standing". The legal resolution of environmental disputes, he noted, thus boils down to resolving conflicts of interests in the sense just defined. "[A] forest as such cannot be aggrieved ... conservationists have interests, but these must be their interests, not the values within a natural process, independent of human interests."[24]

Little appears to have changed since the 1970s. In 2015, a group of high-schoolers in Oregon filed a class-action lawsuit against the US government for failing to protect their future from climate change.[25]

The Mineral King case was significant because of a dissenting opinion delivered by Justice William Douglas. This opinion acknowledged a basis for the legal standing of nature based on its intrinsic value. "The river [running through said wilderness] is the living symbol of all the life it sustains or nourishes ... The river as plaintiff speaks for the ecological unit of life that is part of it. Those people who have a meaningful relation to that body of water—whether it be a fisherman, a canoeist, a zoologist or a logger—must be able to speak for the values which the river represents and which are threatened with destruction."[26]

The discrimination in question arose because the legal system recognizes corporations, for example, as fictional persons, yet it does not recognize ecosystems as the same. As previous chapters have argued, a strong scientific and philosophical basis exists for recognizing ecosystems as autopoietic unities, possessing emergent properties arising from the intrinsic and instrumental values of all their inhabitants, including people. As such, they are tantamount to the 'fictional person' of the corporation. Both corporations and ecosystems are in fact two examples of higher-order natural and social entities exhibiting emergent properties. If one is to have legal standing then, by right, so should the other.

[24] Drengson (1995, pp. 90–91).

[25] *Juliana v. United States.* At the time of writing, the case continues to wend its way through the Ninth Circuit.

[26] William O. Douglas, *Sierra Club v. Morton*, minority opinion. See https://supreme.justia.com/cases/federal/us/405/727/.

This line of reasoning raises a practical problem, which is how does one define such entities? After all, there is no single 'correct' hierarchy of entities, even though there may exist broad ontological agreement on some of them, such as corporations, ecosystems, cities and species.

Justice Douglas' dissenting opinion in the Mineral King case offered a solution. People having a meaningful relation to an entity in question must be able to speak for the values it represents. The entity in question, whether a watershed, a tree, a village or a co-operative, would acquire legal standing through people representing its values. Justice Douglas' use of the word 'values' rather than 'interests' opens the door to the representation of instrumental value or intrinsic value, or both. If upheld, the usage would expand the meaning of 'interest' in a legal sense greatly to encompass not only all possible instrumental values but also intrinsic value.

Another issue arises, though. What is a 'meaningful relation' between a person and a non-human entity unable to speak directly to a court of law? Is there any non-discriminatory way in which it could, or should, be constrained, or could I, for example, insert myself into a legal dispute over a watershed in a faraway country simply because I care for the preservation of watersheds? The question of 'interest' is thorny because the boundary of interests in space and time is not defined. All people and all other organisms on Earth can be said to have had an interest in the outcome of the Mineral King case and countless other environmental disputes, even if that interest were diluted by distance or elapsed time. Wherever one drew a line, it would be arbitrary. Even more thorny, all living things alive today have an interest in all environmental disputes yet to arise. It is a transgenerational problem.

For nature to have standing in any legal case potentially affecting it, at any time, anywhere in the world, it must have agency. The court of law, being a human construct, therefore would presumably have to accommodate a human agent speaking for nature. Until such time as societal norms became aligned with natural processes, however, issues of rights, which may conflict, will present themselves. The reason for providing nature a legal agent—an attorney—is to protect everybody's right to life-support. By protecting the 'rights' of nature in a legal setting, that agent would act in accordance with an agreed-upon, long-term goal of social and economic stability. On the other hand, by not assigning nature the right of due representation as a matter of procedure, society risks nature possibly bringing future injury upon the economy or society for which no legal recourse exists. This is the 'menace to public safety' to which Odum referred.

Providing a mechanism for nature to have legal agency is a different kind of issue from any other rights issue. Civil rights issues, for example, are by definition human. People can articulate the rights they seek. Nature is distinct and unique in that it neither acknowledges the existence of, nor understands, a court of law; yet its exclusion from the human legal process could have negative consequences for people in the long run.

Although the idea of a legal agency for nature may sound far-fetched, so did other rights issues at various times during the past century: votes for women, interracial marriage, non-discrimination in the workplace, and same-sex marriage, to name a few. Recently, a handful of jurisdictions have passed laws recognizing the legal rights of certain natural resources, and granting them personhood before the law.[27] These actions could represent the thin end of a very large wedge. Legal agency for nature could in the long run become the most important rights issue of all, principally because it would benefit everyone. An institutional and legal system providing due process to nature anywhere on Earth would, in and of itself, incorporate an ability to evaluate economic claims upon natural capacity anywhere on Earth. In this way, the institutional and governmental foundations of a large, stable economy would by design incorporate mechanisms for aligning that economy with natural processes.

Although in theory legal agency for nature could become established through the incremental process of case law and national legislation, the point of nature having legal agency is to give it economic agency. What if both could be given in one fell swoop?

The Planetary Sector

To Peter Barnes, who has written extensively on the subject, the whole economy, as presently constructed, is incomplete. "Everyone knows what private wealth is", Barnes wrote, but "common wealth is like the dark matter of the economic universe—it's everywhere, but we don't see it".[28] Although this vast inheritance has always existed, it has never been recognized, nor represented, within the modern economy. The economy

[27] For example, in 2019, in the wake of a series of toxic algal blooms, the US city of Toledo, Ohio, voted to grant Lake Erie legal personhood, giving its citizens legal guardianship over the lake, along with the ability to hold polluters legally accountable (McGraw 2019). Curridabat, a suburb of San José, Costa Rica, recognizes pollinators, trees and native plants as citizens of the town, and has created natural corridors for them (Greenfield 2020).

[28] Barnes (2006, p. 66).

traditionally has been divided roughly into the private sector, which includes households and businesses, and the public sector, which comprises government. The missing 'third leg of the stool' is the commons: all the natural and artificial capacity that enables the other two sectors to function. Conventionally, the commons has been considered external: taken as a given and unpriced. What is missing, Barnes argues, is a new sector: the 'commons' sector, which would safeguard the future against the short-term, profit-driven excesses of the private sector and the electoral myopia of democratic government.[29] I prefer to call it the 'planetary sector' because this term emphasizes not only the instrumental value of everything the economy depends upon but also its existence value.

In a planetary economy, the planetary sector would function as an avatar for this common wealth. It would embed into the economy's very structure the notion of ownership as belonging. Its inherently generative nature would serve as a counterweight to those elements of the private and government sectors that, left unchecked, can become inherently exploitative. The new sector would be brought into existence through a new institution, or institutions. Fortunately, notes Barnes, we do not need to reinvent the wheel for this.

A Planetary Trust

The planetary sector can be created through a trust: a legal institution that has been tried and proven over centuries of use.[30] Many of us are familiar with private trusts that families set up to safeguard inheritances; in all likelihood, you may have one of your own or be the beneficiary of one. Similarly, land trusts and conservancy trusts are widely used to preserve valuable land and historic buildings. The trust furthermore is a common framework for the financial administration of universities. Trusts have intergenerationality built in: they are designed for long-term preservation, which makes them a natural fit for the maintenance of the various forms of capacity on which the economy depends. Its long view acts as a natural counterbalance to financial or electoral short-termism, thereby, as Barnes puts it, "saving capitalism from itself".[31]

[29] Barnes (2001, p. 127); Barnes (2006, p. 87).

[30] Barnes (2006, p. 83).

[31] Barnes (2001, p. 106).

The essence of a trust, as Barnes describes, is a fiduciary relationship. "Neither trusts nor their trustees may ever act in their own self-interest; they are legally obliged to act solely on the behalf of beneficiaries." The trustees must preserve a trust's financial principal, unless instructed otherwise, and they must be transparent in all transactions.[32]

A planetary trust would protect not only Earth's natural capacity but also the economy's artificial capacity: all the commonly-owned assets and infrastructure that enable economic activity, from utilities to the financial system to the internet. A case might be made for natural capacity to be protected separately from artificial capacity, but they could still fall under one trust.

The trust would give every living person one share in natural capacity and one in artificial capacity, analogous to one person, one vote. Like a vote, the share would be created upon birth, would expire upon death, and would not be transferable in the meantime. Alternately, following Barnes, shares could be issued for distinct kinds of capacity, such as 'sky shares' for the atmosphere or 'habitat shares' for ecological communities, each type of share having its own trust.[33] The degree of aggregation of these different forms of capacity would depend in part upon the ease of their measurement. Earth's overall natural capacity is a function of the health of its atmosphere, water systems, soils and biota, but these components cannot be considered in isolation. The value of a share of natural capacity would have to consider them in an integrated way, and could only be arrived at through scientific consensus. Shares in different kinds of artificial capacity, on the other hand, can be distinctly valued, based on the prices that their principal users—corporations—would pay in the market to use them. Ultimately, it might be simplest to establish a single trust for all forms of capacity, which would be itemized on a schedule. As technologies change, so the schedule would be updated.

One of the principal reasons for establishing a planetary trust is to provide a framework for delivering returns to the beneficiaries of natural and artificial capacity: that is, all of us. The other principal reason is to compel the present generation to preserve the ability of this capacity to deliver returns to beneficiaries as yet unborn. How returns from capacity would be disbursed is the subject of future chapters. A variety of mechanisms can

[32] Barnes (2006, p. 83).
[33] Barnes (2001, ch. 7).

be imagined; the point is that the trust would provide a framework for accommodating any and all future instruments that could be dreamed up.

By providing returns from the use of capacity, a planetary trust would deliver a dramatic reduction in economic inequality. It would redirect wealth from high users of capacity to low users of it. Corporations, for instance, use much more capacity than individual people, and wealthy people generally use more capacity than poor people. As Barnes points out, under the current capitalist system, the free use of capacity—both natural and artificial—by corporations earns them a 'commons rent'. If, on the other hand, the rights to use capacity were assigned to a trust, and the trust were authorized to sell those rights to corporations, then the trust would capture that commons rent for the common benefit.[34]

The next chapter describes an institutional framework for a planetary trust, which would form part of a wider planetary governance. The path toward such a planetary trust would begin with the establishment of local, regional and national commons trusts. Government has a key role to play here. As Barnes points out, government is the default trustee unless or until it assigns a commons to another institution. Government alone has the ability to make such an assignment and it has the ability to retrieve into the public domain elements of the commons that may have been shortsightedly privatized.[35] Local or regional commons trusts would be established by legislation to safeguard elements of the natural commons—such as regional ecologies or air quality—or artificial common assets, such as water, power or communications infrastructures. The trusts would be given broad authority to extract commons rent for the use of these assets, returning any benefits to everyone equally.

The trust, as an institution, establishes both legal and economic agency for nature because it formally acknowledges the role of common wealth within the economy, thereby revealing the planetary sector.

Requirements for Legal Institutions

In summary, a planetary economy would have two principal requirements for legal institutions, summarized in Table 4.1.

[34] Barnes (2006, p. 94).
[35] Barnes (2006, pp. 152–153).

Table 4.1 Requirements for legal institutions for a planetary economy

Requirement	*Description*
Legal agency for nature	In order for decision-making to serve long-term economic stability, it must support natural capacity, which means that nature must be given legal agency within the economy.
A planetary sector for trusteeship of the commons	A missing sector of the current economy, the planetary sector, would counterbalance myopia in the private and government sectors with an intergenerational impetus. The institution to accomplish this is a trust which, in a planetary economy, would be a planetary trust, operating in the interests of both natural and artificial capacity. The creation of the planetary sector by the trust also would support widespread prosperity because the sector, by design, would be generative rather than exploitative.

Source: Author's creation

References

Adger, N. (1997). Property Rights, Nature Conservation and Land Reform in South Africa. In F. D. M. Smith (Ed.), *Environmental Sustainability: Practical Global Implications* (pp. 213–232). Boca Raton, FL: St. Lucie Press.

Arrow, K. J., et al. (1995). Economic Growth, Carrying Capacity and the Environment. *Science, 268*(5210), 520–521.

Ashley, C., & Barnes, J. (1997). Wildlife Use for Economic Gain: The Potential for Wildlife to Contribute to Development in Namibia. In F. D. M. Smith (Ed.), *Environmental Sustainability: Practical Global Implications* (pp. 163–192). Boca Raton, FL: St. Lucie Press.

Barnes, P. (2001). *Who Owns the Sky? Our Common Assets and the Future of Capitalism*. Washington, DC: Island Press.

Barnes, P. (2006). *Capitalism 3.0: A Guide to Reclaiming the Commons*. San Francisco, CA: Berrett-Koehler.

Boulding, K. E. (1962). *A Reconstruction of Economics*. Science Editions, New York (reprinting of 1950 edition, John Wiley & Sons, New York).

Capistrano, A. D., et al. (1997). Poverty Alleviation, Empowerment and Sustainable Resource Use: Experiments in Inland Fisheries Management in Bangladesh. In F. D. M. Smith (Ed.), *Environmental Sustainability: Practical Global Implications* (pp. 141–162). Boca Raton, FL: St. Lucie Press.

Costanza, R., et al. (2015). *An Introduction to Ecological Economics* (2nd ed.). Boca Raton, FL: CRC Press.

Drengson, A. (1995). Shifting Paradigms: From Technocrat to Planetary Person. In A. Drengson & Y. Inoue (Eds.), *The Deep Ecology Movement: An Introductory Anthology* (pp. 74–100). Berkeley, CA: North Atlantic Books.

Georgescu-Roegen, N. (1971). *The Entropy Law and the Economic Process*. London: Harvard University Press.

Greenfield, P. (2020). 'Sweet City': The Costa Rica Suburb That Gave Citizenship to Bees, Plants and Trees. *The Guardian*, 29 April.

Hanna, S. S., Folke, C., & Mäler, K.-G. (Eds.). (1996). *Rights to Nature*. Washington, DC: Island Press.

Hardin, G. (1968). The Tragedy of the Commons. *Science, 162*, 1243–1248.

Holling, C. S., & Sanderson, S. (1996). Dynamics of (dis)harmony in Ecological Systems. In S. S. Hanna, C. Folke, & K.-G. Mäler (Eds.), *Rights to Nature*. Washington DC: Island Press.

Kneese, A. V. (1966). Research Goals and Progress Toward Them. In H. Jarrett (Ed.), *Environmental Quality in a Growing Economy* (pp. 69–87). Baltimore, MD: John Hopkins Press.

Lowe, A. (1965). *On Economic Knowledge: Toward a Science of Political Economics*. New York: Harper & Rowe.

Marglin, S. (2008). *The Dismal Science: How Thinking Like an Economist Undermines Community*. London: Harvard University Press.

McGraw, D. (2019). Ohio City Votes to Give Lake Erie Personhood Status Over Algae Blooms. *The Guardian*, 28 February.

McKean, M. A. (1996). Common-property Regimes as a Solution to Problems of Scale and Linkage. In S. S. Hanna, C. Folke, & K.-G. Mäler (Eds.), *Rights to Nature*. Washington DC: Island Press.

Mitchell, R. B. (2013). Oran Young and International Institutions. *International Environmental Agreements: Politics, Law and Economics, 13*, 1–14.

Odum, H. T. (1971). *Environment, Power & Society*. New York: Wiley-Interscience.

Ostrom, E. (1990). *Governing the Commons*. Cambridge: Cambridge University Press.

Ostrom, E., & Schlager, E. (1996). The Formation of Property Rights. In S. S. Hanna, C. Folke, & K.-G. Mäler (Eds.), *Rights to Nature* (pp. 127–156). Washington DC: Island Press.

Rodman, J. (1995). Four Forms of Ecological Consciousness Reconsidered. In A. Drengson & Y. Inoue (Eds.), *The Deep Ecology Movement: An Introductory Anthology* (pp. 242–256). Berkeley, CA: North Atlantic Books.

Sanderson, S. E. (1994). Human Driving Forces: Political-Economic Institutions. In W. B. Meyer & B. L. Turner (Eds.), *Changes to Land Use and Land Cover: A Global Perspective* (pp. 224–252). Cambridge: Cambridge University Press.

Schumpeter, J. E. (1950). *Capitalism, Socialism and Democracy*, 3rd Edition. 2008 reprinting by Harper & Row, New York.
Stiglitz, J. E. (2003). *The Roaring Nineties.* New York: W.W. Norton.
Toledo, V. M. (1997). Sustainable Development at the Village Community Level: A Third World Perspective. In F. D. M. Smith (Ed.), *Environmental Sustainability: Practical Global Implications* (pp. 233–250). Boca Raton, FL: St. Lucie Press.

CHAPTER 5

Political Institutions

Governmental Foundations for a Stable Economy

Political Globalization

Is global governance or global government feasible or desirable as a means to navigate a 'moral extensionist' phase of societal evolution, which we seem to have now entered? This question is politically contentious to those who hold fast to the immutability of the nation-state. Yet it has been around for a long time. Thomas More's friend, the Dutch Renaissance scholar Desiderius Erasmus, dreamed of "a unified empire [with] a sovereign in the image of God".[1] In the eighteenth century, Jeremy Bentham proposed "the common and equal utility of all nations" as the object of "a universal international code".[2] Between the 1770s and the 1830s, writes Mark Mazower, various writers imagined "an alternative international politics, one that acknowledged the diversity of peoples, beliefs and forms of government".[3] The Comte de Saint-Simon, writing in 1814, proposed a federation of the entire continent of Europe, one that would fall under a common institutional organization.[4]

[1] Mazower (2012, p. 14).

[2] Mazower (2012, p. 21). Quite the coiner of words, Bentham was responsible for 'international' and 'codification', as well as the concept of utility, which later became central to the paradigm of neoclassical economics.

[3] Mazower (2012, p. 24).

[4] Quoted in Mazower (2012, p. 96).

F. Murison Smith, *A Planetary Economy*,
https://doi.org/10.1007/978-3-030-49296-0_5

The first international institutions began to form, ironically, during the apogee of the nation-state. The Inter-Parliamentary Union,[5] founded in 1889, arbitrated a number of international disputes that avoided armed confrontation prior to the First World War.[6] Arbitration was seen at the time as the best option in the absence of an overarching authority to enforce international law. Today, the IPU sits as an observer at the United Nations. Its legal successor, the International Court at The Hague, has the ability to try individuals for war crimes, but not to enforce any kind of international law upon any nation-state. It has no jurisdiction over the members of the UN Security Council, nor are they signatories to it. Among nations, it primarily serves regions where judicial systems are weak.[7] The only supranational legal body with binding powers over nations is the European Court of Justice.[8]

The late nineteenth century also saw a proliferation of technical and professional institutions. The Belgian Paul Otlet (1868–1944), who founded the Union of International Associations, believed, like Saint-Simon before him, that societal evolution led eventually to world government, and that diplomacy among nation-states hindered this change. Another of Otlet's initiatives, to transform the world's libraries into information hubs—a quixotic effort predating the internet by about 80 years—ultimately yielded the International Institute of Bibliography and the Universal Decimal Classification.[9]

The first international political institution was the League of Nations, formed in the Versailles Treaty of 1919. The League traced its existence to the ideas of Giuseppe Mazzini (1805–72) and to the former civil servant Leonard Woolf, husband of the writer Virginia Woolf, whose 1916 report, *International Government*, supplied a wealth administrative detail.[10] In 1947, the League was superseded by the United Nations, a supranational organization within which, its architects hoped, all nations could assist one another and resolve differences. Not everyone believed it would work. Albert Einstein and scientists of the Los Alamos National Laboratory issued separate statements calling for unitary world government as a way

[5] www.ipu.org.
[6] Mazower (2012, pp. 88–93).
[7] Mazower (2012, p. 401).
[8] Mazower (2012, p. 411).
[9] Mazower (2012, pp. 108–109).
[10] Quoted in Mazower (2012, p. 125).

to prevent atomic war.[11] Yet, measured by the extent of its membership, the UN has been wildly successful. The only supranational organization with more members is FIFA.[12] Over time, the UN has evolved into a body asserting a 'responsibility to protect' the security and human rights of *individuals*, if necessary, over and above the sovereign rights of member states. The UN operates peacekeeping missions,[13] development programs, human rights programs, humanitarian affairs and international legal efforts. In these efforts it collaborates with a constellation of non-governmental organizations, foundations and businesses.[14] Perhaps it should really be called the United Peoples?

The history of international governance so far has represented a triumph of *realpolitik* over any technocratic dreams of world government. The nation-state itself emerged through the democratization of previously monarchical states, whose borders were rarely constant. The tendency for national borders to cut across cultures, languages and biomes is a "geographical error", according to the geographers Elisée Reclus and Piotr Kropotkin.[15] Today, global governance is little more than a convergence of expectations among sovereign states.[16] These expectations frequently change as administrations come and go.

The US actions on international climate-change agreements represent a case in point. The Kyoto Protocol, which committed countries to reduce greenhouse-gas emissions, was adopted by the United Nations on 11 December 1997. It was signed into effect on behalf of the US by then-Vice President Al Gore on 12 November 1998. However, the US Senate refused to ratify it prior to the 2000 presidential election. In the meantime, 83 other nations signed on. Early in 2001, the newly elected US administration of George W. Bush retracted the US signature to the accord, claiming that other emerging major emitters—particularly China

[11] Mazower (2012, pp. 231–232).

[12] FIFA has 209 members to the UN's 193, notwithstanding the UK's four member regions, each having distinct memberships of FIFA.

[13] Around 100,000 peacekeepers in 16 active operations as of 2014. Source: http://www.un.org/en/aboutun/index.shtml.

[14] This network of partners is known as the UN System. As of the end of 2013, the UN Office for Partnerships had allocated about $1.2 billion in funds to 544 projects and programs in 124 countries through the system.

[15] Mazower (2012, p. 25).

[16] Young (1996, pp. 246–251).

and India—were not subject to mandatory emissions reductions.[17] To the Bush administration, it was an unfair deal for the US; yet, without the US and its 16 percent of global carbon dioxide emissions on board, the strength of the agreement was significantly diminished.[18] Eight years later, incoming President Barack Obama also did not sign the Kyoto accord, instead choosing to develop alternate, bilateral commitments, such as with China. Late in 2015, he signed the Paris accord, which succeeded Kyoto. After the 2016 presidential election, in one of his first acts as President, Donald Trump withdrew the US from the Paris accord. Meanwhile, few if any of the other signatories to Kyoto or Paris are meeting the emissions reduction targets they committed to.

Supranational institutions represent only one facet of institutional globalization, even though they may be the oldest. Modern networked communications technologies have produced new institutions with global reach, ranging from informal online networks to organized affiliations among local and regional governments. Earlier proponents of a monolithic global government failed to acknowledge how diverse actual governments are from one region to another. The one-size-fits-all approach proposed by global government simply would not work in a world as diverse as ours. At the same time, though, 'governance' is not merely international co-ordination, but co-ordination across many geopolitical scales.

Whither the Nation-State?

The governments of the most powerful nation-states may not yet sense or acknowledge an impending change to the hegemony of the nation-state, but it surely will come as the world becomes ever-more interconnected. Many nations, such as the US, jealously guard their 'untrammeled sovereignty', as Mark Mazower puts it, yet this sovereignty might not be as robust as they think.[19] To quote Alexsandr Solzhenitsyn, "no such thing as 'internal affairs' remains on our crowded Earth." [20] Two weeks before a 2015 referendum in Scotland on whether to secede from the United

[17] United Nations Framework on Climate Change, http://unfccc.int/kyoto_protocol/items/2830.php; Borger (2001); Rosenthal (2009). This is an issue Barrett (1996) termed 'leakage'.

[18] Emissions datum from the US Department of Energy, Carbon Dioxide Information Analysis Center (CDIAC).

[19] Mazower (2012, pp. 423–424).

[20] Mazower (2012, p. 321).

Kingdom, Reuters conducted a poll on state secession within the US. Among several thousand respondents around the country, nearly one-quarter 'supported or tended to support' secession of their states. That is a shockingly high number.

Many who polled in favor of secession identified strongly with their state or region. Here, as at the local level, there may be a stronger shared cultural identity than at the national level. A similar sentiment may be at play in Europe, where growing dissatisfaction with the European Union coincided with an upswing in nationalism during the 2010s, leading to Britain's exit from the Union in 2020.

The deep ecologists have been particularly critical of the nation-state. They see it as a severe hindrance to the development of a global, ecologically conscious community. "The national politics of the modern world", wrote Gary Snyder, "are 'states' which maintain their existence by deliberately-fostered craving and fear: monstrous protection rackets."[21] Snyder contrasted the world of nation-states with a concept of 'true community' (sangha) adopted from the Buddhadharma. As a community of all beings, it would signify a 'planetary culture' in the language of the West, one which would follow upon a new understanding of society, leaving behind the nation-state. Whereas "the State is greed made legal", he wrote, "a natural society is familial and cautionary. A natural society is one which 'follows the way', imperfectly but authentically."[22] Wishful thinking, perhaps; but Snyder was looking out a long way.

He was looking about as far out as Steven Sherwood and Matthew Huber, who studied the role of heat stress in limiting people's physiological ability to adapt to a warming climate. Nature, it turns out, is an altogether deeper existential threat to the nation-state than any social ones. Under a scenario of severe global warming, in which the mean surface temperature increases by 7 °C, some parts of the world would be rendered too hot for metabolic heat dissipation. For most practical purposes, a global-mean warming of 3–4 °C would "in some locations halve the margin of safety" between the maximum ambient 'wet-bulb' temperature experienced by a person and the theoretical lethal maximum of 35 °C, this being the temperature of human skin. Under these conditions, people would not be able to move around outside for fear of suffering heatstroke. Livestock would not survive. Although mechanical air

[21] Snyder (1985, p. 251).
[22] Snyder (1985, p. 252).

conditioning could mitigate some of the effects, it would not be available to all, and it would place stresses on power grids. Power outages would become lethal. As the authors point out, the land area lost to uninhabitable heat under a warming scenario of 3–4 °C could dwarf that lost to sea-level rise. The economic effects would be much higher than the 10–30 percent of world GDP that conventional forecasts predict would be lost.[23]

Changes of this magnitude would make a mockery of the nation-state. A global mean average warming of 4 °C is very much within the bounds of likelihood over the coming one to two centuries, according to climate models. The already heat-stressed Middle East could, for all practical purposes, become physiologically uninhabitable, thereby resolving once and for all its ancient, rancorous disputes over land rights. Sherwood and Huber's analysis suggests that large parts of South America, eastern North America and China could become marginally inhabitable. Most of Africa, Australia and India would have to be abandoned. These changes would coincide with the world's population approaching a peak of 11–12 billion. Mass migrations toward the poles would irrevocably alter the geopolitical landscape, consigning many nation-states to the history books.

Sherwood and Huber's analysis does not, of course, consider technological adaptations, nor does it intend to. The effects of technological adaptations are hard to predict. In theory, it might be possible for heat-stressed regions to remain inhabitable by bringing all activities indoors or sending them underground. People in hot regions could live in cities whose buildings were cooled by passive geothermal technologies, whose transportation was all underground, and whose food production was accomplished by industrial-scale indoor growing factories. It would almost be like living on another planet. It would be technologically possible, but would it be desirable?

New geopolitical entities might form, accommodating the expatriates of newly uninhabitable regions. Mass migrations would drive home the imperative for a globally integrated form of governance. The information-connectedness of the twenty-first century would enable this. As far back as 1834, Felix Bodin described a similar structure for a 'Universal Congress' in his book *Le roman de l'avenir* (*The Novel of the Future*).[24]

Current international institutions, such as the World Trade Organization, the World Bank and the International Monetary Fund, "stand in urgent

[23] Sherwood and Huber (2010).
[24] Mazower (2012, p. 23).

need of renovation", writes Mazower. They are "incapable of helping to rectify the global imbalances that threaten the world economy" and "[un]able to coordinate the response to global warming".[25] Yet the nation-state is not going away any time soon, nor should we expect the international organizations to which nations belong to fade away, either. Examples of participatory global governance, in which national governments willingly cede sovereignty, already exist in the form of international standards. The International Standards Organization, for instance, is so widely accepted that even the recalcitrant US recently replaced its old domestic classification of hazardous materials with the global standard.

Another possible avenue toward participatory global governance is through capital markets. The capital markets are so significant in volume that they form a networked institution far more powerful than any sovereign nation-state. If the ten or twenty largest exchanges in the world decided among themselves to play by a new set of rules then no single sovereign government could stop them.

Far from becoming "yesterday's dream",[26] the idea of governing the world is only just beginning to form. As *governance* rather than *government*, it is more than merely a dream for the future; it is a prerequisite for a stable one.

A Global Governance Council

The involvement of entities from multiple levels of social organization within a global governance structure must be participatory, by definition.[27] Feudalism or autocracy may be limited today to relatively few areas of the world, yet the dominant paradigm of democracy, as currently practiced, is itself not participatory enough.

Any institution for governance of global society and the economy would have to support a prime directive of material discipline. It also would have to accommodate a plurality of approaches toward increasing material efficiency and broadening prosperity. The imposition of universal standards, such as for carbon dioxide emissions reductions, is problematic when applied to socioregions in vastly different stages of economic

[25] Mazower (2012, p. 424), a view echoed by Adler and Varoufakis (2019).

[26] Mazower (2012, p. 427).

[27] Prakash and Gupta (1997) come to a similar conclusion through an application of Rawls' 'veil of ignorance' concept.

development. Tangled discussions about fairness ensue among participating jurisdictions. If, on the other hand, the goal to which participants committed were simply to minimize material intensity then a great deal of scope would exist for participants to focus on aspects of their material intensity most critical or meaningful to them, and to develop solutions appropriate to their needs and capabilities. A collection of small, tropical fishing communities, for instance, might focus on stabilizing harvest rates or limiting bycatch, whereas a provincial government in Europe might develop programs to reduce greenhouse gas emissions. In this way, the tricky question of applying an overarching goal upon all jurisdictions would be avoided. Underlying indicators, such as atmospheric CO_2 concentration, then would be used simply to measure progress. Climate researchers and policy analysts acknowledge that the name of the game during the twenty-first century quickly will become one of adaptation to climate change, as much as mitigation of it, if not more so. Although adaptations will vary among socioregions, it can be certain that sustained reductions in material intensity anywhere would move the indicators in the right direction in the long run.

One possible institution for planetary governance is outlined here to illustrate an approach to capturing interests from different kinds of social entities at different levels of organization. It is called a global governance council. The design of this institution employs an approach apparently not widely used within existing Earth system governance literature: hierarchy theory. Subsequent chapters, which outline requirements for policies, standards and instruments, will refer back to this structure as an organizing framework.

A global governance council intentionally would involve entities from several jurisdictional levels within one forum. In a strict, nested hierarchy, a tendency would evolve for top-down control, which has proven ineffective. A global governance structure for the economy's interaction with natural systems would have to represent all levels of social organization, because it is at all levels that problems are identified and it is from all levels that solutions will arise. The structure also would have to be participatory. A global governance council could not impose legislation by majority rule; it would only have jurisdiction over its own policies and over membership. Participation would represent a commitment to one or more prime directives established by the council, such as material discipline and broad prosperity. Membership would be contingent upon acting according to this directive: setting goals and meeting them. The institutional structure itself would provide assistance in this, thereby establishing an incentive to remain. However, legislation committing a jurisdiction to its goals would be left to the jurisdiction itself.

Structure and Membership

Jurisdictions participating in a global governance council can be broadly categorized into municipalities, counties, states or provinces, and nations. Each represents a geopolitical level of organization having certain emergent properties not possessed by entities at lower levels of organization. Nations, for example, are able to enact defense and macroeconomic policy, whereas municipalities and counties administer local services and taxation. At the most fundamental level, the jurisdiction of the individual exists, a jurisdiction unto the self. It has an equal seat at the council of governance along with all the higher-order entities representing agglomerations thereof. The nexus of global governance therefore is the intersection of all these entities, as Fig. 5.1 illustrates.

Since the number of all local, regional and national jurisdictions worldwide is very likely over one million,[28] the council itself would have to consist of a small subset of participating members. In other words, participating members would decide in some way which of them would represent the whole membership in the role of governing members. The structure of the council could be informed by hierarchy theory. Since each organizational level has distinct kinds of interactions with nature, each becomes equally valid in aligning the economy with it. For this reason, each kind of jurisdiction would have equal representation within the council. Participating members would be only institutions; however,

[28] An exact count is not easy to come by. However, the US Census Bureau maintains good data for the US from which an extrapolation may be done. In the 2010 census, there were 3031 counties and 85,073 other local governments, comprising 19,522 municipalities, 16,364 townships, 37,203 special districts (water districts and the like), and 12,884 independent (i.e., non-municipal) school districts. By proportion to the global population, which is about 25 times that of the US, the total number of counties, or local governmental entities having jurisdiction over non-municipal and non-state land, would be about 75,000. The number of other local governments would be about 2.1 million. These values might overestimate the total, as other parts of the world might not have as well developed a system of civic institutions as the US. The actual number likely will fall between one and two million. Source: www.census.gov/newsroom/releases/archives/governments/cb12-161.html.

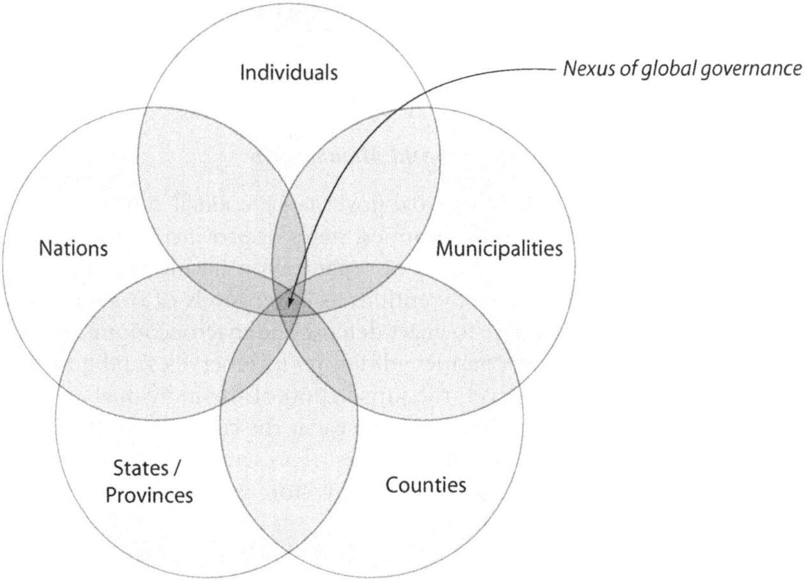

Fig. 5.1 Global governance as a nexus of social entities at various levels of organization. (Source: Author's creation)

individuals not belonging to participating institutions would have an opportunity to represent all individuals on Earth in the council.

One possible structure, attractive for a variety of reasons discussed below, combines elected and randomly chosen members. Within each category, a percentage of the seats—let us say half, for the sake of illustration—could be electable, the other half being randomly selected.

Candidates representing a particular jurisdiction would put themselves up for election, and all participating members of the council, from all categories, would have an opportunity to vote for them. Suppose, for instance, that the number of elected seats on the council were 20, divided among the five institutional types in Fig. 5.1. Then there would be four seats for candidates representing nations, four for provincial government candidates, and so on, down to four seats for candidates running as individuals. In the interest of institutional stability, voting would be staggered over time so that not all seats within each category came up for election at once; perhaps one-quarter or one-half. Terms would be short—say, two years—so that many members would have opportunities to sit on the council.

The randomly chosen seats—whose number would not exceed the number of elected ones—would work rather like jury duty, a process known as sortition. Members within each category, and individuals, would be offered an opportunity to serve, which they could accept or decline. The mechanism would serve to dilute any tendency for elected members to coalesce into party blocs. Council members chosen at random would serve shorter terms than the elected ones—perhaps up to a year—to best preserve their independence.

Having provided an elected or randomly chosen council member, an institution would be ineligible to offer another candidate until a certain minimum proportion of other member institutions had served. An individual having served a term would be ineligible to serve again for life. To ensure that the individual seats were not dominated by populous regions, such as China and India, a provision could be established to render ineligible people from countries who had recently had an individual on the council. Populous countries would become re-eligible to send individuals to the council only after a certain proportion of other countries had done so. In this way, individuals from smaller countries would have a fairer shot.

A handful of additional seats might be open to candidates from anywhere in the world seeking to represent parts of the global commons: Antarctica, bodies of water (oceans, rivers and lakes), the world's soils, the atmosphere and the biosphere. The intent would be to include individuals not necessarily having any political affiliation, who represented the long view, and ensured that this view would be maintained in decision-making. The individuals representing aspects of nature would have special responsibility, as they would act within the council as agents on behalf of nature.

The random selection of individuals from all member entities would present certain technical challenges, particularly in ensuring that the poorest, least-represented, most inaccessible people would be provided such an opportunity. There would also need to exist a procedural training program to bring inductees up to a common level of knowledge and understanding, as well as a mechanism for individuals to be compensated for any income foregone by serving on the council. However, if the number of such seats were relatively small—say, twenty or so—then the associated budgetary requirements would be small also.

The role of the council itself would be to oversee policies and membership. The activities of the council, more broadly, would be overseen by committees staffed by members. Policies would govern overarching goals and directives, as well as the mechanisms by which those goals would be met. Policies also would govern how performance were measured and how membership would be granted or revoked.

The council could meet face-to-face or virtually. In-person meetings of the council itself or of its constituent committees could rotate among participating jurisdictions. Rather than having a single headquarters, like the UN, a global governance council would convene in a different host city each time, selected a year or two in advance by a process of sortition, so as to avoid any potentially biased or expensive bidding process. Council sessions would be guided by an individual serving as a speaker or moderator, chosen through a nomination process and serving a limited term. All participating members would be able to attend council meetings, if not in person then virtually, and able to contribute to deliberations. Individual council members, although drawn from local, regional and national governments, would not represent their particular governments at the council but instead all governments at their respective institutional level. They would be obliged, although not required, to vote according to the wishes of their colleagues. Those wishes might themselves be expressed by vote within sub-councils comprising members from each institutional level. Thus, when a council member convenes at the global council, they already would have received guidance on issues of interest.

Aside from the incorporation of sortition, much of the mechanics of the council, as the foregoing describes, would follow well-established procedures already adopted in assemblies around the world.[29]

Voting, too, could be virtual, given the relatively small number of participating members, compared to public electorates. A secure network would be established for this. To avoid biases introduced by campaign finance, the council would maintain an electoral budget divided evenly among candidates within each institutional level. Budgets may vary across institutional levels, given the likely larger electoral budget requirements of nations or provinces compared to municipalities or individuals.

Membership would be handled by a membership committee comprising diverse members, following policies established by the council. The council would maintain a database of the world's municipal, regional and national governments showing the membership status of each. New members would pledge to make every reasonable effort to fulfil the directives

[29] Robert's (1876) *Rules of Order* is a widely used example.

of the council. Although directives would be non-binding, the council would have procedures for ensuring members' commitments. Foremost among these would be mechanisms to ensure that members acting in good faith are given opportunities to succeed, such as through assistance from other members. Members failing to act in good faith may be subject to progressive warnings or termination, for which an appeals process would be provided. Although termination would be an act of last resort, the council could only function properly if all members demonstrated commitment.

These mechanisms, combined with the benefits of membership, should ensure membership grows over time. At some point, a critical mass may be attained where it would become in the interests of the majority of jurisdictions worldwide to become members.

Functions

Measurement, Verification, Reporting and Communication

A principal purpose of a global governance council would be to enable member jurisdictions to set and meet goals for material efficiency and widespread prosperity. For this, the council would provide services in measurement, verification, reporting and communication (MVRC).

Much of the news coverage on the relationship between the global economy and planetary systems early in the twenty-first century has focused on the urgency to do something to stop various key indicators going off the charts. What is missing from this narrative is an articulation of that 'something'. At the same time, it is hard to understand, without a great deal of digging around, what is actually being done today in different jurisdictions—the success stories, the failures, the plans for the future—and how all these elements connect. An integrated picture is needed, readily available to anybody connected to the world's communications infrastructure. This could be a significant value of the council's work. The more people or institutions see what is being done, the more they will become convinced they can do something, too. Membership of a global governance council would confer economic benefits upon participating institutions through the sharing of knowledge and expertise. A municipality, regional or national government could realize the economic benefits of innovation within a community of jurisdictions offering a degree of protection against first-mover risk. The larger the community becomes, the lower that risk, and the lower the concomitant risk of free-ridership, such as through capital flight to other jurisdictions.

Participating institutions would engage in an iterative process of goal-setting, implementation—with assistance as needed—scoring of performance, reporting to the council and goal-resetting. An off-ramp into membership review would exist for any serial non-performers. The MVRC process itself would be managed by council staff under the oversight of member committees. By providing MVRC functions, the council would act as a kind of standards organization. It would not directly oversee programs established by member jurisdictions but instead would evaluate their performance. Its communications function would share knowledge and information about member programs, thereby enabling members considering new programs to develop ones suiting their needs based on the best available practices of others. By not owning any programs of its own, the council would not require a large annual budget the way United Nations programs do. A global governance council would not need a large bureaucracy, allowing it to be administratively lean and relatively affordable to its members.

One extremely valuable MVRC function would be in providing ratings relative to its prime directives. Since material efficiency is readily quantifiable, it would be straightforward to establish standards, semi-automate much of the MVRC process, and produce a material efficiency rating. MVRC in relation to other directives, such as widening prosperity, may require a degree of judgement. Financial measures of prosperity are easily quantifiable, whereas non-financial ones are not. Nonetheless, a prosperity rating also could be devised. A ratings agency as a division of the council would provide a similar function to the financial ratings agencies already in existence. Unlike many of those agencies, it would be a public-sector institution and more likely to operate impartially. Indeed, as subsequent chapters will argue, the council itself also eventually could acquire the financial ratings role.

The value of a material efficiency rating extends beyond the membership of a global governance council into the business world. In order to place the economy on a path toward material efficiency, markets will need a way to evaluate businesses' material performance relative to one another, particularly within the context of the economic instruments that may come into play. Just as an institution may receive preferential credit terms by maintaining a high credit rating, so it may receive preferential treatment under a variety of economic instruments, including credit, by maintaining a high material efficiency rating. If that rating were provided by a global, public institution with a large and growing membership then it would be expected to enjoy a high degree of trust by businesses worldwide.

Although council staff might manage its MVRC functions, it might not be necessary to in-house them. Many other non-governmental organizations and universities already operate measurement and reporting activities that could be adapted to the council's needs. The council's role would be to collate, store and integrate existing datasets, and to establish new MVRC efforts as needed. Delegating measurement and verification functions to registered NGOs also diversifies the risk of erroneous or dishonest reporting. The council's MVRC programs would be more trustworthy by virtue of originating from multiple sources that can be cross-checked against one another.

Oversight of a Planetary Trust

A global governance council would be a natural home for a planetary trust. In establishing and upholding the directives of material discipline and widespread prosperity, it already will have established the guiding principles of the trust. It also could provide a forum and a mechanism for selection of the trustees. The trustees could be selected in a manner similar to the selection of representatives to the forum itself: through a blend of election and sortition, although a portion of the trustees probably should be appointed from among known experts in their fields. Where political governance will benefit from diversity, so will the trust.

A strong case probably would exist for the trust to operate autonomously from the council, although with reporting responsibilities to it and oversight from it. The trust's fiduciary relationship would be defined as being to the global commons: all people alive and unborn, as well as to all nature on Earth. The global governance council would be the institution through which this fiduciary relationship would run, provided that the council itself represented the global commons. The structure outlined above is designed to ensure this. The council would have the ability to police any mismanagement on the part of trustees but it would not have any direct role in management. The trust, in turn, would be able to report to the council any activities it saw as violating the directives of material discipline and widespread prosperity. The relationship would be analogous to the one between a central bank and a government, except that it would cover common wealth, as distinct from financial wealth.

Relationship to Other Transnational Organizations

A global governance council would not sit in isolation from other transnational organizations. Other TNOs, such as the UN, the EU, the OECD, the World Bank, IMF or any number of others would have close relationships with the council. In fact, there would likely be benefits to their holding 'observer' status at the council, being able to contribute to its deliberations without actually holding a vote. A global governance council would be essentially a civic organization, whereas many existing TNOs are more like clubs, being more or less exclusionary. The UN, for its part, has only nation-states as members. It is an open question whether a global governance council could be a successor organization to the UN.

The civic nature of the council would preclude any direct role for transnational corporations. The council's prime directives—material discipline, widespread prosperity or any other—cannot be clouded by the profit motive. The role for TNCs would be to find creative ways to increase value by increasing material efficiency. They could make policy recommendations to the council in an advisory capacity, although a voting role would undermine the council's independence.

Establishment and Growth

A global governance council thus described would be an organizing nexus of stakeholders in agreed-upon directives drawn from multiple societal levels. By representing a sample of everybody, even down to individuals, it would speak with authority on progress, both local and global, toward its stated goals. Those goals, such as material efficiency and widespread prosperity, would be common goals, not only for member entities but for everybody. By integrating verified information on local, regional and national efforts with global systems indicators provided by a network of NGOs and research institutions, the council would communicate an integrated picture of how we are doing, what is working, what is not working, and what yet might be done.

Early adopters likely will be smaller jurisdictions municipalities and counties—who have practical, local problems to solve and who understand the benefits of sharing knowledge and experience with one another. Larger jurisdictions would come on board over time. However, the structure of the council would ensure that smaller entities retained power even as the membership of larger ones grew. Equal representation across hierarchical levels would empower all participants.

A structure such as this could not be established *de novo* without some prior trialing and refinement. Some of its novel institutional structures and processes could be tested and refined at smaller scales before a global institution were established. Again, local governments are good environments for this kind of development, as they are more nimble than larger governmental organizations and they represent smaller constituencies. Their institutional innovations, where successful, tend to percolate upward.

Two particular institutional innovations of the global governance council are the random selection of representatives and the inclusion of stakeholders from a variety of levels of social organization. The sections below discuss how these innovations could become established.

Direct Participatory Democracy by Sortition

The idea of blending elected seats with randomly selected seats akin to jury duty is not at all new. The committees running the ancient Athenian government were filled by sortition rather than through elections. The random selection of individuals who made themselves available for selection was considered true 'democracy'—in the sense of rule by the people—in a way elections were not, although only men were eligible to serve. Ancient Athenians viewed this 'demarchy' as vastly superior to monarchy or oligarchy. Sortition was later widely used in Italy during the Middle Ages and the Renaissance to choose members of city councils or the committees appointing them. It has been used sporadically in other countries since, such as in Canada, India and Switzerland.

Although sortition's most common modern application is to jury duty, it is not limited to this. A substantial and growing academic literature has examined its application to the composition of modern political assemblies.[30] One of the early recent proponents of the idea was the same Ernest Callenbach who wrote *Ecotopia*. He and Michael Phillips advocated for the US House of Representatives to be formed this way.[31] Other writers, such as John Gastil and Erik Olin Wright, took up the idea of a blended system, proposing one house of a bicameral legislature to be elected and

[30] For example, Burnheim (1985, 2006), Delannoi and Dowlen (2010), and Bouricius (2013).

[31] Callenbach and Phillips (1985).

the other to be sortitioned.[32] This proposal is similar to the blended structure of the global governance council outlined above. Sortition is viewed by its advocates as a means to overcome what Gastil and Wright call the twin "pathologies of legislative elections": the co-option of electoral campaigns by the wealthy or the malfeasant, and the balkanization of legislatures into party-political blocs.[33]

The idea has percolated out of academia in recent years to attract the attention of the popular media, finding a receptive audience among many dissatisfied with electoral politics as usual.[34] When we read in the news of 'citizens' assemblies', it is usually implied that these are formed through sortition. In a 2016 interview, President Barack Obama suggested a future role for citizen participation as a way to integrate what he saw as people's thoughtfulness and open-mindedness into the political process.[35] Organizations promoting sortition have become established during the early years of the twenty-first century to build on the academic research in an attempt to establish pilot projects in actual legislatures. Examples include openDemocracy and the Sortition Foundation.[36]

Sortition forms part of a broader field known alternately as participatory democracy or deliberative governance, concerned primarily with designing institutions for citizen participation.[37] Deliberative democracy is common at the local and regional levels, where assemblies encourage and accommodate public input. County councils and boards of supervisors, as well as regulatory agencies, are examples of deliberative democracy in action. However, the vast majority are not actually sortitioned bodies: the public has a voice but decisions are made by elected officials.

Independent of academic debates about whether legislatures should be wholly or partially sortitioned, the process has certain clear advantages over a purely electoral system. In addition to reining in electioneering and diffusing party politics, it is, by definition, more democratic. By incorporating private citizens serving short terms, sortition should be expected to

[32] Gastil and Wright (2018).

[33] Gastil and Wright (2019).

[34] For example, Davis (2012).

[35] Philip Galanes, The Roles of a Lifetime (Table for Three Series), *The New York Times*, Sunday Styles, May 8, 2016, p. 14.

[36] https://www.opendemocracy.net/ourkingdom/collections/great-charter-convention/constitutional-convention and https://www.sortitionfoundation.org/. The latter is campaigning to replace the UK House of Lords with a House of Citizens.

[37] See, for example, Fung and Wright (2003), Dryzek (2010) and Bachtiger et al. (2018).

reduce the risk of corruption. It gives private citizens who otherwise would not have the means to run for office a chance to wield real political power, albeit for a limited time. People would feel more empowered and engaged with the political process knowing that, even if they themselves might not be selected, others like them are. Legislatures also likely would be more diverse as a result, and more willing to entertain new ideas. Votes cast in a legislature would become votes of conscience to a much greater extent than in today's electoral system, not only among sortitioned representatives but also among elected ones.

Objections to sortition focus on expertise and accountability. If one believes that the country should be run by the best and the brightest, then why would we choose a sample of representatives spanning the entire bell-curve of intelligence, knowledge, capabilities or temperament? Although this is a reasonable objection, the electoral system itself does not select the best or the brightest, either. Sortition is not a viable mechanism for filling administrative roles, where domain knowledge and experience count. The ancient Athenians did not use it to choose their military commanders, for example; it was employed instead for public oversight. In this role, it matters less if a portion of sortitioned representatives occupies the lower end of the bell-curve, provided they are well informed, they vote according to their conscience, and their number is balanced by elected representatives. A wholly sortitioned chamber, even a relatively large one, could be at risk of making ill-conceived decisions, whereas a blended chamber would mitigate that risk significantly. Another potential check against incompetence might be the establishment of a 'fit and proper persons' test to filter out people who were wholly unsuited to the role, although great care would have to be taken to ensure that the test remained independent, incorruptible and fair to the disadvantaged; otherwise a great part of the purpose of sortition would be defeated.

The accountability objection is that people selected at random would not have a popular mandate to govern, meaning their votes somehow would be less legitimate than those of elected representatives. Also, by serving short terms from which they would walk away never to return, sortitioned representatives would have less invested in their decisions than elected ones. How would they be held accountable? These objections are weak. First, the legitimacy of elected representatives within a chamber may be questionable for other reasons, particularly if the electoral process itself is tainted with graft or money, or if legislation is tainted by lobbying. If elected legislatures were as pure as the driven snow then people probably would be satisfied with the legislative process as it is, and there would be no need to consider sortition. Sortition acts as a counterweight to election, not a replacement of it. For this reason, the votes of sortitioned representatives are no more or less legitimate than those of elected representatives. This being the case, sortitioned representatives would be no more or less accountable for their voting records than elected ones. If you were one of the lucky few to be selected, you probably would want to walk away proud of your voting record.

By way of example, suppose that the 435 seats of the US House of Representatives were chosen approximately equally by election and by sortition, where 200 seats were occupied at any one time by private individuals drawn at random from the general population and the remaining 235 seats were elected. Persons chosen by sortition would be able to opt out if they wished. Any who did would be replaced by other randomly selected individuals. Suppose also that all members, elected or sortitioned, had to pass some kind of 'fit and proper persons' test. The test would be applied to prospective electoral candidates as a condition of their initial filing, whereas sortitioned members would take it after being selected, as a condition of assuming their seat. Suppose that the test failed to catch a few percent of the sortitioned representatives who turned out to be 'bad apples' in some way, whether unreliable, incompetent, untruthful or unruly. A few percent would equate to no more than ten out of the whole 435. Since the House already has mechanisms for sanctioning 'bad apples' among its elected representatives, the risk presented by any unfit sortitioned representatives would be statistically negligible.

The global governance council outlined above incorporates sortition as a way to increase representation across hundreds of nations, thousands of regional governments and millions of local governments. There is no reason in principle to exclude individuals: therefore, sortition would apply at the individual level as well.

Assuming sortition were trialed initially at the local level, and it proved successful in increasing the participatory nature of local democracy, then any refinements and improvements would be incorporated over time into successively larger political aggregations, eventually up to a global governance council. In fact, a radiation of sortition through local, regional and national polities actually might enable the emergence of the global governance council itself. The twentieth-century 'top-down' form of democracy, characterized by a periodic electoral mandate, no longer serves decision-making processes affecting everybody, especially in a world where the barriers to entry into elected office are so high. Eliminating big money and big business from politics has long been discussed, to little avail; yet, in addition to electoral and legislative reform, a way to support more direct, participatory democracy—one that does not put too many cooks in the kitchen—could avert a great deal of social unrest in the future.

Cross-representation

Sortition may be beneficial to legislatures in rendering them more democratic but it does not by itself help to connect them politically to one another. From ancient times to the present day, legislatures have been jurisdictions unto themselves, largely closed off from their neighbors. This stovepiping presents an impediment to the emergence of planetary governance. The League of Nations and later the United Nations represented attempts to bring together collections of distinct nation states to manage global affairs. However, they made only a tiny step toward a participatory global democracy. A handful of transnational organizations exist to support local authorities in sharing knowledge and expertise.[38] Valuable though these resources are, any jurisdiction's legislative decisions, whether local, regional or national, has the capacity to affect the lives of people anywhere in the world, directly or indirectly, through its effects upon natural capacity.

A global governance council would manifest participatory governance not only among entities at different levels of organization but also among entities within the same level of organization. Governance as a participatory exercise would be built into the norms and conventions of all of its members, unlike the situation at the UN, whose member states generally regard their sovereignty as inviolate. Participatory governance could even be a prerequisite for membership of a global governance council, allowing for the conventions it embodied to be reflected all the way down the

[38] The International City/County Management Association (https://icma.org) claims more than 11,000 local government members worldwide, whom it serves with proprietary data and resources. The United Council of Local Governments (www.uclg.org), partnered with the European Union and the United Nations, has global operations divided into regional sections representing each continent. It operates a 'Municipal eLibrary' (http://www.municipalelibrary.info) to "[help] knowledge-sharing between local and regional government practitioners around the world". In the US, associations providing resources and national advocacy for local governments have existed for many decades: the Council of Mayors, the National League of Cities, and the National Association of Counties.

organizational hierarchy. The evolution toward a global governance council might in fact entail ever greater degrees of participatory governance among jurisdictions so that, over time, the conditions became established for the emergence of a global institution.

Participatory governance among similar jurisdictions could take a variety of forms. Here, a direct form of participation, called cross-representation, is examined. In cross-representation, a legislative or regulatory body representing a given jurisdiction would include voting members who are external stakeholders in its decisions. External representatives would constitute a small minority relative to the internal representatives, so as to minimize any risk of outside interests dominating the chamber. The intent of cross-representation is to open a legislature, to a modest degree, to allow stakeholder representation to flow across the geopolitical membrane.

To see how the process might work, it is first necessary to understand the sizes of the legislative bodies in question. Below the national level, data on the sizes of legislatures are somewhat spotty. However, Table 5.1 provides a sample indicating the approximate ranges.

Table 5.1 Numbers of representatives in selected local and regional legislative bodies by region

Region(s)	*Min.*	*Avg.*	*Max.*	*Sample*	*Sources*	*Notes*
Local government sizes						
Various	9	56	1,2	35	1	1
US cities	5	6	3–5	35,748	2	2, 3
US counties	3	7	6,7	3098	3	4
Regional government sizes						
Canadian provinces	19	58	125	13	4	
South African provinces	30	48	80	9	5	
US states & territories	13	135	424	56	6	5

Source: (1) Meloche and Kilfoil (2017); (2) National League of Cities, www.nlc.org; (3) National Association of Counties, http://explorer.naco.org/; (4) The Canadian Encyclopedia: http://www.thecanadianencyclopedia.ca/en/article/members-of-provincial-legislatures; (5) Wikipedia: https://en.wikipedia.org/wiki/Provincial_legislature_(South_Africa), (6) Wikipedia: https://en.wikipedia.org/wiki/Comparison_of_U.S._state_governments

Notes: (1) Sample covers municipalities in North America, Europe and Australasia; (2) US 'local government' denotes 'general purpose local government', as distinct from 'special purpose local government', which includes school districts and special districts (US Census Bureau); (3) 'City' includes 19,495 municipal governments and 16,253 township governments (US Census Bureau: https://www.census.gov/data/tables/2017/econ/gus/2017-governments.html, table 2); (4) Min. is an estimate from NaCO map; Includes US overseas territories. (5) Assembly sizes encompass both lower and upper houses

North American local government assemblies tend to be smaller than their counterparts in Europe and Australasia.[39] Regional assemblies, as one might expect, tend to be larger than local ones. National government assemblies are larger still. Among 234 sovereign and non-sovereign assemblies, the average is 210 seats, ranging from a minimum of seven (Cocos Islands) to a maximum of 2987 (China).[40]

Cross-representation might be a challenge to initiate among local assemblies having only a few members, such as in many North American ones. The addition of a single seat to a local council having only half a dozen to begin with could alter the political dynamics significantly, especially if it held a tie-breaking vote. At this level, one elected seat could be added to represent the global commons. That individual would speak for all people and systems outside the municipality potentially affected by the council's decisions, which admittedly is a big responsibility. They could be as partisan or as ill-informed as any other member of the local council; however, the important thing is that the global commons would have a seat and therefore a say in local decisions.

Local assemblies in other regions, such as Europe, having several dozen elected members, could readily accommodate a few additional seats, both for cross-representation and for the global commons. In a local assembly having the average of 56 members in table 4.01, an additional six seats could be accommodated for these purposes. With only 10 percent of the legislature representing external interests, the political balance would not likely be upset by the additions, yet the external voices together would be significant enough to warrant attention. In cross-representation, the effects would be reciprocal, in that residents of the city or county in question would themselves be able to run for cross-represented seats in other jurisdictions.

Exactly what those six additional seats would represent is an open question, one to be addressed by any local government setting out along this path. For example, two of the new seats could be assigned to the global commons: one to a representative of the natural commons and the other to a representative of the social commons. The remaining four could

[39] Meloche and Kilfoil (2017).

[40] CIA World Factbook, 2014: https://www.cia.gov/library/publications/download/download-2014/index.html.

accommodate cross-represented candidates from other jurisdictions. Certain geographical restrictions might be established, such as limiting the pool of available candidates to nationals of the state within which the local authority resides; or, more restrictively, to residents of the subnational region therein, or to neighboring local authorities. Another condition might ensure that no more than one representative would be allowed from any individual cross-represented jurisdiction. If cross-representation were allowed to encompass more than one level of political organization, then it could conceivably allow candidates representing neighboring regional governments or even the national government in a local jurisdiction. This structure would represent a first step toward a hierarchical form of governance.

An advantage of cross-representation is that it is scalable. In principle, two neighboring municipalities could try it out by adding one seat for the purpose. The cross-represented member not only would represent the interests of the other municipality in issues that affected it but also would be able to offer an outside perspective, or suggest novel ideas, on issues in which the other municipality had no especially strong interest. For instance, if one municipality were debating traffic-calming measures, the other municipality might not have a very strong interest in which specific measures were chosen; however, its representative might be able to inform the debate through the experiences of their municipality with the same issue. Although municipalities consult one another all the time, what would be different here is that one municipality would have a direct role in the decision-making of another. This is why the proportion of cross-represented seats should remain small. One of the major benefits of the system would arise where two municipalities were in disagreement over an issue, such as perhaps a zoning issue that affected both of them. Each council chamber would have an opportunity to hear its neighbor's perspective. It would have to take that perspective into account in its decision-making, as its neighbor also would have a vote. The system would tend to incentivate compromise.

Now imagine that the idea caught on, and other neighboring municipalities sought to participate. The system easily could scale from two participants to, say, ten, with each participant offering up to about 10 percent of additional seats to accommodate representatives from other participating municipalities. Not all members would be represented in one another's chambers, of course, but even partial cross-representation would create a networked structure. Even if any given participating municipality did not have representation in any of the other member councils over a given term, it could be guaranteed representation from a sample of the other municipalities in its council.

A variety of voting mechanisms can be imagined, each with certain advantages and disadvantages. A simple one would be for candidates seeking election to another municipality's council to run off against one another within their district of origin. This process would select one candidate to represent that district in elections to the target district. Then voters within the target district would choose among candidates from other districts seeking election to it. A potential disadvantage of this mechanism is that it could serve to reinforce political preferences within target districts, as voters within them would tend to choose like-minded candidates. There is something to be said for incorporating diverse, sometimes contrarian, views within the legislative chamber. A mechanism to accomplish this would comprise a vote among all source districts for candidates to a target district. In other words, the voters within the target district would not have a say in who they received as cross-represented members, although of course they would be able to vote in the same way for their candidates going in the other direction. Rather than declaring a target district at the outset, candidates would be asked to declare it only after voting was complete within their district of residence. Winning candidates from different source districts declaring for the same target district would be selected based on the proportion of votes they secured, rather than the absolute number of votes they secured, thereby controlling for variations in the sizes of local electorates. Any candidate who missed out on their first choice by this mechanism would be able to declare for their second choice. And so the process would iterate until all target seats were filled or all candidates accounted for.

Although the process theoretically could scale to any number of participating municipalities, in practice it would work best for clusters of municipalities forming a metropolitan area, or for rural counties within reasonable traveling distance of one another. Neighboring jurisdictions typically share a high proportion of overlapping issues anyway, and presumably would share a relatively large pool of candidates with knowledge of these issues and similar cultural experiences.

In regional and national legislatures, which are much larger than local councils, several additional elected seats could be available for cross-representation. In the US, for example, every state legislature could add a few seats for individuals representing the other 49 states. It would not matter which particular states they came from, only that they represented all other states. Voters in any particular state would vote for candidates seeking cross-represented seats in other state legislatures. The exact mechanism could take a variety of forms; however, the intent would be to introduce individuals into a state legislature who were elected by voters outside that state. If voters inside the state made that selection then the process would be self-defeating. Inductees could be chosen by sortition or through a blended system. Additionally, each state legislature could add seats representing the global commons, or elements thereof. The total number of seats added for cross-representation thus would be less than ten.

At the national level, a similar structure would provide elected seats within a national legislature for representatives of other regions of the world, as well as for representatives of municipalities and of the global commons. Following the classification outlined above for the global governance council, these external seats would number eleven, except in the very smallest assemblies: one for each of the seven continents and four representing the major natural system types. In this way, somebody from Africa, for example, could run for election to the Africa seat in the US House of Representatives, and conversely somebody from North America could run for election to the North America seat in the Nigerian parliament.

Voting would have to be coordinated at the continental level, a role for which the United Nations would be well-suited. Candidates for global commons seats on national legislatures would have to be nationals of countries outside the one on whose legislature they seek to sit, to avoid bias. This requirement could present constitutional difficulties, not to mention the issue of coordinating global voting.

Since an external representative would speak for an external constituency, they would be responsible for understanding the views or needs of that constituency. How this would be accomplished would depend upon the constituency in question: an external representative of the global commons on a local council obviously would face a different challenge from a representative to the US House of Representatives from Africa.

Evolution of Political Institutions

Existing political institutions tend to be self-reinforcing: any norms they may represent may become more deeply entrenched by the exclusion of a diversity of voices. In an increasingly connected world, they are an anachronism. By allowing external representatives to participate directly in debates with their internal counterparts and to vote, cross-representation would influence decisions in ways that diplomacy or lobbying could not. At the same time, however, an external representative would embody a degree of independence absent among lobbyists. Cross-representation would set up a countervailing force against lobbying which would help to prevent assemblies from becoming too inward-looking. When combined with sortition, an additional level of impartiality would be introduced, as individuals would be much less likely to vote along party-political lines, and their tenures would be relatively short. Yet because their number would be small, cross-represented or sortitioned members would not overcome internal voting power, thereby substantially preserving sovereign control.

Cross-representation and sortition are more likely to emerge at the local level than at regional or national levels. Being trialed there, they could be scaled up to larger legislative bodies. Together, they are mechanisms by which political structures may become incipiently more hierarchical and fully representative. The effect would be a kind of networked legislative process, in which legislative boundaries would become semi-porous, rather like the evolution from single-celled organisms to multicellular life. It hardly needs mentioning that multicellular organisms possess emergent properties not exhibited by any of their constituent cells.

If political representation were to become more networked and hierarchical then it would be essential for any existing legal barriers to be amended or removed. These would vary from region to region. In the US, as in other democracies, it has always been assumed that people would represent only the jurisdictions within which they have legal residency. At the local level, it might be necessary for city or county charters to be amended to accommodate cross-represented candidates. If some form of sortition also were trialed then the charters might have to be amended to accommodate that, also. Amendments such as these need not be permanent. Local authorities have the ability to suspend or amend certain charter provisions for specific periods through legislation. If the trials proved successful then the temporary amendments could be made permanent.

Networked clusters of local governments engaged in some form of cross-representation and sortition would become the spawning grounds for new, increasingly inclusionary, representative and geographically broader forms of governance. The process could take several decades. By mid-century onward, one might expect the model to permeate up into regional assemblies. In this way, a three-level hierarchy would be established. Individuals would have an opportunity through sortition to participate in local or regional assemblies for designated periods. Additionally, both local and regional governments would engage in cross-representation at their respective levels. The system then would evolve further, if technologies developed to support it, by allowing cross-representation across levels of political organization: local into regional and regional into local. Although it may appear complex, the purpose would be to give interests representing different social or political levels an insight, a voice and a direct stake in the activities of other levels.

By century's end, one might see the subnational political fabric transformed from a geopolitical patchwork into a highly enmeshed network. By this time, the nation-state, too, will have become more porous. The next phase of political evolution would see the emerging hierarchical structure extend up to the national level, eventually becoming the kind of hierarchical, representative structure illustrated by the concept of the global governance council.

Such a structure would help to reduce political conflict, even while the population continues to grow and the effects of past material excesses continue to be felt. No longer would it be possible to speak of 'us' and 'them' if one of 'them' held a seat in our assembly—complete with a vote and access to the floor—while one of us held the same privileges reciprocally in theirs. A society embracing cross-representation and sortition would become politically global in a way no society has in the historical past. Any global political institution that emerged, whatever form it took, would have the qualities of a 'meta-parliament': a hierarchically connected, multicellular political organism.

Requirements for Governmental Institutions

Present-day institutions operate on timescales of days to years and on spatial scales from the local to the national. Natural processes, on the other hand, have meaningful effects upon the economy over timescales of decades to millennia and on spatial scales from the regional to the global. Present-day institutions are inadequate to account for the effects of natural processes upon the economy, being too narrowly focused and myopic. They are set up only to react to sudden events in natural systems—sudden in the sense of occurring over days to years—and then only poorly.

Political command-and-control or autocracy is not a viable option. Some studies suggest that when people feel threatened, they seek what Gelfand et al. call 'tighter' social norms and more autocratic governance.[41]

[41] Gelfand et al. (2011).

Yet in an increasingly connected global society, only a much more deeply connected form of democracy than exists today is likely to prevail. The biased, partially disenfranchised, episodic voting system passing for 'democracy' in the majority of countries is limited and probably inadequate for the task of governance toward a stable, prosperous economy in alignment with nature. The social processes forming political representation must operate up to the global scale and must account for effects over long time-frames.

The institutional requirements listed in Table 5.2 apply not only to a future planetary economy but also to the pathway toward that future state.

Table 5.2 Institutional and governance requirements for a planetary economy

Requirement	*Description*
Spatiotemporal alignment of economic governance with natural processes	Governance of the global economy and its relation to natural processes must be global and must account for mutual system effects over many decades. The decision-making horizon thus has to be extended out many decades.
Inclusive, federated economic governance	Such governance should not be confused with the ill-conceived idea of monolithic, global government. Instead, it should be inclusive, representative and federated, with participants both setting and meeting goals. Any jurisdiction, at any level of government, should have a mechanism for participation. Individuals should have a voice as well.
Hierarchical governance to increase representation	The mutual effects of the economy and natural systems upon one another affect all levels of society, from nation states and transnational organizations down to individuals. An institution for steering the global economy toward a stable alignment with nature should include all levels of society: individuals, local governments, regional governments and national governments. The global governance council outlined in this chapter suggests one possible way to accomplish this.
Separation of civic and commercial spheres	Such an institution should not provide votes to commercial or other business entities. This is to ensure that long-term societal aims remain uncompromised by short-term profit or other business motives. Business entities may have observer or advisory roles but could not influence decision-making, financially or otherwise.
Cross-representation at local, regional and national levels	To the extent applicable, local, regional and national jurisdictions should take steps to represent stakeholders outside their geographical boundaries. These stakeholders include other regions and the global commons of natural systems. In this way, nature could have agency at all levels of the legislative process.

Source: Author's creation

This pathway will replace the 'untrammeled sovereignty' of the nation-state, as it came to be understood during the nineteenth and twentieth centuries,[42] with a sovereignty of the 'nation' of Earth, as a confluence of biophysical and geopolitical unities, whose human governance would employ a plurality of institutions across a variety of hierarchical levels.

References

Adler, D., & Varoufakis, Y. (2019). The World Bank and IMF are in Crisis. It's Time to Push a Radical New Vision. *The Guardian*, 31 January.

Bachtiger, A., et al. (Eds.). (2018). *The Oxford Handbook of Deliberative Democracy*. Oxford, UK: Oxford University Press.

Barrett, S. (1996). Building Property Rights for Transboundary Resources. In S. S. Hanna, C. Folke, & K-G. Mäler (Eds.), *Rights to Nature* (pp. 265–284). Washington DC: Island Press.

Borger, J. (2001). Bush Kills Global Warming Treaty. *The Guardian*, 29 March.

Bouricius, T. G. (2013). Democracy through Multi-Body Sortition: Athenian Lessons for the Modern Day. *Journal of Public Deliberation, 9*(1), Article 11.

Burnheim, J. (1985). *Is Democracy Possible? The Alternative to Electoral Democracy* (New ed.). Sydney, Australia: Sydney University Press.

Callenbach, E., & Phillips, M. (1985). *A Citizen Legislature*. Bodega, CA: Banyan Tree Books.

Davis, J. (2012). How Selecting Voters Randomly Can Lead to Better Elections. *Wired Magazine*, 16 May.

Delannoi, G., & Dowlen, O. (2010). *Sortition: Theory and Practice*. Exeter, UK: Imprint Academic.

Dryzek, J. (2010). *Foundations and Frontiers of Deliberative Governance*. Oxford, UK: Oxford University Press.

Fung, A., & Wright, E. O. (2003). *Deepening Democracy: Institutional Innovations in Empowered Participatory Governance* (The Real Utopias Project IV). New York: Verso Books.

Gastil, J., & Wright, E. O. (2018). Legislature by Lot: Envisioning Sortition within a Bicameral System. *Politics and Society, 46*(3), 303–330.

Gastil, J., & Wright, E. O. (Eds.). (2019). *Legislature by Lot: Transformative Designs for Deliberative Governance*. New York: Verso Books.

Gelfand, M., et al. (2011). Differences between Tight and Loose Cultures: a 33-Nation Study. *Science, 332*(6033), 1100–1104.

Mazower, M. (2012). *Governing the World: The History of an Idea*. New York: Penguin Press.

[42] Mazower (2012, p. 423).

Meloche, J.-P., & Kilfoil, P. (2017). *A Sizeable Effect? Municipal Council Size and the Cost of Local Government in Canada*. Canadian Public Administration Publication, ISSN: 0008-4840.

Prakash, A., & Gupta, A. K. (1997). Ecologically Sustainable Institutions. In F. D. M. Smith (Ed.), *Environmental Sustainability: Practical Global Implications* (pp. 47–66). Boca Raton, FL: St. Lucie Press.

Robert, H. M. (1876). *Robert's Rules of Order*. Chicago, IL: S.C. Griggs and Company. (11th ed., 2011, Da Capo Press, Cambridge, MA).

Rosenthal, E. (2009). Obama's Backing Raises Hopes for Climate Pact. *New York Times*, 28 February.

Sherwood, S. C., & Huber, M. (2010). An Adaptability Limit to Climate Change Due to Heat Stress. *Proceedings of the National Academy of Sciences, 107*(21), 9552–9555.

Snyder, G. (1985). Buddhism and the Possibilities of a Planetary Culture. In B. Devall, G. Sessions, & G. Appendix (Eds.), *Deep Ecology: Living as if Nature Mattered* (pp. 251–253). Salt Lake City, UT: Gibbs M. Smith.

Young, O. R. (1996). Rights, Rules and Resources in International Society. In S. S. Hanna, C. Folke, & K.-G. Mäler (Eds.), *Rights to Nature: Ecological, Economic, Cultural and Political Principles of Institutions for the Environment* (pp. 245–264). Washington, DC: Island Press.

CHAPTER 6

Corporate and Financial Institutions

As the previous chapter noted, the same hierarchical structure applied to political institutions of the future also could apply to corporate ones. This chapter discusses an additional institutional requirement for a crowded planet, that a hierarchical model incorporating representation from stakeholders at a variety of organization levels should be considered for the governance of large non-civic institutions, such as transnational corporations (TNCs) or other transnational organizations. These entities, having global reach and influence, would become more adaptable to a changing economic and policy environment by incorporating such representation into decision-making.

Ownership, as Marjorie Kelly points out, "is about who decides everything else in an economy".[1]

Corporate Structure and Ownership

A company, firm or corporation is a group of individuals formed to engage in an activity intended to create value. In economics textbooks, a firm is any economic institution transforming factors of production into goods and services.[2] A corporation is more narrowly defined as a firm owned by stockholders who are not liable for its actions,[3] although this textbook

[1] Kelly (2012, p. 212).
[2] Colander (2010, p. G-4).
[3] Colander (2010, p. G-2).

F. Murison Smith, *A Planetary Economy*,
https://doi.org/10.1007/978-3-030-49296-0_6

definition does not easily apply to non-profit corporations. The value a corporation produces can be measured in a variety of ways, financial and non-financial. To whom the value accrues depends on the corporate structure and purpose.

Businesses are about as old as civilization itself. The growth of family businesses, guilds and other professional organizations resulted in a division of labor mirroring that within society as a whole. Modern corporations can trace their origins back to the *corpora* and *publicani* of ancient Rome. The *corpora*, as the name implies, were considered legally as persons, with all the rights and responsibilities attached thereto. However, they were analogous to today's non-profits, having community or social missions, whereas the *publicani* were associations of private contractors. A key later development was the limitation of liability of the officers of a corporation, which may have originated in France in the fourteenth century. The government oversight of modern corporations can trace its origins to the establishment of corporations by royal charter in medieval Europe, as the Appendix to this chapter describes.

Limited Liability and Corporate Personhood

Although both limited liability and corporate personhood have been established for hundreds of years, these institutional conventions do not necessarily result in corporate behavior well aligned with the maintenance of natural capacity. They tend to benefit the officers of a corporation, who can delegate their legal personhood to the corporation itself, absolving themselves of it. An absence of direct, personal liability on the part of the officers has the potential, in some instances, to tempt recklessness in their actions on behalf of the corporation. In principle, an officer of a limited liability company has a duty to make decisions in the interests of the company, even if those decisions go against their own personal interests. They are personally liable only if they break the law. However, given the way corporate incentive structures operate, an officer's decisions on behalf of a company are almost always in their own interests, as they will benefit personally from any value accrued to the company. There remains ample, legal scope for recklessness on the part of the company, with the potential for negative external effects upon third parties. If a company loses money, then the worst that can happen to its officers is that they lose their jobs. Investors seeking to recover losses cannot 'pierce the corporate veil', as the term goes, and seek compensation from the officers individually.[4] If a company's actions affect third parties, then the officers who decided on those actions would not be liable.

This is how limited liability companies have been able to play a major role in the material expansion of the economy and its emerging coevolution with nature. The specific kind of company with the greatest role has been the limited corporation, which shields its officers from personal liability, and allows a separate class of shareholders to own stock. This structure allows corporate expansion to be financed by investment, in addition to debt. Corporate structures in which officers are liable for a company's actions, such as partnerships, have access to debt financing but generally not to investment.[5] Partnerships are, by definition, owned by the partners: there is no separate class of shareholders. An intermediate structure, known in the Unites States as the limited liability corporation (LLC), limits the liability of officers, and has a similar pass-through tax treatment to a partnership, but it does not allow a separate class of shareholders. Generally, LLCs are not publicly traded, although a few exceptions exist.[6] For a company with no need, or interest, to go public, such as Cargill or Koch, it offers a simpler corporate structure than a convertible corporation and certain tax benefits. Although some LLCs are large, as the Appendix to this chapter details, all the world's largest companies are publicly traded corporations.[7]

It is the separation of share ownership from management that allowed public trading of limited corporations to develop in the first place. Public trading provides a source of capital investment that has allowed some companies to expand to enormous size, with operations around the world. These TNCs are virtual sovereignties in their own right, as the British and Dutch East India Companies became during the seventeenth and eighteenth centuries. Wal-Mart, for instance, has an annual revenue greater than the gross domestic product (GDP) of 169 out of 195 countries for

[4] Bagley and Dauchy (2008, pp. 54–55).

[5] Bagley and Dauchy (2008, p. 60).

[6] Bagley and Dauchy (2008, p. 61). The 'pass-through' in question is where capital gains and tax liabilities accrue not to the company itself but to its partners, or 'members' as they are known in an LLC. An LLC seeking a public offering usually will convert to a C-corporation in advance.

[7] The Appendix also lists the largest partnerships in the world, showing that they are an order of magnitude smaller than the largest corporations.

which the World Bank has compiled GDP data.[8] Its annual revenue is between the 2015 GDP of Norway and that of Austria. ExxonMobil, for its part, is larger than Chile and 158 other countries. Volkswagen Group is larger than Pakistan and 156 other countries.

Having revenue on the scale of some nations, TNCs also are extraordinarily wealthy on a per-capita basis, their employees numbering far fewer than national populations. The largest privately owned firm, Vitol, employs about 7500 people, yet it is only slightly smaller in annual revenue than Finland, whose population is about 5.5 million. Vitol's revenue per capita is more than 700 times Finland's GDP per capita, and Finland is a wealthy country.

There are far more TNCs in the world than transgovernmental organizations, such as the United Nations, the World Bank and the World Trade Organization, as well as global non-profits, such as the International Red Cross/Red Crescent, Oxfam and the World Wildlife Fund. Although the world's TNCs wield more power globally than most states, they are by design non-democratic. This is not a failure necessarily of the corporate structure but rather a reflection of the glaring absence of a counterbalancing global civic institution. Having many offices in many countries, TNCs are able to hedge geopolitical risk through operational redundancy. If one country becomes unstable, operations can be relocated to others. TNCs also have the political power to play national governments off against one another, so as to obtain the sweetest possible deals for themselves. Natural environments and local people may suffer as a consequence. TNCs compete in a global market yet, unlike many national or regional markets, the global market is not comprehensively regulated.

The global market is, in large part, a product of these operations. While it is true that global markets for a large number of products have existed for many decades, the TNCs have been able to create markets across national borders, thereby rendering those borders more diffuse. Their influence over the legislative and regulatory processes of the states in which they operate also further serves to reduce barriers to the movement of capital and resources globally. Indeed, as a 1995 survey of multinationals in *The Economist* found, TNCs tend to prefer "one set of rules everywhere around the world rather than ... different rules in different places". Such standards not only make TNCs' lives administratively easier but also impose the same standards on their competitors. Although a single set of

[8] Source: http://data.worldbank.org/data-catalog/GDP-ranking-table.

rules would result in relatively higher standards across the board, they also would make it harder for TNCs to hide any shady operations.[9]

Jeremy Bentham and his free-trade apostles argued that commerce is a civilizing influence. It is more likely the reverse: commerce is an outcome of civilization. Although commerce has taken a variety of forms—the Mesoamerican civilizations, for example, were based on barter, and in centrally planned economies, commerce was state-controlled—the emergence of corporate personhood more than once in different times and places suggests that it may be a more general outcome of the evolution toward more complex civilizations and trade networks.

Somewhere along the way, the scale of commerce as it grows may begin to have negative effects on civilization, in the sense of introducing social instability as a product of insufficient regulation of competition among TNCs. The twenty-first century could be the period when we see this. Competition is a useful force, but only if steered toward an overarching goal of long-term social stability. In a global economy having measurable effects on natural systems, the TNCs experience an inherent tension between a need many of them recognize to protect natural capacity on the one hand and the competitive nature of their businesses on the other. Royal Dutch Shell, for example, could elect to not extract a new fossil fuel resource in a previously unexploited region, but they have no influence over what the other oil and gas giants do. If Shell does not go and get it, ExxonMobil might. An increasing number of TNCs have called upon national governments to coordinate more effective regulation of common-pool resources. The global governance council outlined in the previous section is one possible institution that could accomplish this. If successful, it would significantly reduce or substantially eliminate the tension in global markets between the exploitation and the conservation of natural capacity.

This is not to suggest that the council would impose financial regulations upon TNCs directly. Mechanisms already exist within many countries to regulate corporate activities at national, regional and local levels. The participation of entities from all those levels within the council would provide a basis for potentially establishing a critical mass of jurisdictions to structure similar regulatory environments in most or all regions where TNCs operate. In so doing, the civic environment would exert a selection pressure upon TNCs to adapt. This is something they are uniquely well

[9]Schmidheiny and Zorraquín (1996, p. 63), citing a 1995 article in *The Economist* by Adrian Wooldridge, "Who wants to be a giant: a survey of multinationals".

equipped to do, many having large reserves of capital along with significant research and development budgets.

A regulatory environment that accounts for natural capacity also would lessen certain negative effects of corporate personhood. One of the prevailing inequities of corporate personhood is that it has allowed corporations to argue legal cases as individuals when it suited them, and yet also protect their executives from personal liability—as individuals—in the face of civil legal action. In the history of environmental litigation, for example, the main reason why plaintiffs brought civil suits against polluting corporations in the first place is because regulations aligning executive action with natural capacity were insufficient or non-existent. If, on the other hand, corporate financial performance was aligned with material discipline, then any material indiscipline on the part of a corporation would reflect negatively on the executives responsible. In this way, the question of whether corporate personhood would help or hinder economic alignment with natural capacity would become moot.

To Profit or Not to Profit?

Limited liability and corporate personhood are institutional characteristics of self-serving companies. Any company acting first and foremost in its own interests is an institutional analog of the self-maximizing individual of the standard economic model. In fact, firms may come closer than individuals to the imaginary self-maximizing entity precisely because they insulate their constituent individuals from liability. An important driver for this behavior is the profit motive, yet it is not the only one. Firms, to varying degrees, have consciences, too. Large firms may emphasize self-perpetuation and market share over profit. And a significant sector of the economy consists of non-profit corporations, whose purpose is mission-based rather than wealth-creation.

On a crowded planet, to what extent should companies be self-serving or mission-driven? What role would market instruments play? Would limited liability and corporate personhood be an impediment to economic alignment with nature, or an asset? And what would be the purpose of the profit motive? These questions are interrelated.

Because Earth is a materially closed system, profit-taking within an economy incurs a loss of natural capacity. The free gift of solar energy allows material to move around and be organized into complex structures. It represents a form of income. 'Profit', in a general sense, is a temporary excess of value. It has a thermodynamic analogue to potential energy, as it tends to dissipate over time. The value in question can be measured any number of ways: energetically, materially or—as is conventional in economics and finance—in currency terms. Yet there exists a tendency toward the liquidation of profit over time, as it is converted into useful processes or outcomes. Profit as a near-term motivator is a means toward the concentration of capital, which in turn enhances an entity's persistence within its system. For example, as a corporation approaches maturity, profit gives way to earnings as a benchmark of performance. Growth gives way to maintenance.

A planetary economy likely would be a low-profit economy. Because the 'principle of conservation of value' articulated in the last chapter states that value cannot be created or destroyed, only moved around, an economy tending toward alignment with natural processes would exhibit a reduction in profit margins over time. The system would become characterized by a blend of competitive and cooperative relationships, as are found within mature ecosystems; whereas, in a growth economy, as has predominated for the past couple of centuries, competition plays a relatively greater role.

In a growth economy, any increasing demand on natural capacity creates a natural capacity debt within the economy. This debt may appear as financial profit if not fully accounted for. If the principle of conservation of value applies, then that profit eventually would be wiped out and would become a deficit. If the deficit became large enough, it could result in economic collapse. Yet significant financial profit can be gained through increases in material efficiency. An economy on an efficiency path would realize internal financial profits through improvements in technologies, both physical and social, until diminishing returns began to take hold. A state of material stability would be characterized by low financial profits and a wider prevalence of non-financial measures of value.

The institutional question is whether this long-term evolution suggests certain changes in business structures. In spite of the worldwide prevalence of non-profits, the vast majority of the present economy's financial wealth flows through for-profit corporations, whose basic structures were formed during the early days of capitalism. The way society draws the "membership boundaries of the enterprise" results in a master-servant relationship between capital and labor, writes Marjorie Kelly. "The ruling oligarch in our economy is capital, for capital is sovereign within enterprise. Only capital has the right to vote inside most publicly traded companies, and only capital has a claim on profits."[10] Indeed, it is ironic that within western society, which prides itself so much on democratic political representation, the vast majority of publicly traded corporations retain many of the trappings of the dictatorship.

And yet, when they get into trouble, they call upon the public for help. General Motors (GM), for instance, was bailed out by the US government in 2009 because the administration feared a deep economic recession would become a full-blown depression if GM were allowed to fold. The bailout ensured that most GM workers were able to keep their jobs; yet, by remaining a limited corporation, GM held onto an institutional structure which insulated—and continues to insulate—its executives from corporate losses, while arguably over-rewarding them when the company does well.[11] In an economy that continues to grow, publicly owned companies can appear as public assets because their stock valuation trends upward. But if a large company over-reaches during a boom cycle, such as by management for short-term gain rather than long-term stability, then it can become a public liability during a downturn. Steer clear of an elephant trying to balance on one leg.

The administration missed an opportunity to restructure the company in a less capitalist, more equitable way, such as a private, employee-owned corporation. In an employee-owned structure, company executives and the board of directors serve at the pleasure of the workforce rather than at the pleasure of absentee investors. One of the reasons employees band together to form unions is because corporations seek profits. Unions do not exist in the non-profit sector, although in theory they could form there for other reasons. Typically, unions act as a counterweight against excessive profit-taking on the part of capital-owners by representing the interests of those who do not own capital. Employees, first and foremost, seek job and wage stability. In a non-profit environment, there is no profit-taking, nor is there a distinction between those who own capital and those who do not. The tension between capital and labor evaporates.

An expectation among the workforce of job and wage stability does not rule out growth, of course, but it rules out unsustainable or unstable growth. If persistence for a corporation becomes more desirable than

[10] Kelly (2012, p. 173).

[11] There is some justification to the argument that the money might have been better spent bailing out the employees of GM and allowing the company to fold. This argument is taken up in the following chapters.

growth, then employee ownership rather than investor ownership may be a preferable institutional structure.

This line of reasoning suggests that a public, investor-owned structure might suit the growth phase of a corporate life-cycle, and that a private, employee-owned structure might apply to the maintenance phase of corporate maturity. Some large companies blend the two. Novo Nordisk, for example, retains 51 percent ownership by a private foundation, the remaining 49 percent of shares being publicly traded. The foundation exercises 'mission-based control', as Kelly describes it, even while a large part of the company's capital trades freely in public markets.[12]

In a growth economy, one should expect a relatively greater prevalence of corporate structures designed for growth; whereas, in a materially stable economy, a greater prevalence of corporate structures would support long-term persistence. In an economy on a path toward long-term alignment with natural capacity, the distinction between 'for-profit' and 'non-profit' may become increasingly artificial and could eventually evaporate. 'Profit', in a planetary economy, may take many forms, financial being only one.

Today, the founder of an enterprise must decide in advance whether to pursue financial profits, or not. This decision has significant implications, as it determines how the enterprise will be organized, where and how it can raise money, and the kinds of taxes it will pay. Investors in a for-profit enterprise expect to make a financial return on their investment through the profits the enterprise generates. It also must pay taxes on those profits. Investors in a non-profit enterprise invest in social outcomes: they expect a social return on their financial investment. No financial profit is made, so no tax is payable on it.

The for-profit/non-profit distinction is convenient for tax purposes, because a for-profit venture has both an inflow and an outflow of money, whereas a non-profit venture has an inflow of money and an outflow of social benefit. However, it is inconvenient with respect to the full range of benefits that ventures provide. For-profit ventures unquestionably deliver social benefits, and certain kinds of non-profits deliver measurable for-profit benefits, business incubators being an example. Many large corporations operate non-profit foundations in their name, understanding that their financial well-being depends at least in part on nurturing their social

[12] Kelly (2012, p. 186).

milieu. And some large TNCs, such as Novo Nordisk and IKEA, are owned and managed by non-profits.

The non-profit structure is a significant portion of the economy. Within the 20 largest US cities, almost 40 percent of the 200 largest organizations are non-profits. 40 million out of about 300 million Americans are served by publicly owned utilities, whose rates average about 11 percent cheaper than their for-profit counterparts. Eighty-three million Americans belong to a credit union, of which there are about 10,000 nationally. 120 million Americans, about 40 percent of the total population, are members of about 48,000 co-ops.[13] Worldwide, the cooperative model has spread to more than 90 nations, encompassing close to a billion members, or about one-eighth of the world's population. Non-profits employ more people than all multinationals combined.[14]

One attempt to bridge the gap early in the twenty-first century has been the establishment of the B-corporation in the US, with variants in a handful of other countries, such as the UK, Italy and France. A B-corporation, where B stands for benefit, can register as a conventional C- or S-corporation in the US, with an ability to raise investment. Directors and executives commit to delivering social benefits in addition to fulfilling their conventional fiduciary duties. While the B-corporation does not take the place of a conventional non-profit, it can be a natural successor to one in situations where a mission-based program, funded perhaps by foundations or government grants, found an opportunity to convert into a venture-backed product or service offering.

The Convertible Corporation

The B-corporation has emerged as a structure to support so-called mission-based ventures within an economic environment that does not yet reward enterprises supporting natural capacity. Yet in a future planetary economy, those rewards would have become well established. Within that environment, profit-taking would serve the growth phase of a company's development, but not necessarily its early development or its maturity. What kind of corporate structure could serve these needs?

One possible model is suggested here: a 'convertible corporation' (not to be confused with a C-corporation). Rather than converting from

[13] Alperovitz (2005, p. 116).
[14] Kelly (2012, pp. 12, 142).

non-profit to for-profit status and back again, a new venture would retain the same institutional structure all the way through. It would circumvent the trap formed by the market economy of the early twenty-first century that precludes an enterprise from seeking grant funding as a non-profit and private investment as a for-profit with an equity structure allowing a return. Early-stage research and development is a risky investment from a venture standpoint, yet it may be a potentially valuable social investment. Most types of pure scientific research are government- or foundation-funded because their *actual* future benefits are impossible to anticipate, whereas their *possible* future benefits are expanded by scientific advances. The scientist in the laboratory, or even the tinkerer in the garage, needs an institutional vehicle able to receive early-stage R&D grant funding. This same vehicle could later receive growth-stage investment if it also incorporated an equity structure. If the corporation grew to maturity then the investors might seek a liquidity event, in the form of acquisition or public offering. The impetus for persistence rather than continued growth on the part of the company would dampen the profit motive, to a point where conversion back to a non-profit status may be desirable.

A convertible corporation would incorporate initially as a non-profit, or as a kind of blended structure able to accept funds not only from grant-giving organizations but also from private investors. At a point when a new technology was ready for commercialization, the corporation would activate a section of its Articles establishing common and preferred stock for the purpose of taking on growth funding. The founders of the company might at this point vest a portion of the common stock. In this way, the company would convert from a non-profit or low-profit state into a mostly profit-seeking state. The non-profit or low-profit part of the company would not be terminated but instead would be maintained alongside the for-profit activity. This would allow mission-driven activities to proceed in parallel with the maximization of shareholder value, without the one cross-contaminating the other. The non-profit part of the business would come back into play during the post-growth phase.

One useful feature that could be built into the Articles of a convertible corporation is a term limit for the growth phase. The for-profit phase would persist for a specific period, thereby providing investors a guaranteed liquidity date, and company management a deadline for achieving breakeven, or whatever metric were designated for liquidity. The provision would be analogous to a financial option. On or before the liquidity date, an investor would exercise their option to sell the asset, recoup their investment and make a return. If a venture failed to perform within the time allowed—many do—then investors would sell at a loss or walk away, and the employees would move onto the next venture.

Investors in for-profit ventures look for a return typically within three to five years, or perhaps as long as seven or eight in technologies with long commercialization cycles. Ventures undertaken during the twenty-first century to develop solutions to aligning economic activity with natural dynamics increasingly will bump up against the conventional venture 'playbook', which is inherently capitalist. That playbook does not consider or acknowledge the wider normative context; venture capitalists have a fiduciary responsibility to deliver a specific return on investment to their institutional clients. Many cleantech ideas, for instance, may have gestation times of ten years or more to commercial viability, yet at the same time, their potential upside may be significant.

A convertible corporation would suit innovating sectors well. Germinating a new technology could take several years, even before it was ready for commercialization. The commercialization cycle itself could be lengthy, as markets tend to take a long time to adopt new ideas or practices. Armed with a guaranteed liquidity date, however, a venture investor might be more motivated to ride with their venture a little longer, rather than pulling out the moment their institutional clients got the jitters from some unrelated crisis in the capital markets.

If a venture makes it through commercialization and its great new technology becomes established in the marketplace, it would become financially self-sufficient and its social mission would help to move the normative needle—even if perhaps only incrementally—in the direction of overall economic alignment with nature. Customers would be buying the solution and they would be buying into the idea behind it. The investors would exercise their liquidity option.

The convertible corporation now would have preferred stock in the hands of institutional investors and common stock in the hands of employees. Profit-seeking will give way to preservation of earnings and maintenance of market share. Conversion back to non-profit status may be desirable if doing so supports the company's social mission or if net profits were heavily taxed. Its Articles already would be set up to accomplish this.

Since a convertible corporation would consist of a non-profit part and a for-profit part, it would have the flexibility to manage the for-profit part in a way that was consistent with the mission of the non-profit part. Whether the liquidity event was a public offering or an acquisition, the non-profit part would be guaranteed a source of funding essentially into perpetuity. The question would be around the degree to which the non-profit had operational control over the business established by the for-profit part. Many options are possible here. One scenario is the Novo Nordisk one, where the non-profit foundation maintains a majority ownership in the for-profit company, effectively bounding its range of possible business activities. Another scenario is more hands-off: the for-profit venture is acquired or goes public, and a percentage of its revenues would fund the operations of the non-profit. In essence, the non-profit would remain a major shareholder in the for-profit business, or in its parent company.

The flexibility assured by the convertible corporation structure is a natural extension of the hybrid structures emerging early in the twenty-first century. In fact, it is a natural replacement for them. It points toward a blurring of the distinction between 'for-profit' and 'non-profit', which itself is a blurring of the distinction between capitalist and collectivist. The future is an amalgam of the two.

The Business Entity of the Far Future

Yet even the concept of the convertible corporation itself is only a stepping-stone toward a more fully integrative concept of the corporation. In the longer term, a more fully generative structure of the corporation may begin to emerge. Gone would be any recognition of 'profit' as a distinguishing feature. Instead, imagine an idea taking shape within a community of stakeholders, some of whom would be the originators of the idea, some of whom represent the global commons in some way, others of whom represent sources of funding. The traditional capitalist model relies upon a private investor, who expects a return on investment over a certain period. The way corporations develop in the far future could be radically different, given that they would all be required to operate consistently within a prime directive of some kind. New ventures would be established to deliver value not only to their founders but also to a wide variety of stakeholders, encompassing employees as shareholders, customers, suppliers, communities and nature.[15]

[15] Kelly (2012, p. 156). Businesses may be taking early steps in this direction. In 2019, the Business Roundtable, an association of CEOs of more than 180 of America's largest companies, announced a new purpose for the corporation which marks a break with the neoliberal

A fully generative business structure would evolve out of something like a convertible corporation into a more deeply networked structure, operating generatively at all points in its life-cycle, yet able to seek the resources needed to grow rapidly if the opportunity presented itself. It would still incorporate a way to reward the people who were instrumental in moving it along its development. Just as the concept of 'value' may broaden over time to encompass not only financial but a variety of non-financial metrics, so the rewards from corporate development might become similarly diversified.

One could imagine a form of corporate governance replicating some of the features of political governance outlined above, such as agency for nature and representation for diverse stakeholders. This kind of inclusionary governance, as Kelly points out, would represent a cultural shift from the exclusionary governance of today's corporations.[16] It implies a variety of forms of investment, financial and otherwise, returns from which would be realized at different phases of the corporate life-cycle. It even suggests different kinds of funding institutions from today. The distinction between a grant-giving foundation and a growth-stage venture investor, for instance, might cease to exist. Instead, early-stage funding would establish the social value of a new enterprise; growth funding would establish its financial security, potentially delivering a liquidity event to growth-stage investors; and then, after that event, the mature business would become established upon a stable platform from which it could more or less self-sufficiently deliver on its various measures of value over a long time.

An inclusionary form of corporate governance also throws into question the applicability of corporate personhood and limited liability. If the business entity of the far future were generative rather than extractive in its purpose, co-operatively owned rather than capitally owned, and inclusionally governed rather than exclusionally governed, then it might be expected to act largely in the interests of external stakeholders rather than against them. This being the case, there might then be little liability to limit in the first place. The boundaries of the corporate 'person' also would be less clear-cut, because of its inclusionary governance. This is not to say that

doctrine of placing primacy upon shareholder value. Under its new principles of corporate governance, shareholder value takes equal place alongside supporting communities, investing in employees and adding value to the supply chain. See https://www.businessroundtable.org/policy-perspectives/corporate-governance.

[16] Kelly (2012, ch. 9).

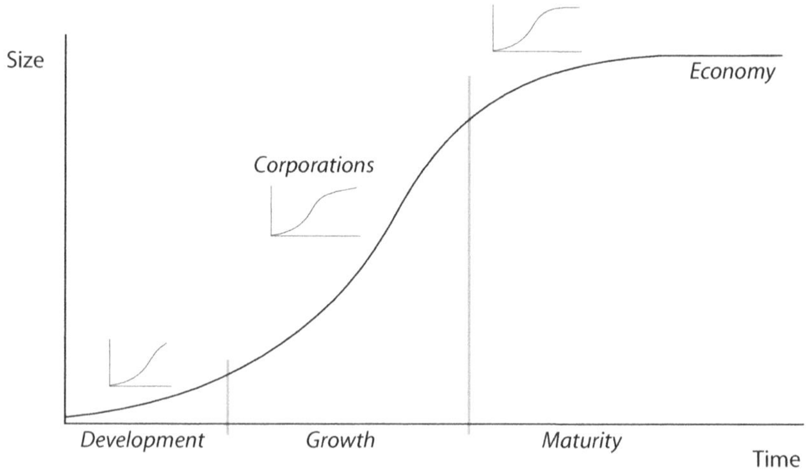

Fig. 6.1 Self-similarity of corporate growth and economic growth over the maturation process of a materially constrained economy. (Source: Author's creation)

corporate personhood and limited liability would become long-lost conventions, but they could evolve into legal provisions significantly different from today's.

The self-similarity between the corporate life-cycle and the long-term economic one should by now be apparent. Figure 6.1 illustrates it. In an economy's early exponential development, such as during early industrialization, natural capacity is not particularly constraining. Corporate culture, correspondingly, is highly extractive, based on an assumption of boundless natural wealth. This culture is illustrated by the small graph in the bottom-left of the figure in which growth does not level off. During the growth phase of the economy, which has occupied much the twentieth century, corporate culture continues to assume the possibility for unlimited growth, even while acknowledging that growth slows as a corporation approaches maturity. This culture is shown by the small graph in the middle section. During the maturation phase of the economy, which is yet to come, the limitations of natural capacity and the leveling of population cause growth to slow, with the material size of the economy reaching a plateau. In this future economy, corporate culture is no longer a growth culture but, like the economy itself, is a maintenance culture. Corporate growth is perceived as an intermediate phase of the corporate life-cycle rather than an end in itself, illustrated by the small graph, whose shape replicates the large one.

Although this future economy may be materially large, its material intensity would be significantly lower than today's economy, by virtue of recirculating most of its material. Material stability would represent a low-growth or zero-growth state, in material terms, although, as argued earlier, technological innovation still could produce growth in economic wealth. Even so, the profit motive would be significantly weaker than today, particularly for businesses that had reached a mature state. Motives for persistence and social value would be correspondingly stronger. Such a régime might render largely unnecessary an institutional or legal distinction between 'for-profit' and 'non-profit' in the sense we use the terms today.

The deep ecologists argued that a prerequisite for an ecologically sustainable economy would be a refocus of economic activity onto small, local entities, with a view toward local self-sufficiency. Keeping supply chains local certainly is an admirable way to reduce material intensity, yet this does not necessarily mean the demise of the transnational corporation. TNCs may have a useful role to play within a globally connected economy on a pathway toward alignment with nature, provided that their governance and culture morph from the extractive and exclusive toward the generative and inclusive.

Given the present economy remains very much within the growth phase—although already past the maximum growth point—significant returns may be possible over the coming decades from investments in ventures delivering material efficiency or improvements in natural capacity more generally. These returns are not only financial, of course, but also social and cultural. Although the financial markets themselves may realize decades of growth ahead by investing in material discipline, they too eventually will see diminishing returns as the economy plateaus and the profit motive fades. In a future planetary economy, the absentee investor may have become largely a relic of the old growth economy.

Role of Publicly Owned Corporations

Although much of the foregoing has focused on generative and extractive ownership in relation to the interests of direct stakeholders in corporations, it has not addressed the broader public interest. Some corporations, particularly utilities, form natural monopolies, having control over critical infrastructure, such as water, power, transit networks or telecommunications. These infrastructures, it is often argued, serve a public interest above and beyond the corporate interest, by virtue of their monopolistic nature. Public ownership would accomplish not only elected accountability for their safe operation and maintenance, but also convergence with a broader social principle or prime directive of some kind. Taking utilities as an example, there exists an innate conflict of interest between the motivations of passive shareholders for short-term profit and the provision of water or electricity for maximum material efficiency and system redundancy.

For this reason, investor-owned utilities generally cannot be expected to work in the public interest. For instance, allowing customer-generators to sell power back onto the grid eats into utilities' profits. The solution, argues Kate Aronoff, is to take investors' thirst for short-term profits out of the equation.[17] Non-profit structures, such as co-operatives and publicly owned utilities, represent the model to follow, because their operations can be tuned to align with broader public imperatives. One of the facets of corporate evolution we may see as the twenty-first century unfolds is an acknowledgment of the public importance of critical infrastructure—particularly in the face of climate change—and a growing motivation for public or non-profit control over that infrastructure. This does not necessarily imply rampant nationalization but it might entail a reworking of the investment model from the passive profit-seeking stockholder to the publicly accountable stakeholder.

Financial Institutions for a Planetary Economy

In a planetary economy, the volume of trading on the stock market probably would be significantly lower than in the growth economy of the twentieth and twenty-first centuries. The market itself likely would be less volatile as well, although certainly not approaching an equilibrium as suggested by standard theory. The inherent nonlinearities of the market would continue to provide room for arbitrage and speculation. A principal difference in this market compared to today's would be a much-reduced tendency toward boom-and-bust dynamics. An economy in stable alignment with nature would be less subject than the current one to stock-market crashes, whereas at the same time upsides would be limited by the pervading low-growth environment. Excessive speculation would be

[17] Aronoff (2017).

virtually impossible. A stock market whose total capitalized value is neither particularly variable nor growing significantly over time would offer relatively lower returns and lower risk. If a crash did occur, its wider economic effects would be relatively weak. Fundamentally, the stock market would not be the dominant economic influencer it is today.

The question then arises what would be required to reach that future state, given today's starting point of a growth economy characterized by a large, volatile stock market? Long-term economic stability will depend upon long-term stability of the financial markets. Therefore, reducing cyclical volatility in the financial markets will become a key requirement for establishing a pathway toward economic alignment with nature.

The institutional challenge is one of curtailing irresponsible financial practices rewarding short-term gain at the expense of long-term objectives. Financial speculation, perhaps justifiably, has taken on certain negative connotations in the wake of the speculative bubbles of the 1980s, late 1990s and mid-2000s. Yet speculation in and of itself is not what causes financial and economic instability: it is excessive speculation. Where in the 1970s the US financial sector traded an amount far less than the US gross national product (GNP) in a year, by the 1990s it traded a sum in currencies, stocks, bonds, futures and derivatives greater than the annual GNP of the entire US *within a single month*.[18] Only about 3 percent of that staggering volume ($800 billion per day) went into trade and real investment. The remaining 97 percent was speculation. By the early 2000s, the volume of trades had risen to $2 *trillion* per day, of which less than 2 percent was for trade and investment.[19] As of 2013, this figure had grown to $59 trillion per day, which was nearly 3.5 times the US annual GDP of $16.8 trillion. And this volume was *down* from a high of about $175 trillion per day in 2008, immediately preceding the economic recession of 2009 to 2013.[20]

The economic historian Fernand Braudel referred to this kind of speculation as the 'financial economy' to distinguish it from the 'real economy' of tangible assets.[21] Hernando de Soto described financial assets as leading an "invisible parallel life" alongside their tangible equivalents.[22] To Nigel Dodd they were not invisible at all but a "mutation of money into a

[18] Phillips (1994) cited in Dowd (2004, p. 184).

[19] Phillips (1994) cited in Dowd (2004, p. 185).

[20] Figures from the World Bank, World Development Indicators: http://data.worldbank.org/country/united-states.

[21] Braudel (1992).

[22] de Soto (2000).

cancerous state":[23] what Susan Strange referred to as 'casino capitalism'.[24] Many commentators on the era of financialization observe that "money itself appears to have been financialized".[25]

Transnational corporations, needing to do business in many different currencies, are among the major speculators. Holding assets in many currencies, they can earn billions of dollars a day, or hedge against risk in certain currencies, by buying and selling in large volumes on the world's currency markets. While technology has been a major enabler of financial deregulation and speculation, changes in the ownership structures of major investment banks—many of which went from being partnerships to publicly traded companies since the 1980s—also are responsible. The recursive chains of obligations among parties speculating in the financial sector are 'phantom assets', to use Marjorie Kelly's term, being only distantly tied to physical assets. Yet when the total obligations become multiples of real GDP, the value of all physical assets could not possibly repay them.[26] Such a network would be doomed to collapse, given enough time.

In the US, the Glass-Steagall Act of 1933, which separated commercial banking from investment banking following the stock market crash of 1929, was repealed in 1999 by the Gramm-Leach-Bliley Act. Prior to the passage of the Gramm-Leach-Bliley Act, regulations in the US, including Glass-Steagall, were more restrictive than those in eighteen European Union and G-10 countries. Yet, by the time the act was passed, many financial institutions already were all-but-operating under a commingling of commercial and investment activities through a Glass-Steagall loophole allowing affiliations among commercial and investment institutions.[27] The repeal of Glass-Steagall may have been a major contributing factor to the financial meltdown and recession of the late 2000s.

In 2010, the US passed the Dodd-Frank Act, a package of broad financial reforms aimed to improve the stability, accountability and transparency of the financial system in the wake of the market meltdown of 2008–09. Among its numerous provisions was the so-called Volcker Rule, named after the former Chairman of the Federal Reserve, Paul Volcker, limiting banking entities from holding no more than 3 percent of their

[23] Dodd (2014, p. 201).
[24] Strange (1994).
[25] Dodd (2011) cited in Dodd (2014, p. 245).
[26] Kelly (2012, ch. 5).
[27] Barth et al. (2000).

assets in hedge funds or private equity funds. The rule also protects against conflicts of interest in dealings between banks and private funds, and requires financial institutions to hold relatively larger capital assets during periods of expansion.[28] The jury remains out on whether Dodd-Frank has curtailed cyclical instability in the US financial markets, as questions have lingered about the robustness of enforcement. Large-scale instability remains a concern.

All investment is a form of speculation. Even grant-giving by government bodies and private foundations, as a form of social investment, is speculation: a bet that this or that project will improve people's lives. The commingling of commercial banking, investment banking and insurance services following the repeal of Glass-Steagall would not have resulted in financial meltdown had the total volume of speculation remained small, even if it might have raised legal or ethical conflict-of-interest issues. The voluminous financial speculation of the 2000s on risky investment vehicles, such as credit default swaps, took place because the prevailing regulations allowed it to. As Dodd writes, "securitization [such as of home loans] did not so much involve diversifying risk as its wholesale *transfer*: first to counterparties ... and second..., through government, to *society*" (italics his). The point at which government stepped in to take on both liquidity and credit risks, these 'private' risks became socialized.[29]

Within the current growth phase of the economy, an institutional structure for the financial markets in support of long-term investment will become essential. The continuing growth of the economy provides ample opportunity for speculation; yet the present financial system is not structured to invest in material discipline. Chapter 11 describes financial reforms that would bring financial speculation into line with the value of financial institutions' real assets, as distinct from their leveraged ones. Specifically, certain firewalls would be erected between depository functions, lending functions and financial brokerage, thereby rendering it virtually impossible to use leverage financial assets for short-term speculation.

In support of long-term investment, the institutional challenge extends beyond financial institutions. Institutional investors operate over much shorter time-horizons than the one facing the whole economy in its

[28] Wikipedia: Dodd-Frank: https://en.wikipedia.org/wiki/Dodd%E2%80%93Frank_Wall_Street_Reform_and_Consumer_Protection_Act.

[29] Dodd (2014, p. 124).

relation to nature. The need to align the economy with natural processes is not first and foremost a financial need, but a social one. The financial system will play an important role but the investment in this outcome must come from social institutions, in particular, from government. Chapter 11 outlines monetary and financial reforms that would allow governments to make the necessary long-term investments, harnessing the wealth-multiplication properties of the financial sector.

Banks in particular have a critical role to play in the stability of the financial markets and of the economy as a whole. As successive financial crises over the past century have shown, the establishment of pyramid schemes by large banks—mostly within the letter of the law, it should be added—has probably been the single-largest contributor to global financial instability, with significant negative effects upon the economy as a whole. Fractional-reserve banking, for example, played a major role in the subprime mortgage crisis of the mid-2000s, the bankruptcy of Lehman Brothers, and the deep worldwide economic recession that followed. The global interconnectedness of the financial system ensures that banking crises now unfold on a global scale rather than the local or regional ones of the historical past.[30]

The problem, Nathan Schneider argues, may not be the large banks being too big but being too selfish. "What if," he asks, "a bank seeking new borrowers had to bring them on as co-owners, with a vote in the boardroom?"[31] The institutional structure for a business entity of the far future outlined in this chapter applies equally well to banks as to any other business, representing an extension of Schneider's idea. Banks, he argues then would become too *good* to fail rather than too big to fail. They would be restrained from careening off into predatory behavior by the very stakeholders who otherwise might be vulnerable to it.

Reinhart and Rogoff argue for an international financial regulatory institution, which would place controls upon fractional reserve banking and other forms of leverage. Such an institution would impose common financial reporting requirements upon banks operating anywhere in the world, seeking also to ensure transparency in financial reporting. It would be politically insulated from lobbying, being transnational in scope.[32] Such

[30] Reinhart and Rogoff (2009, p. 156).

[31] Schneider (2016).

[32] Reinhart and Rogoff (2009, pp. 281–282).

an institution could be an adjunct to the global governance council outlined earlier. Yet it is probably a way off. Simpler, more direct institutional reforms, which can be implemented at the national level—most importantly in the large economies—probably are needed in the near term to stabilize the financial system. These, and other, wider overarching institutional questions relating to monetary and financial governance, are covered further in Chap. 10.

Appendix: Corporations from Early History to the Present Day

A Brief History of the Corporation

The existence of distinct institutions for conducting business is as old as civilization and its accompanying division of labor. In early civilized societies, the emergence of functional specialization among individuals gave rise to a new institution: the family business. In ancient Sumer, as far back as 5000 years ago, temples made agricultural loans to peasant families, and extended credit nominated in silver to merchants to finance trade expeditions with neighboring populations. Ancient China operated a kind of manorial system similar to the one that grew up in Europe during the Middle Ages, 2000 years later. Records of family businesses and individually owned enterprises existed in ancient India as far back as 2800 years ago. There were guild-like organizations, including *nigama* (merchants), *pani* (craftspeople) and *sreni* (artisans). The *sreni* in particular shared certain characteristics with modern corporations. A *sreni* was able to hold property separately from its owners, make its own rules of conduct, and sue or be sued in its own name. In other words, it embodied the concept of corporate personhood familiar today in the institution of the limited corporation.[33]

By Roman times, it had become customary for various kinds of business, such as manufacturing, trade or agriculture, to be conducted through private institutions having a degree of legal standing before the prevailing

[33] Sources: Wikipedia: (1) Sumer https://en.wikipedia.org/wiki/Sumer; (2) Economic history of China before 1912, https://en.wikipedia.org/wiki/Economic_history_of_China_before_1912; (3) Economic history of India https://en.wikipedia.org/wiki/Economic_history_of_India.

governments of the time. The term 'corporation' is itself Roman, deriving from the Latin *corpora*, for 'body', the name implying the treatment of the collective as equivalent to a person. Whereas Roman corporations typically served community or religious purposes, similar to today's non-profit corporations, the for-profit business entities in Roman society were known as *publicani*. Initially, these were associations of contractors formed to bid on contracts. Over time, they evolved into permanent companies with investors and managers. Some of the larger ones employed thousands of workers, no doubt exhibiting a division of labor reflecting the broader division of labor within society, a characteristic of modern corporations.[34] There is evidence that some of the *publicani* received corporate status, granting limited liability to investors.[35]

The corporate form that has come to dominate the global economy over the past 400 years can trace its roots to the *publicani*. The modern limited corporation in its various forms is structured so that investors' liability for losses is limited. However, for more than 1000 years after the fall of the Roman empire, most private businesses were family-run concerns. The oldest known surviving business is Kongo Gumi of Japan. It was founded in 578 to build the Buddhist Shitenoji Temple, which still stands. Other notable old businesses include Stiftskeller St. Peter, a restaurant inside St. Peter's Abbey in Salzburg, which was recommended as a dining spot by Alcuin, a follower of Charlemagne in 803. Staffelter Hof of Germany is considered the oldest surviving winery, dating from 862. Two of the oldest known pubs are Sean's Bar in Ireland (ca. 900) and the Bingley Arms in England (953). The oldest surviving brewery is thought to be Weihenstephan of Freising, Germany, founded in 1040. Offerton Mill in England is the oldest operating mill, dating from 1068, and the Halydean Corporation, which today is an agricultural holding company headquartered in Wisconsin, traces its roots all the way back to an English church-owned chartered corporation holding land and livestock beginning in 1128.[36]

[34] Galbraith (1973, pp. 81–83).

[35] Source: Risk Encyclopedia: History of Corporations, http://www.riskencyclopedia.com/articles/corporation/.

[36] Sources: (1) Wikipedia: List of oldest companies, https://en.wikipedia.org/wiki/List_of_oldest_companies; (2) Business Insider: How 16 of the oldest companies on Earth have

The evolution of the Halydean Corporation is itself a microcosm of the evolution of the private corporation over the past several centuries. Most entities having corporate status after the fall of Rome were public or non-profit entities, such as cities, guilds, universities and monasteries. Corporate status was granted typically by royal charter; hence, they were known as chartered corporations. Private businesses, being family affairs, typically passed from one family generation to the next. In Japan a system of *mukoyoshi* arose, in which a business owner with no male heirs, or with male heirs considered unfit to inherit the business, could legally adopt another heir.[37] This allowed Japanese companies to preserve their cherished 'family-run' status, at least legally; yet it is not a huge leap from the *mukoyoshi* system to one where many unrelated individuals hold a stake in a private concern. In Halydean's case, ownership was appropriated by the crown in 1545, the company becoming rechartered as a Royal Burgh. In 1602, it was assigned to private owners as a Proprietorship of Barony and Lordship. It continued to operate as a chartered private corporation in the UK until 2014, when it was acquired by new owners and reorganized as a Delaware Corporation.[38]

Two well-known early chartered corporations were the British and Dutch East India Companies, founded in 1600 and 1602, respectively. They were given royal charters to exclusively operate trade routes to the East Indies, yet were privately owned and privately funded. Their often predatory operations in those regions over many decades resulted in the establishment of British colonies in the Indian subcontinent and Dutch colonies in what is now Indonesia. As the governments of the two nations took direct control of those colonies, the companies themselves were dissolved: the Dutch in 1799 and the British in 1874.[39] In chartering these trading companies, the English and Dutch crowns had drawn upon the

been making money for centuries, http://www.businessinsider.com/oldest-companies-on-earth-2014-8.

[37] Source: Priceonomics: Why are so many of the world's oldest business in Japan? http://priceonomics.com/why-are-so-many-of-the-worlds-oldest-businesses-in/.

[38] Source: Wikipedia: Halydean Corporation https://en.wikipedia.org/wiki/Halydean_Corporation.

[39] Sources: Wikipedia: (1) East India Company, https://en.wikipedia.org/wiki/East_India_Company; (2) Dutch East India Company, https://en.wikipedia.org/wiki/Dutch_East_India_Company.

long-established tradition of granting royal charter to guilds. Guilds were formed to create monopolies in certain kinds of business or certain geographical regions. They were established, among other things, to protect trade routes. Guild members competed with one another for business as sole proprietors or as family-run affairs, yet competition from outside was prohibited.[40] What the guilds lacked that modern corporations have was limited liability.

While the British and Dutch East India Companies are cited as two of the earliest examples of the limited liability structure, there may be a case for the Bazacle Mill in Toulouse, France, to be the oldest. In 1369, the mill owners, who shared a perpetual lease on a portion of the Garonne River where the mill is situated, signed a profit-sharing agreement. They were sued in 1372 by a merchant to recover a loan made to one of their members ten years earlier. The concept of limiting liability to shareholders, according to the French author Germain Sicard, can trace its roots to this lawsuit, which prompted the establishment of a new type of ownership contract known as *pariage*.[41] Prior to pariage, all the owners of the Bazacle Mill were at financial risk from the lawsuit, in the same way that the partners in a partnership would be today.

The practice of forming a corporation by charter continues in modern times, only not usually by royal grant any more. Municipalities are formed by public charter, which governs the terms and conditions under which a city or county does business. The Articles of Incorporation of a private company are typically filed with a local or regional governmental entity having jurisdiction over it, that entity then granting the corporation a license to operate. The Articles of Incorporation name the shareholders and officers of the company, as well as its terms and conditions of operation. Upon adoption, they serve the same role as the charter of olden times.

World's Largest Corporations by Category

Tables 6.1, 6.2, 6.3, and 6.4 list the largest public, private, partnership and non-profit corporations in the world, as of 2014 to 2015.

[40] Source: Risk Encyclopedia: History of Corporations.

[41] Sicard (1954).

Table 6.1 Ten largest public or state corporations in the world as of 2014 to 2015, showing type, ownership, exchanges where listed, headquarters, annual revenue and effective period

Rank	*Name*	*Type*	*Ownership*	*Public exchange listing(s)*	*HQ*	*Revenue (US$ bil.)*	*As of*
1	Wal-Mart Stores, Inc.	C-corp. (US)	Family/ public[a]	New York Stock Exchange (NYSE)	Arkansas, USA	$482	Jan. 2016
2	Sinopec Group	Corporation	State-owned	Hong Kong and Shanghai	Beijing	$455	2014
3	China National Petroleum Corp.	Corporation	State-owned	NYSE, Hong Kong and Shanghai	Beijing	$428	2014
4	Saudi Aramco	Corporation	State-owned	None	Dhahran, Saudi Arabia	$338	2014
5	State Grid of China	Corporation	State-owned	None	Beijing	$333	2014
6	Samsung	Conglomerate[b]	Mixed[c]	Korean, London and Luxembourg[d]	Suwon, South Korea	$305	2014
7	Royal Dutch Shell	Public limited company (UK)	Public	London, Euronext, NYSE	The Hague and London	$273	2014
8	ExxonMobil	C-corp. (US)	Public	NYSE	Irving, Texas	$268	2015
9	Kuwait Petroleum Corp.	Conglomerate	State-owned	None	Shuwaikh, Kuwait	$252	2014
10	Volkswagen Group, AG	Corporation	Public	Frankfurt and others[e]	Wolfsburg, Germany	$245	2014

Source: Author's creation, from various sources, including Wikipedia, Forbes, Fortune, Bloomberg and company websites

[a]The Walton family, which founded the company, holds a greater than 51 percent stake in Wal-Mart. The remainder is publicly owned in the sense of being traded on public exchanges

[b]Samsung owns or part-owns many subsidiaries in many countries, having a variety of corporate structures. Some are publicly traded, others not

[c]Ownership is part private, part public. For example, the major shareholders of Samsung Electronics include the company itself (in the form of treasury stock), Samsung Life Insurance, the Korean National Pension Service, Samsung C&T Corporation, and the company chairman

[d]Exchanges listing stock in Samsung Electronics

[e]VW Group's primary listing is on the Frankfurt Stock Exchange, with secondary listings on the London, Luxembourg, Swiss and New York Stock Exchanges

Table 6.2 Ten of the largest privately held companies reporting revenue, showing type of corporate structure

Rank	*Name*	*Type*	*Industry*	*HQ*	*Revenue (US$ bil.)*	*As of*
1	Vitol	Corporation	Commodity trading	Rotterdam and Geneva	$270	2015
2	Cargill	LLC	Agribusiness	Minnetonka, MN, USA	$120	2015
3	Koch Industries	LLC	Conglomerate	Wichita, KN, USA	$115	2015
4	Trafigura Group	Corporation	Commodity trading	Singapore[a]	$97	2015
5	Schwarz Gruppe, GmbH	Conglomerate	Retail	Neckarsulm, Germany	$90	2015
6	Robert Bosch GmbH	Corporation	Manufacturing	Gerlingen, Germany	$80	2015
7	Aldi Einkauf GmbH	Corporation	Retail	Essen and Mülheim, Germany	$60	2010
8	Dell	LLC	Computers	Round Rock, TX, USA	$59	2015
9	Albertsons	LLC	Retail	Boise, ID, USA	$58	2015
10	INEOS Group	Limited company	Chemicals	Rolle, Switzerland	$54	2014

Source: Author's creation, from various sources, including Wikipedia, Forbes, Fortune, Bloomberg and company websites

[a]Trafigura's 2015 annual report lists Trafigure Group Pte. Ltd. As the new legal consolidating entity for Trafigura Group businesses, taking over from the Dutch-registered Trafigura Beheer BV

Table 6.3 World's largest five firms having a partnership structure (LLP or general partnership)

Rank	*Name*	*Type*	*Industry*	*Ownership*	*HQ*	*Revenue (US$ bil.)*	*As of*
1	Pricewaterhouse Coopers	Network[1]	Professional services	Private	London	$35	2015
2	Deloitte	Network	Professional services	Private	New York	$35	2015
3	Ernst & Young	Network	Professional services	Private	London	$27	2015
4	KPMG	Swiss Cooperative	Professional services	Private	Amstelveen, Netherlands	$24	2015
5	BDO International	Network	Professional services	Private	Brussels	$7	2014

Source: Author's creation, from various sources, including Wikipedia and company websites; https://en.wikipedia.org/wiki/List_of_wealthiest_organizations

Table 6.4 World's largest non-profit organizations by value of assets held

Rank	*Name*	*Type*	*Industry*	*Ownership*	*HQ*	*Assets (US$ bil.)*	*As of*
1	Catholic Church	Roman Curia[a]	Religion	Sovereign[b]	Vatican City	$100s[c]	–
2	Church of Jesus Christ of Latter-Day Saints	Non-profit[d]	Religion	Private	Salt Lake City, UT, US	$40–150[e]	2012
3	Stichting INGKA Foundation	Dutch non-profit	Philanthropy, retail[f]	Private	Liechten-stein	$59	2018
4	Bill and Melinda Gates Foundation	US non-profit corporation	Philanthropy	Private	Seattle, WA, US	$42	2015
5	Harvard University	University endowment	Higher education	Private	Harvard, MA, USA	$38	2015

Source: Author's creation, from various sources, including Wikipedia and organization websites; Business Financial Post, 31 January 2018

[a]Akin to a royal court, acting under the direction of the Pope

[b]The Holy See is the sovereign jurisdiction of the Catholic Church, recognized under international law

[c]Because the Catholic Church is not a unitary organization but a collection of dioceses, its total assets are not fully known. Estimates range into the hundreds of billions of dollars

[d]Incorporation may vary from country to country. In the US, the LDS Church is an unincorporated entity operating a variety of non-profit and for-profit subsidiaries

[e]A 2012 article by Bloomberg Businessweek placed the net worth of the LDS Church at $40 billion. Other estimates, which include other real property and investment funds, vary up to $150 billion

[f]Stichting INGKA, also known as the IKEA foundation, is the holding company of IKEA

References

Alperovitz, G. (2005). *America beyond Capitalism*. Hoboken, NJ: Wiley.

Aronoff, K. (2017). Utilities Companies Won't Let You Sell Your Own Solar Power. Why Not? *The Guardian*, 1 August.

Bagley, C. E., & Dauchy, C. E. (2008). *The Entrepreneur's Guide to Business Law* (2nd ed.). Mason, OH: Thomson West.

Barth, J. R., Brumbaugh, R. D., Jr., & Wilcox, J. A. (2000). The Repeal of Glass-Steagall and the Advent of Broad Banking. *Journal of Economic Perspectives, 14*(2), 191–204.

Braudel, F. (1992). *Civilization and Capitalism* (Vol. 1–3). Berkeley: University of California Press.

Colander, D. C. (2010). *Economics* (8th ed.). New York: McGraw-Hill.

de Soto, H. (2000). *The Mystery of Capital: Why Capitalism Triumphs in the West and Fails Everywhere Else*. New York: Basic Books.

Dodd, N. (2011). 'Strange Money': Risk, Finance and Socialized Debt. *British Journal of Sociology, 62*(1), 175–194.

Dodd, N. (2014). *The Social Life of Money*. Princeton University Press.

Dowd, D. (2004). *Capitalism and Its Economics: A Critical History* (2nd ed.). London: Pluto Press.

Galbraith, J. K. (1973). *Economics and the Public Purpose*. Boston, MA: Houghton Mifflin.

Kelly, M. (2012). *Owning Our Future: The Emerging Ownership Revolution*. San Francisco, CA: Berrett-Koehler.

Phillips, K. (1994). *Arrogant Capital: Washington, Wall Street and the Frustration of American Politics*. New York: Harper Perennial.

Reinhart, C. M., & Rogoff, K. S. (2009). *This Time is Different: Eight Centuries of Financial Folly*. Princeton University Press.

Schmidheiny, S., & Zorraquín, F. J. (1996). *Financing Change: The Financial Community, Eco-Efficiency and Sustainable Development. With World Business Council for Sustainable Development*. Cambridge, MA: MIT Press.

Schneider, N. (2016). Breaking Up the Banks Might Make Things Worse. Instead, Let's Take Ownership. *The Guardian*, 14 April.

Sicard, G. (1954). *The Origins of Corporations: The Mills of Toulouse in the Middle Ages*. English Trans. M. Landry and Ed. W. N. Goetzmann, Yale University Press (2015).

Strange, S. (1994). *Casino Capitalism*. Manchester, UK: Manchester University Press.

CHAPTER 7

Policy Development

Policies are formed by institutions. An institution, being a social group having a certain structure and organization, convenes to decide the nature of a problem, what to do about it, and how. The what and the how is policy. The process by which an institution decides these things is policy development.

If the norms and institutions that served modern society in the past are not altogether well-suited to supporting a planetary economy, then the same must be true of today's policies and the ways they are developed. To continue as usual would be like "requiring a man to wear the coat which fitted him when a boy", to quote Thomas Jefferson. As norms change, so do institutions, and therefore policies. "Changes in laws and constitutions", Jefferson continued, "must go hand in hand with the progress of the human mind."[1] To quote another American politician, Al Gore, writing two centuries after Jefferson, "science thrives on uncertainty and politics is paralyzed by it."[2]

Policy Development in the Face of Uncertainty

Policy development in the face of radical uncertainty cannot follow the mechanical paradigm of twentieth-century economics but instead must be evolutionary and adaptable. A brief example shows why. Julian LeGrand

[1] Meacham (2012, pp. 467–468).
[2] Gore (2006, p. 260).

F. Murison Smith, *A Planetary Economy*,
https://doi.org/10.1007/978-3-030-49296-0_7

developed a concept of 'production substitutability' by applying the mechanical formalism of Pareto optimality to the allocation of limited resources, a common policy problem. "An allocation of resources is efficient", he wrote, "if it is impossible to move towards the attainment of one social objective without moving away from the attainment of another objective",[3] this being an adaptation of the Pareto rule. In LeGrand's formulation, individual utilities in the standard Pareto analysis were replaced with aggregate objectives, presumably competing with one another for fulfilment. In principle, then, one could trade off any number, *N*, of social objectives, creating an *N*-dimensional policy space. He added a caveat that this formulation is possible only if it has the same mathematical structure as the conventional social welfare function, implicitly assuming that the objective of society is to raise individual utilities.

"There are two forms of trade-off to be distinguished", LeGrand stated: "the value trade-off and the production trade-off. They are distinct not only in conception but also in the methods necessary to establish their existence and magnitude."[4] While social efficiency theoretically is a state attained by trading off among social objectives, the objectives themselves are determined by consensus among people: that is, they are driven by values. "It would not be possible", LeGrand continued, "to find a definition of efficiency that was value-free; for efficiency can only be defined in relation to more fundamental objectives, acceptance of which is dictated by values."[5]

LeGrand was correct about the normative basis of economic 'efficiency'; however, his approach had three problems. First, he merely showed that under certain assumptions one can extend the standard microeconomic concept of Pareto efficiency to a group level, which does not tell us very much about how policies actually are developed. Second, and more fundamental, a policy system with many degrees of freedom—that is, where the number of social objectives is large—may not allow a single optimal allocation. Third, and most fundamental, such a policy system is not really a mechanical analogue anyway. Just as real markets are evolutionary systems, so the same is true of policy systems. If it really were possible to optimize across many policy objectives, such an optimum probably would have been found and put into action.

[3] LeGrand (1991, p. 27).
[4] LeGrand (1991, p. 27).
[5] LeGrand (1991, p. 15).

Policy development needs adaptive, learning-based approaches, which mirror the trial-and-error of evolution. Policy objectives, it must be understood, are never actually achieved, only approached, because of the constantly shifting social and natural environment within which they are developed. Just as technology development has moved away from rigidly linear methods inherited from military command structures to so-called agile methods which are opportunistic and adaptable, so policy development would benefit from a rejection of linear methods inherited from standard, twentieth-century economics in favor of 'agile' approaches.

The globally interconnected nature of living and non-living systems on Earth presents a particularly wicked policy problem to a materially large economy on an increasingly crowded planet. By the time the coevolution between the economy and nature became recognized, it had snuck up on a constellation of institutions systematically unprepared for it. The structure and organization of governments, businesses and international institutions were remnants of a bygone age in which the economy was, or was perceived to be, insignificant within an apparently boundless natural world.

Cognitively, people are adapted to respond decisively to sudden threats, such as the appearance of a voracious predator. Our primal instincts probably will never catch up in time to modern-day threats, such as climate change, which are diffuse and stealthy. We will have to rely on our conscious intellects. Science and technology have changed the human perception of nature and of society's relation to it in ways that people living 500 years ago could scarcely have imagined. At the same time, what was a barely perceptible threat at the dawn of the industrial revolution itself has gathered pace, so that now, early in the twenty-first century, decisive action is being demanded.

The standard economic response has been to offer ways to 'internalize the externalities' into the price mechanism of the market. Aside from the methodological error of treating the market as a mechanical analogue, this response is epistemologically flawed in that it does not account for radical uncertainty. It cannot be assumed that the market will internalize all unanticipated future externalities. Even if it could, a measurement problem would remain of converting natural values into financial ones. The radical unpredictability of nature renders the forward projection of natural values into present market values an exercise in futility.

Ecological economists and others offered a partial workaround in advocating a 'precautionary principle' in economic policymaking. Economic policy, like the Hippocratic oath taken by doctors, should aim to 'do no harm' to nature. In practice, harm would be minimized rather than eliminated altogether.[6] Yet precautionary thinking does not address how economic policy would be formed, nor the kinds of economic policies that would be instituted.

In the meantime, governments, who could have begun passing legislation as early as the 1970s, have been plagued by an 'after you' syndrome. Presented with the professional uncertainty of natural scientists—a natural by-product of scientific caution—and the relative certainty of economic advisers—a by-product of their theoretical orthodoxy—no national government dared be the first to step out and address the growing threat of nature's coevolution with the economy, for fear of losing any macroeconomic advantage. On climate change, for instance, it took about 40 years for natural scientists to remove sufficient uncertainty about whether it was actual rather than a statistical artefact, and to confirm it as largely caused by people. This is one reason why resource taxes or carbon taxes have languished in legislatures around the world since they were first floated in the early 1990s.[7] During that time, economists' advice to policymakers reflected a *certainty* (in their minds) about the benefits of establishing market-based solutions to social and economic problems, and the benefits of market-based economic growth. Policymakers can work with advice offered from a basis of certainty because it is politically defensible. Finally, by the late 2010s, political momentum to legislate on climate change has intensified, because the language used by the natural scientists to urge action has become much more direct and unequivocal.[8]

Legislation is the only realistic way forward, J.K. Galbraith argued. The state, rather than the market, is supposed to be the arbiter and protector of the public interest,[9] even though it may often fall short of this ideal. Absent a global legislature, it falls to national, regional and local governments to legislate; also to supranational bodies in certain areas where they have jurisdiction. Transnational corporations have edged along in the absence of government action but they, too, have been hampered by the same 'after you' hesitation as nation-states.[10] Rapid, large-scale change requires some form of global governance. It requires legislatures to

[6] Costanza et al. (2015, pp. 121, 171).
[7] See, for example, Schmidheiny and Zorraquín (1996, p. 26).
[8] IPCC (2018), USGCRP (2018).
[9] Galbraith (1973, p. 289).
[10] See in Schmidheiny and Zorraquín (1996, p. 27).

recognize, in concert, that the "public damage [of not taking action] … [would be] greater than the public enjoyment of a service or product".[11]

The key to policymaking for an economy facing an existential threat is to act now, based on an identifiable future that people want. That future can be expressed as a set of policy requirements, such as those outlined in this volume. Uncertainty, however, is inherent to the process, as Joseph Stiglitz pointed out, "[and] good policies recognize this".[12] The implication is that if policymaking must perforce be iterative and evolutionary, resembling the systems it seeks to influence, then twentieth-century economic orthodoxy would no longer have any place in the committee room or in the executive briefing room.

In a planetary economy, the global interconnectedness with nature would no longer be seen as a policy problem but as the policy *solution*. Because it would be normal to view the economy and nature as structurally coupled subsets of a whole system, it would be normal to develop policies to improve natural capacity, thereby returning benefits to the economy. Somewhere along the path from the present economy to the future one, a shift would take place from an 'after you' mentality to an 'altogether now' mentality. An 'altogether now' mentality would signal an emergent holistic, or whole-systems understanding, a kind of normative takeover within modern society, allowing the establishment of institutions and policymaking processes for economic alignment with nature.

A first step along this path would be the recognition of first-mover advantage. The macroeconomic disadvantage that many national governments have perceived in the past about moving ahead on policies for material efficiency, if it exists at all, would actually be short-lived. A large carbon dioxide emitter who unilaterally reduced its emissions by, say, 20 percent over a ten-year period might find itself at a competitive disadvantage during that period. But then its fortunes would change. Five to ten years of technological innovation in emissions reductions would find its way into increased economic productivity, and potentially a head start in new industries, resulting in long-term comparative advantage.

[11] Galbraith (1973, p. 318).
[12] Stiglitz (2003, p. 199).

Social Selection as Driving Force for Policy

The normative and cultural shifts of the historical past represent social 'speciation events', to borrow from the language of evolutionary biology. Thinking of them this way, the question becomes, how might policymaking within the present economy encourage such a speciation event in the future? The process would be one of selection upon the variation in policy effectiveness. This is not altogether a new idea. Studies of policymaking as an iterative process of trial and error go back at least as far as the 1970s.[13]

Up to the present, most attempts to develop economic policies accounting for natural processes have drawn upon single-point valuation studies, such as of ecosystem services,[14] whose underlying assumptions are contestable, even if they may be reasonable. Attempts have been made also to voluntarily incorporate such valuations into corporate accounts and best practices, although some commentators have called for it to be required by legislation.[15] The problem with single-point estimates is that they are not actionable within real markets because they do not send price signals. All they can do is give policymakers a rough idea of the value of maintaining natural capacity.

In the spirit of what Robert May called 'robust, self-correcting strategies',[16] what is needed is price signals which act as indicators of natural capacity. By allowing the value of instruments as indicators of material intensity to evolve over time, it becomes possible to allow the market itself to discover these values.

Any given instrument operating within a market is the result of a policy. It was created through legislation or regulation, reflecting a process of policy development. In an economy on a path toward material efficiency, policies that reduced material efficiency would be selected against within the prevailing normative and institutional environment. Consequently, the mix of market instruments within the economy at any point along the path would reflect the prevailing selective environment. It is impossible to say in advance what those instruments would be or how they would work together. However, the same process of selection that would maintain a stable policy environment for a planetary economy also would guide the present economy toward that future state.

[13] For example, Majone and Wildavsky (1978), cited in Nelson and Winter (1982, p. 384).
[14] For example, Costanza et al. (1997).
[15] See, for example, Juniper (2013, p. 281).
[16] May et al. (1978).

A policy is implemented through instruments, such as quotas, incentives, taxes and moratoria. Some instruments operate within markets, others outside them. The extent to which a policy spreads through the legislative and regulatory environment depends upon the performance of the instruments it brings forth. In order for an instrument and its underlying policy to persist within the social system, it must, at a minimum, not violate one or more prime directives. Even if the normative standing of a prime directive for material discipline were weak, as it is today, any instrument that enhanced both material efficiency and prosperity would carry a selective advantage, and therefore should be expected to spread.

Policy development for a planetary economy, then, is not about trying to accurately measure the value of 'environmental externalities' for the calibration of market instruments but is instead about trial and error. As Beinhocker describes, selection acts upon a 'portfolio of options', a basket of different kinds of policies. It is impossible to know in advance which options will perform best, or when. However, if the selection process begins with a 'diverse collection of small bets' then some of them may pay out. Beinhocker argues for companies to foster greater experimentation by enabling managers to make many small-scale bets, thereby fostering greater adaptability.[17] The same applies to any kind of institution, and to policymaking in general, particularly in the face of radical uncertainty. Many bets will not pay off: the failures are an integral part of the process. Policymakers, and the people who elect them, should not view failed small bets as failures of performance. "We may not be able to predict or direct economic evolution", Beinhocker writes, "but we can design our institutions and societies to be better or worse evolvers."[18] The evolution of policy for a planetary economy thus becomes an iterative cycle of adaptation, diversification, selection and amplification.

If policy development mirrors how natural systems evolve anyway then it holds promise for long-term success. Yet this way of thinking about it still may be a conceptual leap for today's policymakers. Employing trial and error might seem tantamount to an admission of failure, yet there exists a degree of trial and error anyway in policymaking that is not always admitted. The challenge is to embrace the idea that trial and error is preferred.

[17] Beinhocker (2006, p. 359).
[18] Beinhocker (2006, p. 324).

As new policies spread through the policy 'meme pool', so old, maladapted policies would be replaced. The social and political environment determine which policies, at which times, become prevalent. A policy moving the needle on material efficiency but also perceived as draconian—for example, limiting the operation of internal combustion vehicles to one day per week per owner—might not spread within a society, like today's, on the cusp of significant climate change, but it could become established within the same society later if things became really dire.

One of the great benefits of trial and error is that it produces surprises. Within complex, adaptive systems, most surprises turn out to be evolutionary dead-ends. Within the policy environment, happy accidents can occur when a piece of legislation or regulation produces unintended beneficial consequences. Yet the reason why some policies have unintended consequences is the same reason why some apparently well-reasoned policies fail in the policy meme-pool. It is impossible to predict how a new idea will fare within the complex system that is human society. It is impossible, therefore, to predict how a new proposal to increase the material efficiency of the economy actually will influence natural capacity. When policymaking is an exercise in trial and error using a portfolio of small bets, not only will a majority of such bets fail but also some will succeed, perhaps in surprising ways.

Start Small

If the process consists of seeding the legislative and regulatory environment with a variety of small bets, then the best place to begin might be at a small scale. The previous chapter made a case for proving and refining a participatory institutional structure first at the local level, then, once it was sufficiently developed, taking it up to the regional level. In this way, eventually a global, hierarchical governance structure could emerge. The same reasoning applies to policymaking by social selection.

Local governments are able to pass all manner of ordinances and resolutions to promote material efficiency within their jurisdictions, such as recycling and composting programs, restrictions on disposable plastics, and energy efficiency incentives for buildings and vehicles. An example of a beneficial unintended consequence of local legislation is the effect upon air quality of pedestrianizing town centers, which took off in Europe in the 1970s. The original intent was to reduce congestion; the air-quality benefits were not widely articulated at the time.

A municipality beginning with none of these kinds of measures, and with a limited budget, might wonder where to start. Yet it could readily discover which option might be best adapted to its particular circumstances by placing a series of small bets in the form of neighborhood pilot programs. One neighborhood gets bicycle lanes on a trial basis, another gets a park and ride scheme, another gets a recycling and composting scheme, for instance. After a year or two, the pilots can be compared in terms of their material performance as well as their rate of uptake and their popularity. The municipality then would be in a position to select which ones to prioritize for roll-out across their entire jurisdiction. A degree of competition among neighborhoods for selection even might emerge, which of course would increase overall participation, as well as the demand for permanent programs.

Grow from There

Going a level up, it is possible to see how programs established at the municipal level would become pilots at the regional level. Within a region where some municipalities have established programs and others have not, a regional government could fund the establishment of pilot programs to discover, through a process of selection, which programs could be established at the regional level. For example, measures to reduce the consumption of non-compostable plastics have taken hold in a number of cities and counties within California during the early 2010s, although the state government has yet to consider statewide measures. Its evaluation of the various municipal programs would weigh material efficiency against financial costs in considering whether or how to offer financial incentives for the adoption of compostable or recyclable materials.

There is, of course, nothing new about piloting ideas before implementing them. However, the difference in the way the legislative or regulatory process would unfold within a selective environment lies in the way a jurisdiction would make a commitment to a program. Decisions to embark on pilot programs conventionally are made at the staff level, usually with approval from a departmental commission. Municipal councils or regional assemblies rarely are involved at this stage. They become involved only once a department has completed a pilot and seeks legislation to establish a permanent program. Such legislation becomes necessary particularly where municipal revenues, such as from taxes, would be used. The legislation typically is written to adopt and implement a single program. A pilot managed by a department does not represent a commitment by a local or regional government to adopt a program. In contrast, a policymaking approach adopting a selective mechanism would build into the legislation itself a commitment to select among a portfolio of options having first been trialed through a pilot phase. The role of local participatory governance in this process is clear.

The difference may appear superficially trivial, yet it is important in two ways. First, on a practical level, it is a commitment to actually do something rather than to simply investigate or evaluate. Second, on a more fundamental level, it represents a tacit acknowledgement of the superiority of a selective, adaptive approach over a monolithic approach. When legislation is written to adopt a single program, a legislature makes a single, large bet. Consequently, it assumes a bigger risk of failure. This is one reason why departmental program managers can encounter resistance on the part of elected officials to put new or unfamiliar ideas into action. A staff member's job probably will not depend on the success of a pilot program but an elected official's seat may well depend on the success of a permanent program. By viewing legislation as a commitment to choose among a portfolio of options, rather than to adopt a single one and hope for the best, a policymaker can mitigate electoral risk while at the same time being seen to get things done.

Once a policy meme reaches a national legislature through a selective process, policymakers should have a reasonably good idea what to expect from different options. To provide an illustrative example, a measure that has gained increasing attention during the 2000s and 2010s is a 'carbon tax' to disincentivize carbon intensity, thereby reducing atmospheric emissions. In theory, a carbon tax would be placed upon goods and services having a high carbon intensity. However, the measurement problem again rears its head, because parties could argue endlessly about the validity of various measures of carbon intensity. The estimated value of carbon intensity of any given good or service would depend significantly upon the boundaries of analysis. Some stakeholders might try to make a case for exempting certain goods or services altogether, such as those whose demand is relatively price-inelastic. Then there is the politically unpalatable prospect of increasing the overall tax burden on individuals or businesses, necessitating some kind of mitigating measure in the form of financial relief elsewhere. There also could be equity implications for low-income people who pay little or no income tax.

This is the kind of policymaking tangle to which a selective approach is well suited. Rather than seeking to determine in advance which particular configuration of a carbon tax package to implement through legislation, a national legislature could farm out the discovery process to regional or local jurisdictions. They would be tasked with implementing carbon taxes on a pilot basis. They would be given free rein to determine which goods and services are subject to the tax. They would be able to decide whether to shift the tax burden from elsewhere, such as from state income tax or local sales tax, and they would have discretion to provide relief to low-income taxpayers, as needed. They would be required to report on carbon throughput abated by the tax, as well as on the degree of support for it among taxpayers. Running the pilot over a few years would allow regional or local governments to adjust their programs annually based on the performance of neighboring programs or on input from taxpayers and other stakeholders.

A mechanism such as this would allow for the discovery of successful programs through selection. By the time such a program became considered for national implementation, a national government then would know more or less what to expect from it. A national program itself need not be a 'one-size-fits-all' program. It could instruct regional legislatures to implement a carbon tax to meet certain carbon-reduction goals determined during the pilot phase, offering an offsetting reduction in national income tax or other financial benefit. It would be up to regional governments to determine what to tax, and by how much, in order to meet the goals in question.

The whole process, from local implementation to national implementation, could take as little as ten years. The alternative, a national carbon tax established without any selective discovery at smaller scales, may still be justified given the urgency of addressing climate change; however, it might not get things right the first time, as the French government discovered in 2018 when it tried to raise fuel taxes nationally without any countervailing financial relief. The measure prompted widespread social unrest from a populace frustrated with high costs of living.

A process of selective discovery can in theory be applied to any kind of policymaking, not only to policies designed to reduce material intensity. It will bring forth the most effective and most productive combinations of instruments for the job. Policymaking thus becomes a process of establishing a 'market' for policy ideas and then allowing the competitive, selective process of the market to produce the best-adapted packages. As Beinhocker writes, "there needs to be a willingness to experiment, collect feedback, and change course, none of which is politically easy",[19] at least, perhaps until such time as it becomes the norm.

Role of the Market in Economic Policymaking

For the sake of illustration, let us suppose that a national legislature establishes a carbon tax program, or any other kind of tax on material intensity, through a process of selective discovery, and that the program incorporates a tax-shifting mechanism to render the majority of taxpayers financially indifferent to the carbon tax. This means that whatever extra you may spend on motor fuel, heating fuel, electricity or food likely would be offset by lower taxes elsewhere, such as on income. Taxpayers have little control over the amount of tax they must pay, except occasionally at the ballot box, and even then only indirectly. Yet taxpayers have great control over the choices they make in the market for goods and services. Those choices are manifested in selections made among alternate technologies for energy delivery, and among alternate products embodying varying degrees of material intensity.

The price-elasticity of demand for carbon-intense products can vary widely. Heating fuel is a carbon-intense product whose demand is relatively price-inelastic. Monopoly conditions often prevail in this market, in part because of the sunk costs of existing infrastructure, and because few substitutes exist. A carbon tax on heating fuel then would increase the financial burden upon taxpayers while delivering more revenue to the providers of heating fuel. On the other hand, such conditions are less prevalent in the market for motor fuel because of the emergence of electric vehicles. A carbon tax on motor fuel would shift demand toward electricity. It would also shift demand onto cleaner forms of electrical generation. A carbon tax on electrical generation, in turn, would amplify the shift toward lower-carbon electrical generation. Overall, carbon taxes on energy supply would create demand for products using more energy-efficient technologies.

At the store, products having higher material intensity would cost relatively more, thereby creating demand for products having lower material intensity. For example, meat and processed foods would become more expensive. Products containing a high percentage of materials derived

[19] Beinhocker (2006, p. 427).

from petrochemicals, such as artificial plastics, would become more expensive. Manufacturers and food producers would change the way they packaged their products as a result. Produce grown by industrialized agriculture would become more expensive, shifting demand toward lower-carbon or other less materially intense growing practices, such as dry farming, polyculture and urban farming.

Initially, consumers probably would pay more at the store and more for their energy needs than before. Balanced against this, they would pay less tax, whether on income, sales or property. Now, however, the consumer has control over their material intensity. They would have an ability to realize net savings through their selections in the marketplace. As demand drove the introduction of materially efficient or low-carbon options, so over time those products would compete on price with their materially intense or high-carbon counterparts.

A major purpose of developing policy through a process of social selection becomes clear. It allows the market of policy ideas—the policy 'meme pool'—to calibrate the most socially acceptable rate of a carbon tax for the economic market to meet a specified goal of carbon reduction. Selection also allows the policy meme pool to establish such a goal in the first place. If the bar were set too low then the carbon tax would be small and the economy could, over time, experience selection pressures from nature to increase it. If the bar were set too high, then too few consumers could afford to make the shift and the economy would be unable to adapt. The process of adaptation and selection should find a bar height that could be cleared by the economy—in the sense that a large enough number of consumers would be indifferent to the net effect, or otherwise consider it acceptable—while being high enough to produce the largest possible reductions in material intensity.

If this sounds like a process of optimization, it is not. Evolutionary systems do not produce 'optimum' outcomes, like a marble rolling down to the bottom of a bowl. A better analogy is an apple rolling down a hill into a pond whose surface is disturbed by a breeze. The apple will bob around on the water, never settling into one place. This is the adequate solution of an evolutionary system, as distinct from the perfect one of a reversible system.

A carbon tax, or any tax on material intensity, once established, would evolve through the same process of selection. Legislation would increase the carbon tax rate and shift the burden further away from other forms of taxation, or establish other countervailing financial measures. The values of these adjustments can be discovered by trial and error just as the initial values were. Through successive adjustments, the economy would evolve away from a high-carbon state toward a low-carbon state.

Any economic policy for material efficiency must be able to convert between material units—the units of the natural world—and the currency units of the economy. In this regard, a policy of this kind will have two parts: a material part, such as a target improvement in material efficiency, and a currency part to distribute financial costs and benefits. The material part of a policy for material efficiency is the non-market part of the policy. The financial part of a policy for material efficiency is the part that exists within a market. The two parts form an integrated whole, addressing both natural system dynamics and economic or social dynamics. The policy can be operational only with both parts in place because it affects both the economy's material intensity upon nature and the flows of currency within economic markets.

Markets therefore have a central role to play in translating material goals into changes in behavior and preferences. Organic foods, for example, have gained a foothold in grocery stores because consumer preferences have become to some extent driven by norms associated with personal health and ecosystem health. Yet organic foods remain for the most part more expensive than non-organic foods, and therefore accessible only to consumers with the requisite disposable income and an inclination to spend it. If organic food, on the other hand, were cheaper than non-organic food, it would fairly quickly come to dominate the supermarket. This kind of large-scale change is what markets are good at. A carbon tax alone would nudge the system some way in this direction because virtually all the fertilizers and pesticides used in conventional agriculture would become more expensive.[20] The shifting of government subsidies from non-organic production to organic production would move the system further. It is not hard to see how a portfolio of instruments, operating upon different points in the supply chain from grower to market, could invert the price-comparison, tipping the scales for organic foods.

If one traces the decisions made by economic actors far enough up the supply chain then, at the point of primary extraction, one finds decisions being made about quantities of extraction being influenced by expectations about currency flows within downstream markets. How much ore is

[20] These have been blamed for recent precipitous declines in insect abundance, threatening the very pollination basis of agriculture itself. See Sánchez-Bayo and Wyckhuys (2019).

mined, lumber felled, fish harvested or crops grown is a decision influenced by an expectation of how much can be sold in the markets for those resources.

At the other end of the material pathway through the economy lies the point of consumption. Consumption represents an exit point for material flows from the economy. The currency used to pay for material consumption is cycled back upstream to a vendor. Here, consumers' decisions are influenced partly by price, partly by a perception of relative social status, and—depending on the individual—by a product's perceived desirability for personal or natural health. Will the product be bad for me? Could it be bad for natural systems? Do I care?

Almost all material pathways through the economy travel through one or more markets. It is within markets that currency is used to pay for a material flow: in fact, such an exchange pretty much defines a market. Markets, then, play a central role in the mediation of material flows through the economy. It follows, then, that markets will play a central role in the *alteration* of material flows through the economy so as to evolve in the direction of material efficiency. Economic policies for material efficiency thus are definable *only* within the context of natural systems, the material and financial parts being two essential components of the whole, each acting as a point of connection to the system it represents. Policies for material efficiency necessarily will be policies affecting markets or, to be more precise, the behavior and preferences of economic actors within markets. This is not to say that all policies for material efficiency will have market components, only that market mechanisms will be the predominant mechanisms in such policies.

It is not that markets *should* play a central role in policymaking for a planetary economy, it is that markets *will* play a central role, because of their fundamental status as a class of complex, adaptive systems. Markets likely will be just as central to a planetary economy a couple of hundred years hence as they are within today's growth economy. Therefore, policies to steer the current economy in that direction not only will use markets but also may be most effectively developed through a market of ideas.

Legal and Regulatory Processes

Policy is an agreement among members of a society about how norms should be enacted. It is a statement to which members of a group subscribe which articulates behaviors to be encouraged and others to be discouraged, as well as reasons why. In communities too large for effective verbal communication, policy typically is conveyed in written form. When adopted by a designated institution, it takes the form of legislation, which creates and amends statute. A statute is the body of laws applied within a geographical jurisdiction.

Laws are enforced through regulation. Legislation often is written in such a way as to leave the details of implementation to the regulators. Legislation thus can be broader, less detailed and therefore shorter. While a piece of legislation might remain unchanged for years, the regulations it enacts could change from one year to the next as programs are established, monitored and updated. Regulation reflects a kind of clamor among various stakeholder interests for agreement of the regulators—albeit couched in dry legalese. In this way, regulation tends to be more participatory than legislation. Regulators must ensure that any regulation they enact stays within the letter and the spirit of its enabling legislation, and that it integrates stakeholder interests in such a way as to create a program that most effectively promotes the intent of the legislation. The process is inexact, which is a strength, because it leaves room for adaptation and evolution.

A large economy such as ours can only exist with government regulation. A synergy of government regulation with market processes will be critical to a successful transition toward material stability and widespread prosperity.

Government Regulation

"Markets", observe Krugman and Wells, "are a remarkably effective way of organizing an economy."[21] They are in fact a kind of organization toward which all human economies above a certain size seem naturally to gravitate. Yet the organization of an economy does not end with markets. Markets are called into existence by institutions and regulation, which define their scope and set their rules. This is as much true of today's globally interconnected markets as it was of the small, highly dispersed markets of the historical past.

Ultimately, of course, the economy is regulated by nature. The economy as we know it conceivably could be destroyed if its deleterious outputs produced sufficiently inhospitable conditions for people on Earth. It would seem to be government's role to intercede on behalf of nature, that

[21] Krugman and Wells (2009, p. 111).

is, to regulate the economy as a proxy for nature—for the economy's own protection. Government regulation of the economy would seek to maintain long-term economic stability by, among other things, regulating the volume of its interchanges with nature so that they can be sustained. This concept is not foreign to economics, as evidenced by the theory of maximum sustainable yield. However, because historically governments were in the habit of regulating economies that were small in volume relative to natural processes, the regulation largely ignored natural constraints, focusing instead on raising revenue through taxation, and on social issues, such as income distribution. Despite an increasing number of regulations in recent decades designed to protect natural systems, to limit consumption of natural resources and to reduce waste streams, there is not yet an overarching mindset within society treating government regulation as protecting the economy from nature rather than the other way around.

Government regulation is as old as the economy itself—at least any having a division of labor. The cuneiform tablets of ancient Sumer recorded harvests, documented trades and confirmed the ownership of property, all hallmarks of a society with established administrative roles governing the division of labor.[22] From ancient Sumer through the centuries to Renaissance Europe, the prevailing economic orthodoxy was one of government regulation over commerce. Governments raised revenue through taxation; they levied duties on goods traded across borders; they even restricted in some situations the kinds of goods and services that could be offered within their jurisdictions. However, a feature of almost all these economies, one which persisted through centuries of government oversight, was the presence of markets. Indeed, without the presence of markets, there would have been little for the ruling classes to oversee. As the Renaissance gave way to the age of European exploration, some of the peoples encountered by European explorers were found to operate markets, whereas others did not. Societies having high population density, such as in Mesoamerica, South Asia and parts of West Africa, traded internally or with neighbors through markets. Regions of low population density or geographical isolation had few if any markets.

An essential precondition for a market is proximity. In the historical past, this meant geographical proximity, whereas in the early twenty-first century it can mean virtual proximity. Billions of transactions occur online every day, between parties who may be on opposite sides of the world.

[22] MacGregor (2010, ch. 15).

The industrial revolution brought with it a change in how markets were viewed. The *laissez-faire* philosophy of Adam Smith and his followers led to a curious inversion of economic orthodoxy. Instead of government calling the shots over economic activity and the operation of the markets, the new market orthodoxy portrayed government regulation as 'intervention' in otherwise perfect market operation. The more freely a market could operate, the argument went, the more wealth it could generate. Government 'meddling' in the markets was considered 'grossly inefficient' because it was assumed to diminish aggregate welfare.

It was a curious inversion of economic orthodoxy because, taken to its logical conclusion, it implied the abolition of government altogether. Modern economists acknowledge the institutional context for markets, in that it defines their rules of operation although, for the convenience of market models, it is generally regarded as an unchanging background. When in fact the regulatory context changes, the orthodoxy may regard it as a violation of *laissez-faire*.

Regulation does not necessarily mean 'big government' or government 'meddling', although it sometimes can be. Fundamentally, regulation should capture those influences which cannot be internalized into a market economy, and represent them within markets in ways they can process. What is regulated by government would include all the bounding conditions within which a market operates, and their conversion into market units.

Government Regulation as Economic Guidance

In practice, governments regulate some, but not all, of these bounding conditions. Many bounding conditions imposed by nature, for instance, are left unregulated. This is a problem not for nature but for the economy. It illuminates the central role of government regulation for a planetary economy. Where the market is an engine of economic change, government regulation takes on the role of guidance system. Under the market orthodoxy of the twentieth century, it was assumed that the economy could drift in any direction, or no direction in particular, because its long-term trajectory was not governed by any overarching norms. It was a kind of political Brownian motion. The emerging material coevolution between the economy and Earth's natural systems creates a normative impulse for market guidance. Economic policy has to reflect a prime directive for alignment with natural processes.

As the guidance system of the economy, the role of government is to interpret the signals coming to it from the economy's environment and to steer the economy in the desired direction. If the guidance system falters—that is, if insufficient normative support for the prime directive existed within society—then the vessel would run the risk of crashing. If the guidance system functions well then the vessel eventually would make a landing within the target zone.

Fundamentally, a guidance system translates environmental inputs into machine units. If an airplane pilot encounters a headwind, they may increase power. Wind speed translates into fuel flow. If the pilot encounters a storm cloud then he or she may divert around it or over it, changing direction or increasing power. In both cases, they may watch the fuel gauge to make sure enough fuel remains to reach the destination. The engine itself is blind to the plane's vector and position, as are the airfoils and flaps. The engine responds only to fuel flow; the flaps only to electrical signals to their servomotors.

Most of the global economy exists as a network of many interlinked markets. Each of these potentially is an economic engine for material efficiency. Today, many of them pull in conflicting directions, some more materially efficient than others. Yet, when harnessed together, they would be able to move the entire global economy in a more or less materially efficient direction. Where an airplane engine is a metaphor for the market, the airplane itself represents the whole economy. The pilot, as the guidance system, represents government regulation. A flight plan to travel from one place to another represents policy. The destination point is the prime directive.

Like the engine on an airplane, a market delivers economic thrust. This thrust is directionless in the absence of a normative guidance system. This is another way of stating that markets are 'amoral'.[23] In large societies where anonymity is common, moral principles become codified into regulations. Small communities, being better able to police themselves, need less in the way of exogenous regulation, in part because they are normatively more uniform.[24] With global average population density set to increase by perhaps another 50 percent over the next century, regulation will remain part of the social landscape for centuries to come.

[23] For example, Diamond (2005, p. 485).
[24] Naess (1989, p. 159).

Government regulation for material discipline must encompass regulation of both material flows and financial capital. Material intensity must be translated into currency values, a process which is most effective when performed adaptively. The adjustments a pilot makes to the flight path are analogous to legislative and regulatory alterations to maintain course on policy.

If government regulation were the only controlling influence over the engine of the market, then the issue would boil down to political will. However, other non-governmental forces also attempt to exert control over markets, transnational corporations being an example. Market prices, argues Peter Barnes, are influenced by four factors: supply, demand, market power and political power. "The first two determine what might be called fair market value", he writes; "the last two determine ... rent."[25] Entities having market power use their dominance over markets to impose rents, thereby increasing prices. Entities having market power also have political power, enabling them to influence the legislative process. Legislation that otherwise might be unfavorable to them can become diluted or filled with loopholes through which they then jump. The failure of national cap-and-trade legislation in the US during the 2000s is an example.[26]

It is as if there were several pilots in the cockpit, not all agreeing on the flight path, or even on the final destination.

In a materially large economy, a proxy is required for natural constraints, exerting a counterbalance of power against overconcentration. Limits on the market power and political power of large players would translate into limits on the volume of natural resources going through the economy. Market processes therefore would become more influenced by the forces of supply and demand than by power. Political decision-making would reflect the interests of all to a relatively greater extent, which is why institutional structure is so important. In our aircraft analogy, the designated pilot, which is government regulation, must find ways to keep imposters off the controls. This is not always easy, because an imposter may offer the pilot a large amount of money to alter the flight plan. Flight plans driven by lobbying or bribery tend not to end well.

For long-term economic stability on a crowded planet, an entity uninfluenced by money needs to have the ability to regulate material flows

[25] Barnes (2014, pp. 51–52).

[26] Barnes (2014, ch. 8).

through the economy. Government is the logical choice, provided the influence of non-governmental money were removed. A planetary economy would be regulated by governments who acted first and foremost as proxies for natural capacity, adjusting the controls on material flows and on the values of goods and services within the economy so as to maintain or improve economic prosperity for minimal resource intensity.

Government regulation attending first and foremost to material efficiency would still deliver wealth over time, yet now along a defined trajectory. The modern economy, Jared Diamond reminds us, is a luxury that can be afforded only while its natural container is able to support it. This perspective is the inverse of the anthropocentric one, which portrays environmental concerns as a luxury. In the anthropocentric view, environmental problems left unsolved are perceived as a money-saving device. The reality is the opposite, as Diamond illustrated with reference to the US Clean Air Act of 1970. That act is estimated to have yielded health savings of about $1 trillion per year in equivalent dollars, compared to an annual cost of premature deaths due to air pollution in the US of around $650 billion per year.[27]

Diamond was not arguing—nor am I—for government command-and-control. As Robert Frank pointed out, by the late 1990s, environmental regulators had shifted away from command-and-control toward policies employing economic instruments. "The results", he wrote, "have been exactly as predicted: the costs of a given level of pollution abatement are lower … under the new incentive-based systems."[28] That said, the most important measure of success is not the *cost* of pollution abatement but the *degree* of pollution abatement: the physical amount removed from the effluent stream. If the amount of pollution abatement is insufficient to move the needle on natural capacity improvement, then its financial efficiency is unimportant.

It is not necessary for government regulation to be based on an accurate understanding of natural capacity, even if such an understanding could be attained. As *Economics of a Crowded Planet* describes, the minimization of the economy's material intensity will improve the ability of nature to sustain the economy. This is because nature deals in material units. The role of government regulation, then, is to guide the economy toward a more materially efficient state. Success in attaining long-term

[27] Diamond (2005, pp. 503–504).
[28] Frank (1999, p. 210).

stability in alignment with nature would reveal, through a process of selective discovery, an understanding of what improves and maintains natural capacity.

This line of thinking also suggests that government regulation can and should be relatively simple. Government statutes have ballooned in size since the middle of the twentieth century as lawmakers have attempted to regulate ever-more complex and globally interconnected economies. As Adam Levitin argues, governments do not have any real control over their regulations when the statute is thousands of pages long. The larger the statute, the more loopholes exist for market players to exploit. Yet, if government regulation were guided by an overarching principle—a prime directive, such as material efficiency—then existing regulations and potential new ones would be evaluated against this principle. In many situations, it should be possible to collapse existing regulations into a relatively small number of simple rules. Much of today's environmental regulation would be rendered unnecessary in an economy operating under, or moving towards, a prime directive of material discipline, having the requisite economic controls in place to steer market activity in that direction.

If government codes were on the order of a few hundred pages rather than tens of thousands, then at least they could be comprehended by an individual person. Participatory democracy would be easier to institute. People would have an ability to engage in the legislative and regulatory processes on an intelligible level, and legislators and regulators themselves would enjoy greater clarity about what it is they are trying to manage. Legislation, or even perhaps a constitutional amendment, to impose strict limits on the size of government statutes would ensure that policy development would be geared toward winnowing existing language and replacing it with something new rather than the endless legislative and regulatory sedimentation seen today.

Synergy of Government and Markets

The bounding conditions established by government regulation over economic activity are analogous to the fence around a corral within which livestock can roam where they please.

This position is distinct from the philosophy of *laissez-faire* which, if taken to its logical conclusion, would do away with the fence, thereby essentially doing away with the livestock. Countries with weak government institutions, writes Eric Beinhocker, often have economies "in a low-cooperation, low-competition dead-end",[29] characterized by widespread poverty and highly skewed distributions of income. Nelson and Winter's seminal work on evolutionary economics gave theoretical support to the empirical observation that market structure matters.[30]

Economic policy for a crowded planet is not so much the antithesis of *laissez-faire* as a constraint upon it: *laissez faire sous règles dures* ('let do under hard rules'). Although the rules themselves may not necessarily be hard to meet, they would constitute a clear and firm boundary, like the fence around a corral. Guiding the economy from a materially inefficient state toward a materially efficient one then would become a process of incrementally moving the fence in a particular direction, thereby shifting the location of the corral. The occupants of the corral should be indifferent to its location provided the quality of their grazing remains unaffected. This indifference requires the maintenance of adequate distribution of grazing access among individuals, the 'adequacy' itself being determined by the members of the herd. Excessive inequality would violate the indifference requirement. A variety of mechanisms exist to accomplish this, to be reviewed in the coming chapters.

The issue, then, is not "states versus markets", as Beinhocker argues; "it is how to combine states and markets to create an effective evolutionary system." Consistent with a prevailing norm of strong reciprocity, which has been found by behavioral psychologists to be widespread across societies and cultures, "the state has an obligation to ensure that all its citizens have an equal opportunity to participate in the economic system, and to provide a basic level of support". Markets, on the other hand, generate new physical and social technologies through a process of discovery, differentiation and selective amplification, thereby creating wealth, the selection pressures being applied through policy.[31]

In the early-twenty-first-century economy, however, governments are influenced at least in part by some of the market participants whose activities they are responsible to bound. In the corral analogy, there exists a degree of influence by members of the herd over the positioning of the fence around it. In a world whose economy were materially small, the desirability of such influence would be merely a moral question. Where the economy is materially large relative to its natural container, the

[29] Beinhocker (2006, p. 425).

[30] Nelson and Winter (1982, ch. 14).

[31] Beinhocker (2006, p. 427).

question takes on a 'meta-political' significance, one encompassing the long-term stability of the whole economic system. In this world, the influence of market participants upon the boundaries of the market could distract from a longer-term imperative of moving the market in a materially efficient direction. The government's primary responsibility would become the material efficiency of the economy, accomplished through the guidance of market activity, unhindered by the influence of market players. The prevalence of institutions and laws enabling or preventing the capture of political power by large market players has a profound influence over government's ability to wield power in the collective good.

A properly functioning process would represent a synergy of government and markets. Government power would be manifested in the form of hard, absolute caps on certain kinds of material flows. This is the sense in which government regulation would represent the long-term directive of material discipline. However, such caps alone would not have the ability to change the economic system. For that, we need the market, since the market is far and away the most important mechanism for translating individual attitudes and decisions into aggregate economic effects. "Markets", writes Beinhocker, "are the best social technology devised yet for integrating the views of large numbers of people to put prices on complex assets, and to allocate capital."[32] It is principally through the ability of major players to move the market, and through the forces of supply and demand, that markets, and consequently the economy as a whole, transform over time. Yet market power and supply and demand are themselves shaped by cost and benefit structures established through government regulation, and ultimately by the limits to market activity, established by government. In a synergistic system, the center of gravity of the market should move in the same direction that government is moving the boundaries of the market. Government regulation, argues Beinhocker, should "shape the fitness environment, leaving business plan selection and amplification to market mechanisms".[33]

Regulation, then, boils down to two major components. One is outward-looking, toward nature, and is materially based. It creates the boundary conditions within which markets operate. Its units are in tonnes or any other equivalent material measure. It establishes a path toward material efficiency. The other component is inward-looking, toward the

[32] Beinhocker (2006, p. 403).
[33] Beinhocker (2006, p. 426).

economy, and is currency-based. It creates the incentive structures that harness the power of the market toward the material objectives. These structures guide market behavior toward materially desirable attributes and away from undesirable ones. Some of these structures may have indirect effects upon material efficiency, such as by helping to stabilize markets.[34]

Incentive structures do not need to represent a direct measure of material flows. All they need to do is to nudge the material flows toward greater efficiency, a process that is adaptive and evolutionary. While specific species of incentive may come and go over time, it is essential for government to retain control over the selection pressures placed upon markets. If major market players acquire an ability to influence these selection pressures then the pace and direction of the economy's evolution would become influenced by their interests, which could be orthogonal to a prime directive or simply short-sighted.[35]

Where major market players can—and should—have influence is in the development of the incentive structures themselves. All market players, in fact, are stakeholders in this issue. The development of market incentives for material efficiency, which ultimately become manifested through legislation and regulation, must involve the very stakeholders it will affect. As the entity responsible for establishing regulations that will implement market incentives, government is also responsible for incorporating stakeholder input into those incentives. Typically, this occurs through the regulatory rulemaking process. For a synergy of government and markets to be successful, however, the major market players need to stay out of the legislative process itself. It is there that the material bounding conditions for markets are established. For a stable, materially efficient economy to be attained, those bounding conditions must be informed primarily by the material influences of nature.

Until a governance structure to represent the whole planet becomes established, it is the duty of national governments to establish bounding conditions over the scope of markets, and to establish incentive structures to guide market behavior. If any such bounding conditions and incentive

[34] See, for example, Levitin (2016).

[35] Oil and gas companies, who buried their knowledge of the climate threat in the 1980s, did so because the market in which they operated allowed them to. Although a moral case for transnational corporations to act in the public interest can be made, as Bill McKibben has argued in multiple news articles during the 2010s, the ultimate responsibility lies with government for erecting the system in the first place.

structures were to place the electorate in long-term danger then, strictly speaking, the government would be in dereliction of its duty.

As the coming chapters describe, an altering of the selection pressures upon the economy likely will produce what paleontologists call an adaptive radiation. In economic evolution, it would represent a mass 'speciation' of new technologies, new corporations, new jobs, new wealth generation. Holding back such a speciation event today are the very government incentive structures allowing materially unsustainable corporate practices to persist, as well as the continuing influence of many of the same corporations over the legislative process.

References

Barnes, P. (2014). *With Liberty and Dividends for All.* San Francisco, CA: Berrett-Koehler.

Beinhocker, E. D. (2006). *The Origin of Wealth: Evolution, Complexity, and the Radical Remaking of Economics.* Boston, MA: Harvard Business School Press.

Costanza, R., et al. (1997). The Value of the World's Ecosystem Services and Natural Capital. *Nature, 387,* 253–260.

Costanza, R., et al. (2015). *An Introduction to Ecological Economics* (2nd ed.). Boca Raton, FL: CRC Press.

Diamond, J. (2005). *Collapse: How Societies Choose to Fail Or Survive.* Allen Lane and London: Penguin.

Frank, R. H. (1999). *Luxury Fever: Why Money Fails to Satisfy in an Era of Excess.* New York: The Free Press, Simon & Schuster.

Galbraith, J. K. (1973). *Economics and the Public Purpose.* Boston, MA: Houghton Mifflin.

Gore, A. (2006). *An Inconvenient Truth: The Planetary Emergency of Global Warming and What We Can Do About It.* Rodale, Emmaus, PA.

IPCC. (2018). *Global Warming of 1.5°C: An IPCC Special Report on the Impacts of Global Warming of 1.5°C Above Pre-industrial Levels and Related Global Greenhouse Gas Emission Pathways, in the Context of Strengthening the Global Response to the Threat of Climate Change, Sustainable Development, and Efforts to Eradicate Poverty* (V. Masson-Delmotte et al., Eds.). Geneva, Switzerland: World Meteorological Organization.

Juniper, T. (2013). *What Has Nature Ever Done for Us? How Money Really Does Grow on Trees.* London: Profile Books.

Krugman, P., & Wells, R. (2009). *Microeconomics.* New York: Worth Publishers.

LeGrand, J. (1991). *Equity and Choice: An Essay in Economics and Applied Philosophy.* London: HarperCollins Academic.

Levitin, A. J. (2016). Safe Banking: Finance and Democracy. *University of Chicago Law Review, 83*(1), 357–455.
MacGregor, N. (2010). *A History of the World in 100 Objects*. London: Allen Lane.
Majone, G., & Wildavsky, A. (1978). Implementation as Evolution. In H. Freeman (Ed.), *Policy Studies Annual Review* (Vol. 2). Beverly Hills, CA: Sage Publications.
May, R. M., et al. (1978). Exploiting Natural Populations in an Uncertain World. *Mathematical Biosciences, 42*, 219–252.
Meacham, J. (2012). *Thomas Jefferson: The Art of Power*. New York: Random House.
Naess, A. (1989). *Ecology, Community and Lifestyle* (D. Rothenberg, Trans. and Ed.). Cambridge University Press.
Nelson, R. R., & Winter, S. G. (1982). *An Evolutionary Theory of Economic Change*. Cambridge, MA: Belknap Press.
Sánchez-Bayo, F., & Wyckhuys, K. A. G. (2019). Worldwide Decline of the Entomofauna: A Review of Its Drivers. *Biological Conservation, 232*, 8–27.
Schmidheiny, S., & Zorraquín, F. J. (1996). *Financing Change: The Financial Community, Eco-Efficiency and Sustainable Development*. With World Business Council for Sustainable Development. Cambridge, MA: MIT Press.
Stiglitz, J. E. (2003). *The Roaring Nineties*. New York: W.W. Norton.
USGCRP. (2018). *Impacts, Risks and Adaptation in the United States*. Fourth National Climate Assessment, Volume II, edited by D. R. Reidmiller et al. U.S. Global Change Research Program, Washington, DC, USA.

CHAPTER 8

Requirements for Economic Policies

A 2013 survey by the Public Policy Institute of California reported that "66 percent of residents … would like the state to lead the way on fighting global warming, including making policies separate from the federal government …" Yet only a fraction of respondents reported changing their own ways to the same end. "It is easy to ask someone else to fix a problem," wrote the *San Francisco Examiner*, "but much more difficult to make personal changes."[1]

Yet fixing the problem at the state level inescapably entails people making personal changes. The difficulty individuals have in doing so arises from an instrumental environment that does not encourage everyone to do so. People's apparent reticence to make changes in their own lives represents a reticence to make unilateral changes. A principle of 'strong reciprocity' is at work. People will make changes to their lives if others also do so, because then it is perceived as fair.[2] Frustration with a general lack of progress toward a more environmentally aligned economy may be widespread, but progress in the population at large has to occur on a level playing field. This is accomplished through economic instruments, which in turn emerge from policy. The respondents to the survey were not asking someone else to fix the problem; they were asking the state to make new rules under which everybody would work.

[1] *San Francisco Examiner*, August 5, 2013.

[2] See, for example, in Beinhocker (2006).

F. Murison Smith, *A Planetary Economy*,
https://doi.org/10.1007/978-3-030-49296-0_8

The problem with the current economic system is that it places the onus to align with nature upon the consumer, who is least equipped to do so. The consumer is, by definition, a reactive entity, responding to choices before them in the market. They have little influence over the range of choices presented. That range is determined upstream in the supply chain. Consumers need a range of choices enabling alignment with nature, as well as the ability to fulfil them.

The solution is to place the onus to align with nature primarily upon production, not consumption. Assuming higher costs would be passed down the supply chain, the solution also must allow consumers to afford those costs. If these two conditions are satisfied then other incentives and disincentives can be placed upon consumption, such as incentives to recirculate material and disincentives to waste it.

This chapter outlines requirements for policies that could guide the economy toward a long-term material alignment with nature and widespread prosperity. The policies reflect the current state of physical and social technologies early in the twenty-first century. As technologies change, so will the policy options. The requirements outlined here therefore can be taken as a starting-point.

Summary of Policy Requirements

A planetary economy—one that is materially large, stable, aligned with nature, and broadly prosperous—would have to satisfy the following requirements, discussed the sections and chapters to follow:

- Government programs would maintain or improve natural capacity. Any that diminish it would be disfavored.
- Material throughput between the economy and nature would be expensive, whereas materials recirculated within the economy would be comparatively cheap.
- Deleterious or unacceptable material flows for which no price can act as a disincentive would be banned or limited directly.[3]
- Ownership of real property would bear a responsibility to maintain natural capacity.
- Any business or government entity undergoing legal incorporation would be required to apply for and receive a material efficiency

[3] Contemporary examples include chlorofluorocarbons and ivory.

rating. Ratings would be under public jurisdiction, overseen and coordinated by a global governance institution.

- Economic entities, such as businesses and government agencies, would be required to report both their financial performance and their material performance.
- Productive labor would be minimally taxed.
- Government tax revenue would come primarily from social bads, such as material intensity, rather than from social goods, such as productive labor.
- Everyone alive would receive one share in natural capacity under a planetary trust.
- Everybody participating in the economy, anywhere in the world, would be assured a usable minimum living standard.
- Personal income would have no upper bound, provided it does not come at the expense of everyone else's income, meaning that the minimum income would remain guaranteed and usable. The distribution of incomes would remain within a range having broad social acceptance.
- People would be assured access to all physical and social technologies enabling them to act upon the prevailing incentives within the economy, thereby ensuring substantially full participation in material discipline and widespread prosperity.
- Financial speculation and arbitrage would be limited to ensure relative long-term stability of the financial system.
- Government regulation would be as simple as possible. A clear prime directive should lead to a clear and simple set of policies. These in turn should translate into minimally complex regulations.

Subsequent chapters go into details on material reporting, money, and instruments to implement these policies. This chapter discusses the rationale for the requirements and their relations to one another.

Policies for Natural Capacity

Policies to maintain or improve natural capacity are organized into three sections: stewardship of real property, government operations and individual and corporate actions. Property is addressed first because it is held by both government and private entities.

Stewardship of Real Property

The transition from a hunter-gatherer existence to a life of settled agriculture brought a profound change to the way people view the natural resources that sustain us. As Chap. 2 described, hunter-gatherer societies, being resource-constrained, held that people belong to their surroundings, a participatory worldview that has carried down to modern-day indigenous peoples. Its wide geographical prevalence suggests it is archetypical.

Once people began cultivating crops, the perception flipped. Individuals or families delineated plots of land over which they had responsibility for cultivation. While they may have helped one another in certain aspects of their agriculture, such as in harvest, their neighbors' fields were not their responsibility. The field, and all it yielded, belonged to the individual or family tending it. The words 'mine' and 'yours' became attached to the land. With this, a new, anthropic concept of property emerged.

Agriculture created a degree of human control over nature, both perceived and actual. No longer were people at the mercy of natural ecosystems for food, although in cultivating crops they remained at the mercy of weather, soil conditions or pest outbreaks. Agriculture expanded the food supply, which in turn allowed the population to expand. With a growing population, inevitably conflicts arose over land ownership. After the advent of writing, people began owning titles to land, which were recorded by government bureaucracies. The concept of property expanded to encompass not only farmland but also structures, equipment and other possessions.

The concept of property extended to the group as well. A chieftain of an early settled community not only held personal property but also was responsible for the management of community property. Roads, grain storehouses and armories were assets of the civilized environment owned not by individual people but by the community as a whole, through its government. Just as plots of land titled to individuals became demarcated and recorded, so early governments began to demarcate and record community boundaries. All land or water contained within those boundaries were described as 'belonging to' the community. Here, we had the origin of the city-state which, through mergers—some by marriage, some peaceful, some violent—produced larger geopolitical entities: principalities, kingdoms, nation-states and empires. All were characterized by a concept of ownership over the regions they controlled.

Although this exploitative concept of property may have served a materially small economy, it has become problematic on an increasingly crowded planet because it ignores the global commons. During civilized history, it was rare to conceive of the 'property' of the whole planet or of all humanity. Now, however, the concept of property comes full circle, back to its original participatory meaning. "The task of redesigning ownership", writes Marjorie Kelly, "is part of the larger task of bringing human civilization into harmony with the earth."[4] The challenge is to reconcile short-term, exploitative impulses of private property ownership with long-term, generative prescriptions of common ownership. Howard Barnett suggested that government should own all the land and lease it to private parties for their use, thereby improving aggregate welfare. However, there may be another, less draconian, more flexible way to resolve the issue.[5]

Agency for Nature Through Property Retitling

What does it mean to 'own property', in a world where real property is in increasing demand, and the natural capacity it hosts is becoming increasingly critical to the long-term stability of the economy? 'Ownership' in this world does not confer complete freedom of decision-making, as it might under frontier conditions. Governments, particularly in countries with well-developed legal and judicial systems, have become increasingly diligent—some might argue intrusive—in protecting against local or regional negative externalities of property owners' decisions. Barnett's hypothesis was correct: these regulations, for the most part, have improved aggregate welfare over the past 50 years. However, whether an individual or an institution were to own title to a property or lease it from the government would not alter their obligations under the law regarding its management.

Society overlays the concept of 'property' onto natural ecosystems. The Ricardian concept of 'land' is an example of terminological laziness that led generations of economists astray. David Ricardo conflated land, which can be titled, with natural capacity. The ownership of land in and of itself carries no value unless the land hosts flora, fauna, soil and water useful to people. Further, neither Ricardian 'land' nor any existing legal conventions regarding entitlement to private property incorporate the air above

[4] Kelly (2012, p. 15).
[5] Barnett (1966, p. 18).

it or the geological strata below it.[6] There is no recognition of the ecological and physical interconnection of a plot of land with its biophysical milieu. Absent civilization, what would remain is simply an ecosystem. While titles to property refer to the *land*; they do not refer to the *ecosystem*. This disconnect allows property owners to do more or less what they want on their land, without regard for any potentially negative effects upon natural capacity.

The disconnect is Cartesian. The 'Enlightenment' mindset of dominion over nature placed the human subject, who owned title to a plot of land, separate and distinct from the natural objects upon it. Existing concepts of property, being Cartesian in nature, may well be fundamentally inconsistent with long-term economic stability on a crowded planet. They represent what Marjorie Kelly calls 'extractive ownership'.[7] The indigenous concept of ownership, in contrast, is participatory and generative. It is ownership as belonging.

Any plot of land can be defined as a quadrant. Ecologists use quadrants to sample biota within a defined area as a way to build a species profile of a biological community. Quadrants can be as small as a one-square-meter metal frame tossed onto a substrate of interest or as large as a marked sampling area of a hectare or more. It is taken that ecological processes operating within a quadrant are connected to ecological processes outside it, since the quadrant itself is a purely arbitrary delineation.

Property boundaries are equally arbitrary from an ecological point of view, yet the same premise applies: ecological processes within one property are connected to those outside it. Borrowing from Maturana and Varela, one can think of real properties as ecological unities in structured coupling with one other, as well as with wider physical systems. Real property thus is not only a legal entity but also an ecological one. The concept can be logically extended up the hierarchical scale to consider the whole integrated system of nature on Earth as one big piece of property 'owned' by everybody.

This line of thinking begs the question whether a property owner who has title to some land also has title to the ecological and physical systems residing upon it, passing over it and lying beneath it. The legal assumption

[6] A rare exception is the international legal convention of 'airspace', which pertains not to private property but to national jurisdictions. Even this concept came into effect only with the advent of air travel.

[7] Kelly (2012, pp. 11–12).

has always been yes, at least for the terrestrial part. Hypothetically, if the ecology, geology and atmosphere were removed from consideration then there would be no 'land' as such to title. Yet a title, as a legal document, does not recognize the connected nature of these systems: it simply defines a quadrant.

Because the economy is in measurable coevolution with nature, everybody, not only a property owner, has a stake in decisions made on their property that potentially affect natural processes. Even someone on the other side of the world is a stakeholder. Prior to the modern age, when much of the land on Earth was not yet titled, the implications of any given property owner's decisions regarding the natural processes on their property were that much smaller. In the modern world, there is nary a scrap of land untitled or unclaimed. Although Antarctica is not titled, even it is divided into segments 'claimed' by some of the participating nations to the Antarctic Treaty of 1961. The open ocean is designated as 'international waters' under the terms of the 1958 United Nations Convention on the High Seas. As such, it represents a common resource. The atmosphere has the same status in practice, even though there exists no equivalent treaty governing it.[8]

Given that virtually all terrestrial ecosystems and many aquatic ones are now titled to private property owners or to governments, the act of 'owning' property becomes an act of owning a stake in everyone else's property as well as your own. Instead of being isolated dominions, properties are biophysically connected units having distinct stewardship. This perspective on property is systemic rather than anthropocentric. It becomes necessary because the actions of millions of property owners—taken, as many of them have been, without consideration of the actions of other property owners—have resulted in the present material coevolution between the economy and nature.

Although the term 'stewardship' has come in for criticism, particularly from the deep ecologists, it is appropriate in this context. Stewardship is analogous to safeguarding a family heirloom for the next generation. In the case of real property, the 'family' in question is everybody alive today and yet to be born. A libertarian complaint about stewardship is that it implies a responsibility curtailing individual freedom. This it does, yet

[8] Since 2010, a good deal of litigation in the US and elsewhere has declared the atmosphere a public trust (Kelly 2012, p. 4) and has sought recourse from public entities failing to protect it. This idea was explored by Peter Barnes in his 2001 book, *Who Owns the Sky?*

there is no such thing as unfettered individual freedom in any society. Furthermore, on a crowded planet, the Cartesian idea of 'dominion' over nature is hubris. The selective pressure of nature upon humanity demands what Marjorie Kelly calls 'generative ownership'.[9]

A mindset of stewardship should not necessitate any significant change in the day-to-day practices of responsible property owners, since those owners would act as stewards to varying degrees already. They do not seek to liquidate natural assets within their property boundaries but rather to maximize returns on those assets.

A concept of stewardship applies to all real property, not only to land, aquatic or maritime assets. Buildings, for instance, account for about one-third of greenhouse gas emissions worldwide.[10] A growing movement exists, particularly within higher-income countries, to reward measures reducing the energy and material intensities of the built environment, particularly in new construction. Even a condominium in a high-rise will have certain measurable or estimable effects upon its physical surroundings.

Although most property titles are legally recorded by local governments, their format and content are governed commonly through regional and national legislation. National governments also have de facto responsibility for maintaining natural capacity, at least until such time as some form of global governance may emerge. Therefore, national governments have an ability to pass legislation embedding within the titling of private property the responsibility of a property owner to maintain any and all natural capacity falling within the property line. Such maintenance would become a condition for retaining title. The negligence of this responsibility would become grounds for relinquishing title. The concept extends naturally from private property also to public property, thereby placing upon public entities the same responsibilities as private ones.

From there, it would be up to the legal process to determine what constitutes 'maintenance' of natural capacity within the bounds of a property, and what constitutes 'negligence' of the responsibility to maintain. Yet now there would exist a mechanism for the engagement of stakeholders in the natural capacity of any geographical area whose ownership is defined. This mechanism is particularly important in a world in which not everybody owns title to real property. Property, traditionally taken as a form of capital has been able to "hire, organize and direct labor" to the benefit of

[9] Kelly (2012, pp. 11–12).

[10] See, for example, United Nations Environment Programme (2009).

the property owner.[11] That similar benefits do not accrue to labor creates the well-documented inequities between the 'haves' and the 'have-nots'. Yet the literature has ignored the other, arguably more important, quality of real property, which is its natural *capacity*. Conferring title to property conditional upon the maintenance of natural capacity corrects some of the distributional imbalance, as it forces property owners to consider the planetary effects of their property management, including upon people without title to property.

One kind of property ownership partly resembling this concept already exists: the conservation easement. Title to a property remains in the hands of a private owner, while rights to develop it are transferred to a separate institution, such as a conservation organization or a government agency. In return, the property owner is given certain tax exemptions.[12] Conservation easements are purely voluntary, and they do not apply to all kinds of real property. The concept presented above goes further, in that it builds into the very process of titling a responsibility for natural capacity.

In countries having a well-developed system of property rights and legal procedures, a simple change to the obligations on a title to real property could have profound and beneficial long-term consequences, not only to the maintenance of natural capacity in a practical sense but also more broadly in shaping attitudes toward property ownership. However, in places where legal structures are weak and property rights are ill-defined or unenforceable, such a revision would have little effect. Much of the deforestation in the Amazon rainforest, for instance, is illegal under Brazilian national and state law. Yet "about 53% of land in the Legal Amazon has uncertain property rights", according to the World Resources Institute.[13] Illegal deforestation in the Amazon is difficult to enforce not only for this reason but also because of the remoteness of many affected areas and the wide dispersion of the perpetrators. The situation is even more intractable in other, less developed countries. In these regions, other, more fundamental measures would be necessary before any amendments to property entitlement could be made.

[11] Daly (1991, p. 158).

[12] Kelly (2012, pp. 105–106).

[13] http://www.wri.org/our-work/project/governance-forests-initiative.

Government Operations for Natural Capacity

Governments have the unique ability to develop and maintain infrastructure in the public interest. They also have the singular responsibility of protecting natural systems in the public interest. Governments at all levels, from the local to the national, fund these activities in a variety of ways, such as through taxes and bonds. Publicly developed infrastructure and publicly protected natural systems are, by definition, inclusionary. Governments cannot exclude access to them, except in rare cases where security takes precedence. The rationale for funding public works through such mechanisms as taxes or bonds is that since everyone benefits, everyone should chip in.

For a planetary economy, the policy requirement for government projects is that they must improve or maintain natural capacity. Public works diminishing natural capacity would constitute a dereliction of duty. Any government entity that signed up to a prime directive of material discipline would make an official commitment to this policy requirement. Failure to meet it would expose the entity to liability.

A government operating under a prime directive of material discipline, or under an even broader commitment to improve natural capacity, would evaluate projects differently from today. First, the kinds of projects the government would consider would be different. A larger proportion of public works would address ecosystem health, clean air, clean water, soil regeneration and the like. Public lands and aquatic systems would be managed in such a way as to enhance these qualities. Second, the way public projects were undertaken would differ. Conventional infrastructure projects, such as roads, railways and utilities, would maximize material efficiency, minimize resource extraction, and minimize effluent flows. These practices would go beyond the conventional environmental review undertaken in many jurisdictions, although they are a natural extension of it. Project costs would be higher than today. Essentially, the public would pay more for a greater benefit.

Public projects that have no hope of improving or maintaining natural capacity would be decommissioned. Under a policy régime for natural capacity, publicly owned electrical generation, for instance, would move away from fossil fuels toward clean energy sources, such as wind, hydropower and solar.

Although public projects conventionally are funded through the public budget process which, in turn, is endowed through taxes and bonds, another funding option exists for public projects in support of natural capacity. This is the hypothecated tax, which is used to fund only a designated activity, not anything else. Income taxes, in contrast, are not hypothecated: they are spent by government on whatever is decided in the budget process. Hypothecated taxes have been examined as mechanisms to fund healthcare: the healthier the population, on average, the lower the tax.[14] The same approach can apply to natural system health. To ensure that all have access to the support of nature, and that the economy remains stable within natural constraints, a 'natural capacity' tax would be levied on all, irrespective of their risk perception with respect to economic collapse. The revenues from this tax would be used by governments to maintain and improve the health of natural systems.

There may be a case for the tax to be somewhat progressive: for example, by exempting people on low incomes and transferring this burden onto high earners. Corporations also would pay a version of this tax, scaled perhaps by headcount. To balance the overall tax burden, other taxes could be commensurably reduced to accommodate it.

Policies for Individuals and Corporations

Policy Options for Material Discipline

Government spending its money in support of natural capacity will accomplish only so much if individuals, households and corporations are not also doing the same. Whereas government projects can have direct benefits on natural capacity by improving the health of natural systems, private actors within the economy have much less influence, on average. Only large corporations may be able to move the needle by changing their operations, and they operate within competitive markets. For this reason, a measurable proxy is needed for natural capacity. Material intensity is a natural choice. Because private actors operate mostly within markets, it is market operations themselves that should be addressed. Consequently, policy options for private actors in support of natural capacity should use markets to address material intensity.

If markets do not encourage material intensity then, as Schmidheiny and Zorraquín pointed out, this is not a failure of markets. The markets are operating as expected. It is a policy failure. Schmidheiny and Zorraquín

[14] LeGrand (1991, pp. 114–123).

identified three broad policy requirements to guide market behavior toward greater resource efficiency:

1. Remove subsidies encouraging overuse of material and energy inputs to the economy.
2. Introduce market instruments encouraging pricing according to scarcity.
3. Introduce market instruments to price, or otherwise regulate, use of the planet's absorptive capacity for anthropogenic outputs.

These changes would be "simple but not easy".[15] They would increase the costs of goods and services within the economy. Implementing them politically would become an uphill struggle unless countervailing measures were put in place. Such measures include incentives and subsidies for materially efficient products and services, and taxation adjustments to shift the tax burden from 'goods', such as productive labor, to 'bads', such as material intensity.

During the transition toward a policy régime supporting natural capacity, some markets may have to be terminated altogether. Moratoria exist for this purpose. A moratorium is a non-market policy option. It operates from outside the market, either upon activities not having markets in the first place, or to shut down existing, undesirable markets. California, for instance, imposes a moratorium on wood-burning on 'Spare the Air' days, wood-burning being a non-market activity. An example of a moratorium on a market activity, in contrast, is the Montreal Protocol. No significant quantity of chlorofluorocarbon (CFC) production can be supported by a market because CFCs are so deleterious. Consequently, the Protocol all but eliminated the market.

Another kind of policy option is the bond, which serves as a form of insurance. It has been proposed by the ecological economists as preferable to the 'polluter pays' principle, which attempts to apply monetary penalties to polluters *ex post facto*. Decades of environmental litigation have demonstrated that the 'polluter pays' principle is not very successful in the courtroom. Courts place the burden of proof upon the injured party, requiring demonstration of a direct causal relationship between pollution and damage which, within the milieu of complex, overlapping systems, is very hard to prove beyond reasonable doubt.

[15] Schmidheiny and Zorraquín (1996, p. 170).

An environmental performance bond, on the other hand, would place the burden of proof upon a prospective polluter. It conforms to what the ecological economists call a 'precautionary polluter pays principle'. It has legal precedent in the construction trades, where contractors are required to post a bond guaranteeing performance, which can be forfeit if performance is not achieved. Forfeiture carries direct financial penalties to the bond-holder, as well as indirect ones in the form of higher insurance premiums. The ecological economists argue that an environmental performance bond is preferable to the 'polluter pays' approach because a prospective polluter would be required to post a bond in advance of commencing operations. "Portions of the bond", Costanza et al. write, "would be returned if and when an agent could demonstrate that the suspected worst-case damages had not occurred, or would be less than originally assessed."[16]

How long would an entity posting such a bond have to wait? Hundreds, maybe thousands of years? The environmental performance bond, well-intentioned though it may be in shifting the burden of proof, suffers the same measurement problems in the face of radical uncertainty that the traditional 'polluter pays' argument suffered in the courtroom. Costanza et al. point to how "farsighted firms have already started to protect themselves against possible future lawsuits and damaging claims by setting aside funds",[17] yet this behavior is not the same as posting a bond, it is simply hedging bets.

Far-sightedness on the part of firms or any other institutions is a norm to be encouraged. However, in today's growth economy, under radical uncertainty from selective pressures imposed by nature, the simple act of producing the same or more economic value with less material intensity will increase its chances of subsequently attaining a materially stable, persistence-driven state. This reasoning would seem to argue in favor of market mechanisms to drive material discipline, supplemented with moratoria where needed.

Commons Rent Recycling Through a Planetary Trust

The planetary trust introduced in Chap. 5 would act as a framework for recycling commons rent. This is the rent extracted by the trust from users of it for using shared capacity. As Peter Barnes describes, everybody uses

[16] Costanza et al. (2015, pp. 252–253).

[17] Costanza et al. (2015, p. 255).

common wealth, whether it is provided by nature or by artificial forms of capacity, such as the financial system or the internet. Some use more of it than others: corporations use more than the wealthy who generally use more than the poor. Under the present system, high users of shared capacity essentially get to use it for free. In the case of pollution, for instance, a polluting corporation captures commons rent in the form of uncharged pollution upon nature. Even if pollution is priced and constrained under a cap-and-trade or similar system then the corporation still would benefit, because of higher prices of pollution-laden products. None of this commons rent would go to consumers; in fact, all else being equal, they would lose commons rent due to the higher prices.[18]

If the rights to shared capacity were held by a trust in the name of all people alive and unborn, and the trust had the ability charge users for shared capacity, then it would be able to distribute the commons rent equally to all shareholders. Everybody pays in and everybody is paid out; however, because the trust would issue one share in natural capacity to every living person, everybody would benefit equally, while not everybody would pay in equally. Low users of shared capacity would come out with a net benefit, while high users of shared capacity would incur a loss. This system would provide an incentive to use shared capacity efficiently. In the case of pollution, limiting of the supply of available permits by the trust would drive prices up, increasing the overall commons rent—at least up to a point of vanishing marginal returns.

A variety of instruments could be used within this framework. They are the subject of Chap. 13.

Tax-Shifting

Tax-shifting is the shifting of a tax burden from one taxable activity to another. Shifting the tax burden from such social 'goods' as income onto such social 'bads' as material intensity not only will drive the economy toward greater material efficiency but also will support the companion goal of broadening prosperity.

Discussions on this variety of tax-shifting emerged through work in environmental and ecological economics during the 1980s.[19] By the mid-1990s, the idea had received considerable support within the business

[18] Barnes (2006, pp. 94–96).

[19] See, for example, in Costanza et al. (2015, p. 247).

community.[20] Because economic actors have differential access to the means of production, an uneven distribution of wealth is created at the point of production. The textbook concept of the distribution of wealth as slices of a pie *ex post facto* is an artifice: a "blind statistical aggregation" of what Kenneth Boulding called many smaller tarts—some bigger than others.[21] Taxing resource intensity and simultaneously applying countervailing tax relief to productive labor would reward low-intensity producers by shifting wealth to them at the expense of high-intensity producers.

Imagine, for example, if personal income tax rates were one-tenth of their current values and motor fuel were five to ten times as expensive. Your disposable income could be markedly enhanced by ditching the gas-guzzler for an energy-efficient vehicle. Tax-shifting is not a magic bullet, though. It fails to benefit economic actors without the means to increase their material efficiency. The distributional issue relates not only to the distribution of financial wealth but also to the distribution of physical and social technologies for material efficiency. Without access to these technologies, an economic actor would have no way of benefiting from tax-shifting. They would therefore have no net incentive to increase material efficiency. Tax-shifting can be successful only if it is part of a broader suite of measures to ensure maximum participation.

Tax-shifting gives economic actors control over the amount of tax they pay through their decisions in the market. Although this is a desirable feature of the policy, it also ensures that governments lose a similar amount of control over their revenue stream. That tax-shifting has not been enacted sooner might be because legislatures fear it might deprive them of essential revenue. Yet this is not a reason to delay tax-shifting. As later chapters describe, taxes would be shifted gradually rather than in one fell swoop, and governments would have time to adjust. By the time the economy approached a realm of stable alignment with nature, the overall cost of government regulation could be significantly lower than in today's growth economy, meaning that government budgetary needs might also be lower.

[20] Schmidheiny and Zorraquín (1996, p. 172).
[21] Daly (2010).

Policies for Widespread Prosperity

Distributional Equity, Wealth and Prosperity

An increasing body of research has established a connection between natural capacity and the distribution of wealth within an economy. National or regional economies exhibiting relatively more equitable distributions of wealth tend to take better care of the natural systems within their borders.[22] They also are less competitive as societies. Competition is fueled by inequality: people desire the lifestyles or status they do not have. This so-called status competition in turn fuels economic growth. In the US, 1998 was a boom year for economic growth; yet in that same year, one family in 68 filed for bankruptcy. Bankruptcies were higher in those parts of the US where income inequality was the greatest.[23] By 2002, the average unpaid consumer credit card debt in the US was $9000, out of an average per-capita income of about $32,000 before tax.[24] This unsustainable debt-to-income ratio led in part to the financial crises that followed in the late 2000s.

In the wealthier countries, economic growth no longer delivers overall improvements in quality of life, compared to countries where rates of absolute poverty remain high. It merely results in greater material consumption at a cost of higher income inequality and concomitantly reduced societal quality. Reducing inequality, Richard Wilkinson and Kate Pickett argue, would "[shift] the balance from divisive, self-interested consumerism driven by status competition, towards a more socially integrated and affiliative society".[25] Such a society might exhibit a very distinct use of language from today's society, due to a difference in norms and attitudes. For instance, many current marketing and advertising campaigns promote the notion of exclusivity. You can have this or that as an 'exclusive' offer because, so the shtick goes, our marketing department has figured out that you are somehow 'special'—or the database has selected you at random to be. Exclusivity drives status competition. The cycle set up by these forces hinders the emergence of a more co-operative, inclusive and less exploitative culture.

[22] Costanza et al. (2015, ch. 3) provide a review of recent literature.

[23] Frank (1999).

[24] Source: US Department of Commerce, Bureau of Economic Analysis, Survey of Current Business: http://www.bea.gov/newsreleases/relsarchivespi.htm.

[25] Wilkinson and Pickett (2009, p. 233).

Although the term 'wealth' often is conflated with financial income or capital, a nation's natural systems also are ineluctably part of its wealth. Wealth, whether at the national level or the individual level, is a multivariate concept, having both quantitative and qualitative aspects. The quantitative aspects of wealth can be measured a variety of different ways, from the financial (money) to the social (such as crime rates) to the natural (such as biodiversity).

A future planetary economy will have achieved its material stability by becoming also more socially and financially stable. Social or financial instability are impediments to aligning the economy with nature. The only way the economy can approach a future stable state is if just about everybody is engaged in economic activities toward material discipline: that is, participating in it. For that to happen, people need access to tools, technologies and options. The transition toward alignment with nature must also be a transition toward widespread economic prosperity.

If more egalitarian societies tend to take better care of their natural resources, then it should follow that policies improving distributional equity without directly addressing natural capacity should nonetheless have a small positive effect upon it. Policies for distributional equity should enhance policies for natural capacity: the effects should be synergistic.

Yet there remains a nagging objection that policies to reduce income inequality would curtail individual liberties. Why not give the wealthy a *choice* whether to 'give back' to society? The experience in many OECD countries since the 1970s indicates that generally they do not. The emergence of a 'trickle-down' taxation policy from the new market fundamentalism was predicated on the assumption that reducing income taxes on high earners would generate more wealth in the wider economy through higher consumption rates. The wealthy did indeed consume more but most of the wealth did not trickle down. From the late 1980s to the late 2000s, 56 percent of income growth in the US went to the top one percent, while median earnings, adjusted for inflation, actually fell.[26] Prior to the 1980s, income inequality worldwide had been falling since medieval times.[27]

During the 2010s, it became popular to conclude that all this was a conspiracy on the part of political and corporate elites. Just as the

[26] Kelly (2012, p. 92), citing studies by Robert Reich, Lawrence Mishel and Heidi Shierholz.

[27] https://ourworldindata.org/income-inequality/#redistribution-as-a-solution.

neoclassical economists asserted that a market solution exists for any social ill, so the market fundamentalists proposed that the market would redistribute wealth. In fact, agent-based economic models have shown that the distribution of resources within a simulated economy will tend to become highly skewed, absent any form of regulation, with a few actors controlling most of the available resources. The market itself does not produce a socially preferred distribution of wealth; in fact, quite the opposite. The skewed distribution of wealth resulting from unregulated market activity would result in social conflict.

On a crowded planet, policy cannot be driven by abstract economic theory, only by an economic theory grounded in concrete reality. Pareto-optimal distributions of welfare would not be adjusted for social desirability, as the neoclassicists would have it. Instead, social desirability would be decided ex ante, and the market is then employed to discover solutions within the boundaries thus defined. This means that, at some minimal level, individual liberties would have to be curtailed for the collective good. In a finite world, assuming all members of society have basic equal rights to its riches, it follows that the protection of those rights will necessitate a distribution of resources narrower than in the unregulated computer model. Yet this requirement does not automatically mean a big-brother state imposing high income taxes.

Absolute Poverty and Relative Wealth

Policies for widespread prosperity need to address two distinct but related problems. One is absolute poverty and the other is people's perceptions of relative wealth. Widespread prosperity does not result from the elimination of absolute poverty alone. It can only emerge with the conquering of 'relative poverty', as it is known, which is a perceptual issue.

Absolute poverty is where an individual is unable to meet their basic living needs, such as nutrition, shelter and healthcare. Financially, it is a state of non-participation or minimal participation in the money economy. Absolute poverty also is a state of low material intensity. Anybody rising out of absolute poverty generally then finds themselves a participant in the money economy and therefore subject to the aspirations of a better lifestyle. In a world of skewed income distributions, they would now find themselves in a state of relative poverty.

Because many on lower incomes aspire to the lifestyles of those on higher incomes, an economy where high incomes are many multiples of low ones will have a wide consumption gap. Where incomes are more evenly distributed, the consumption potential will be lower. Absent any countervailing incentives to increase material efficiency within the economy, an increase in relative poverty would result in higher aggregate material intensity of the economy.

Although economic globalization has reduced rates of absolute poverty overall, the expansion of the money economy to almost all parts of the world has coincided with a significant increase in financial disparity between the poorest and the wealthiest. To put the present global income inequality into perspective, the wealthiest few individuals alive early in the twenty-first century have a net worth on the order of $100 billion each, which is more than the estimated total wealth of Uruguay and each of about 120 other countries.[28]

Globalization has enabled financial capital to become significantly more mobile. In some places, human capital has been left stranded. Industrialized economies have seen middle- and lower-income jobs outsourced to emerging economies, while many in poorer countries have become economic migrants into wealthier ones. Money is allowed to migrate virtually anywhere, whereas people and jobs find it much harder. People generally do not want to uproot themselves. They would rather stay put if they had a way to make ends meet: that is, if they could increase their incomes.

In addition to the issue of geographical mobility, Wilkinson and Pickett show that the ability to move from one income stratum to another within an economy—so-called social mobility—is correlated with distributional equity. The more evenly incomes are distributed, the easier people find it to move from one income level to another. In societies with wide gaps between rich and poor, social mobility is significantly lower.[29]

If lower-income individuals living within a more equal setting can be sure that society will catch them from falling into destitution, then they may be more inclined to take vocational risks. This inclination would be particularly prevalent among young people, whose incomes are, on average, lower anyway than those in mid-career. Further, lower-income people will perceive the upper income brackets within society as relatively more attainable, or at least not completely out of reach. Knowing that their downside risk was low and their upside potential relatively more attainable, lower-income individuals also would have greater latitude to shift

[28] Data from Credit Suisse (2018).

[29] Wilkinson and Pickett (2009, ch. 12).

their patterns of consumption in response to incentives to increase material efficiency. Additionally, knowing that the higher echelons of income are not many multiples more than theirs, people on lower incomes would more likely accept that all echelons of society are doing their part toward a prime directive.

From the point of view of higher-income individuals, greater income equality would reduce their perception of free-riding on the part of those on lower incomes. Being able to take vocational risks and adjust their consumption patterns in the marketplace, lower-income people would become full economic participants. No longer would most of them need or want to 'sponge off the welfare state', with all its attendant social stigmas. They would have their dignity, and high-income individuals would not feel resentful to be funding 'a collection of slackers' through their income taxes. The principle of strong reciprocity suggests that a sense of fairness should exist at all income levels. This is important because, unless we feel all are doing something, few will likely do anything.

The material stability of a future economy would be predicated upon a relative absence of demand to consume more. The range of material consumption would be narrower than today, and so would be the distribution of incomes. Starting from today's growth economy, the challenge of increasing material efficiency becomes also the dual challenges of lifting people out of absolute poverty into the money economy, and reducing the demand for more income among those already in it. Greater equality, write Wilkinson and Pickett, is "the gateway to a society capable of improving the quality of life for all … and [developing] a sustainable economic system"[30] It necessitates moving from the purely individualistic norm of modernity toward a collective norm. Harnessing the power of the market toward greater material efficiency then becomes equivalent to harnessing individual self-interest toward collective prosperity.

Raising the Income Floor

Arguing from a Rawlsian perspective, Matt Miller asks "if we did not know anything about our draw in the birth lottery, what kind of system would we want?" The intent of the question is to design a system ex ante of the birth lottery, one that would be politically and ideologically neutral. Among the policies he advocates for this, which include universal health

[30] Wilkinson and Pickett (2009, p. 237).

coverage and well-funded public education, Miller proposes to place a floor under people's incomes via a minimum living wage or universal basic income scheme. He estimates that these initiatives would cost around 2 percent of GDP in the US.[31]

Basic income sometimes is referred to as a form of 'pre-distribution' to distinguish it from redistribution. As pre-wage income, it avoids the politically contentious issue of taxing the wage-earnings of people on higher incomes to redistribute to those on lower incomes.

A guaranteed minimum income, according to Roope Mokka and Katariina Rantanen, enables "the capacity to produce and thus to participate actively in society".[32] By ensuring participation in the economy, it ensures participation in the drive toward material discipline and long-term economic stability. In contrast, people who rely solely on wage labor for access to sustenance—which is most of us today—have neither the time nor the resources to fully participate in this endeavor. Yet most of the economy's useful economic production probably can be accomplished with a fraction of the people-hours actually expended.

"Money", writes Nigel Dodd, "diminishes our capacity to appreciate qualitative aspects of value because of its unremitting orientation to *quantity*" (italics his). "[It] stifles individual excellence: its very existence encourages us to ... erase individual distinction. Money is the great leveler."[33] All the more reason, then, to use money as a great liberator, by creating a level basis for all. Not only would basic income enable broad participation toward material efficiency but also it would stimulate an unimaginable collection of other innovations, enabled by freeing individual creativity. The individualism of modernity is a self-maximizing individualism. The individualism of a planetary society is a self-fulfilling one.

Variants on the basic income concept within Western thought date back 500 years. Johannes Ludovicus Vives (1492–1540), a close associate of Sir Thomas More referred to it as 'public assistance'. Eighteenth-century writers on a 'social dividend' included the French mathematician and political activist Antoine Caritat, Marquis de Condorcet (1743–94), the American revolutionary Thomas Paine (1737–1809) and the English radical Thomas Spence (1750–1814). The dividend was conceived as a natural endowment rather than as actual income. The idea was taken up in the

[31] Miller (2003), cited in Beinhocker (2006, pp. 445–446).

[32] Mokka and Rantonen (2017).

[33] Dodd (2014, pp. 137–138).

nineteenth century by the French writers Charles Fourier (1772–1837) and Joseph Charlier (1816–96), the latter becoming the first to formulate a basic income scheme as it is understood today, which he called a 'territorial dividend'. John Stuart Mill also was supportive of the concept.

Twentieth-century proponents included Clifford Hugh Douglas, John Commons, Bertrand Russell, Friedrich Hayek, James Tobin, Paul Samuelson, J.K. Galbraith and Erich Fromm. Douglas proposed an idea of 'debt-free credit' to be created through a national dividend, a minimum payment to all workers.[34] Commons, writing in the midst of the Great Depression, proposed a scheme in which government would create new money, paid directly to the unemployed rather than inserted into the banking system.[35] Hayek recognized the plight of those unable to acquire income through participation in market activity: "the sick, the old, the physically or mentally defective, the widows and orphans." A society that has reached a certain level of wealth, he argued, can afford to provide for these people. "The assurance of a certain minimum income for everyone, or a sort of floor below which nobody need fall even when he is unable to provide for himself, appears not only to be a wholly legitimate protection against a risk common to all, but a necessary part of the Great Society in which the individual no longer has specific claims on the members of the particular small group into which he was born."[36]

Erich Fromm, writing in 1976, argued for the abandonment of economic growth to stave off economic disaster, instead creating working conditions fostering "nonmaterial satisfactions and nonmonetary motivations". To close the gap between poor and rich, an issue he viewed as urgent, he proposed a universal guaranteed income based on a universal right to a minimum standard of living. "In this way, social life itself might be elevated to an end in itself."[37]

Various sources of funding were proposed for basic income and its variants. Many of the early writers favored a policy of taxing land value, on the basis that the economic value derived from land belongs to all, as distinct from any value people produce themselves. This concept sometimes is referred to as Georgism, after the nineteenth-century economist Henry

[34] Dodd (2014, pp. 355–356), citing Douglas (1920, 1924).

[35] Striner (2015, p. 44).

[36] Hayek (1979, pp. 54–55).

[37] Dodd (2014, pp. 338–339), citing Fromm (1976). Fromm also, by the way, proposed more open and participatory forms of governance through more effective dissemination of information, a theme explored in the previous chapter.

George, who developed the first systematic treatment, although the concept dates back further. Other economists later proposed funding basic income through income taxes or, more precisely, negative income taxes. Milton Friedman argued in 1962 that incomes below a certain level could be increased through a negative tax rate, which would result in a government disbursement, paid for by higher tax rates on the wealthy. In 1967, James Tobin et al. refined the idea to propose an automatic payment to be made to all citizens, funded through negative income taxes.[38] Tobin et al.'s analysis influenced political thinking at the time. A basic income scheme known as the Family Assistance Plan was introduced in 1970 by President Nixon. It passed in the House, but ultimately was rejected in the Senate.[39] Basic income came very close to becoming law in the US.

In 1993, James Meade proposed a scheme in which the government, through the taxation of large-scale wealth, would invest in corporate stock, the dividends from which would be distributed equally to every citizen.[40] The size of the dividend, of course, would be subject to the whims of the financial markets. In the late 1990s and early 2000s, Thomas Pogge argued for a 'global resources dividend' to reduce global poverty by levying taxes on nations' resource use. However, the idea came in for criticism for being impractical and potentially exacerbating existing distributional inequities.[41]

Since 2000, basic income has received much more attention, in economic research, policy circles and the popular press. Philippe Van Parijs presented a case for it in 2000, prompting a flurry of other commentary. Gar Alperovitz, for instance, highlighted the need to look beyond the taxation system as a means to fund basic income, suggesting, as Van Parijs did, to examine the Alaska Permanent Fund as a model. Anne Alstott emphasized breaking the link between social-welfare benefits and paid work, while also highlighting the potential synergies between a basic income scheme and consumption taxes or environmental taxes. Robert Goodin argued alternatively for paid 'workfare' to expand to encompass "people doing socially useful unpaid work". This proposal, however, ignores the people Hayek identified as incapable of doing any kind of work

[38] Tobin et al. (1967).

[39] Adapted from *History of Basic Income*, Basic Income Earth Network, www.basicincome.org.

[40] Meade (1993), cited in Alperovitz (2005, pp. 45–46).

[41] See Pogge (1998) and Tim Hayward's (2005) critique.

at all, such as the very young, the very old or the infirm—the people arguably in greatest need.[42]

Academic research on basic income has built significantly on this earlier work.[43] A non-profit society, the Basic Income Earth Network, has been formed, and the field has established its own dedicated scholarly journal, *Basic Income Studies*.[44] This academic work has caught the eye of policymakers, resulting in pilot schemes in a number of places, including Manchester, Scotland, Finland, Oakland, Barcelona, Shenzhen and the Netherlands. These in turn have led to increasing coverage in the popular press.[45]

Relatively little of the basic income literature or policy has considered the possibility of funding it through sources other than taxation. One exception comes from the advocates of sovereign or public money creation, who have proposed for governments to create new basic income or 'citizens' dividends' from scratch, as Chap. 10 will elaborate. Peter Barnes makes a case for charging fees to corporations to use common assets, such as nature, the electromagnetic spectrum and the internet. Rights to these assets would be held by a commons trust, or planetary trust, having the ability to collect them and disburse dividends from them. The dividends would form a basic income for all citizens. The rationale for such a scheme is that the privilege to use shared capacity carries a cost, in the form of

[42] See Van Parijs (2000) and responses therein.

[43] A Palgrave Macmillan series edited by Karl Widerquist incorporates a variety of theoretical work and case studies: http://www.palgrave.com/gb/series/14981. Guy Standing (2014) argues for a basic income as part of his 'precariat charter'. Widerquist et al. (2013) provide an anthology of papers related to minimum income going back to Milton Friedman. Along more popular lines, Philippe Van Parijs and Yannick Vanderborght (2017) provide a historical context for an idea which they argue will shake up welfare economics over the coming decades. Nick Srnicek and Alex Williams advocate for the same idea in their 2016book, *Inventing the Future: Postcapitalism and a World Without Work.*

[44] http://basicincome.org/research/basic-income-studies/.

[45] In 2017 alone, *Forbes* published three articles on basic income: 'Top economists endorse universal basic income' (Frances Coppola, August 31, 2017), 'The IMF gives a cautious welcome to universal basic income' (Frances Coppola, October 15, 2017), and 'In Silicon Valley, basic income gets serious' (Gregory Ferenstein, October 20, 2017). *The Daily Telegraph*'s recent coverage includes: 'Will all Scots get free money from Nicola Sturgeon? What is a universal basic income and how is it funded?' (Helen Chandler-Wilde, September 5, 2017), 'How Universal Basic Income may hurt those it seeks to help, (Anna Isaac, October 23, 2017), 'Berlusconi woos voters with tax breaks for pet owners and a basic income for all Italians' (John Phillips, December 28, 2017), and '"Universal basic income": proposals to give £10,000 to everyone under 55' (Sophie Christie, February 16, 2018).

rent. The flow of funds would bypass the national tax authority, as it would not be counted as income. It would therefore be straightforward to administer.[46] Barnes estimates that even relatively modest fees would produce an annual dividend to every adult and child in the US of about $4000. From birth, this amount would yield $72,000 before interest to a young person turning eighteen, a fund with which they could begin their adult life.[47] Chris Hughes recently extended the idea to include charging fees on the data that corporations collect on customers, a neat simultaneous solution to data privacy concerns.[48]

The benefits of using small fees levied on corporations to fund basic income are that downside financial risks to all would be mitigated at negligible upside cost to the few, and individual income tax rates would not be affected. The idea is inspired by the Alaska Permanent Fund, an oil dividend established in 1976, paid by the state of Alaska to all Alaskan residents. Oil companies pay the state to access Alaska's oil resources, and the state redistributes this income to its citizens, who are its primary stakeholders. The dividend, which averages around $2000 per person per year, is not taxable income.[49]

Most of the research and development on basic income has assumed that payments would be made in national currency. This does not necessarily have to be the case, as David Orrell and Roman Chlupatý point out. "If basic income were delivered using a complementary currency", they write, "it would be more clear that the funds represent a kind of birthright." Hoarding of this complementary currency would be discouraged by levying a small account fee, proportional to the amount held, which also would cover administration costs. In this way, the currency would be released back into circulation.[50] The currency would have to be legal tender, of course, accepted widely as payment. Although the idea raises questions about whether or how the currency would be convertible into national currency, or other forms of money, for that matter, it merits further investigation.

Basic income is not the only way to raise the income floor. A family of alternate proposals centers on establishing what could be called a starter

[46] Barnes (2001, 2006, 2014).
[47] Barnes (2014, p. 94).
[48] Hughes (2018).
[49] Barnes (2014, ch. 6).
[50] Orrell and Chlupatý (2016, p. 195).

fund. They would provide every individual an annual income from birth that can be accessed as a lump sum upon adulthood. In this way, rather than receiving a Social Security pension—or perhaps in addition to it—an individual would receive much-needed financing at the beginning of their working life.[51] Although many of us looking back on our early adulthood would agree that such a payment could have been very useful, an advantage of basic income is that it would continue uninterrupted through life.

Could a basic income system result in perverse incentives? Possibly. For a start, it could encourage people to have larger families. Each additional child would render a family eligible for an additional annual unit of basic income. One way to counteract this tendency would be to limit basic income to the replacement rate of two children per family. Another possible effect would be a downward pressure on wages. Employers might be tempted to freeze wage increases or to actually reduce them, on the basis that part of an employee's income would be met through basic income. However, two counter-effects also would operate. First, being relatively less dependent upon wage-based income, people would be correspondingly less dependent on wage-based employment. The demand for employment would go down along with any demand for wages. If this were the case then, second, employees' bargaining power to maintain wages would be enhanced. Any employer reducing wages in the face of a basic income scheme might find itself losing employees to a competitor who did not. As a result, wages might continue to depend on the overall employment rate in the economy, as they do today. Modeling techniques, such as game theory, would become useful to explore the potential effects of these forces.

Finally, it should be noted that financial income is only one of potentially many measures of prosperity. Chapter 13 proposes expanding the idea into a 'basic living program' including also health care and education.

Elective Redistribution

Basic income schemes are on the right track. Yet they do not, in and of themselves, wholly resolve the issue of people's relative perception of

[51] See, for instance, Kuttner (1998), Ackerman and Alstott (1999), Brown (2001) and Boshara (2003), cited in Alperovitz (2005, p. 44). Alperovitz also cites attempts at Congressional legislation in the US to establish funds of this kind, such as the KidSave Account.

wealth which, if widely disparate, can feed excessive consumption and undesirable material intensity. At the other end of the distribution spectrum is a uniform distribution, which is what the communists advocated. That does not work because it stifles competition and assumes everybody is equal in their abilities to create wealth. There is likely a 'Goldilocks zone' of income distributions that conforms to the principle of strong reciprocity: everyone should have an opportunity for reward but everyone also should be treated fairly.

One way to estimate the bounds of such a Goldilocks zone is to examine the income distributions of different countries. An index exists for this purpose, the Gini coefficient, named after the Italian statistician who invented it. The Gini coefficient is a measure, between zero and one, of inequality among values in a frequency distribution. Although the coefficient does not capture the shape of a specific income distribution, it allows comparison of income inequality among countries. By way of illustration, if the top one percent of a population accounted for 50 percent of all the income then its Gini coefficient would be 0.49.

In the US, whose Gini coefficient is 0.42, the top 10 percent of incomes is just under four times the median income. The Gini indices of relatively egalitarian economies, such as Norway, Slovenia, Sweden and Japan, are in the range 0.25–0.3, corresponding to the top 10 percent being around twice the median.[52] Suppose for the sake of illustration that a broadly agreed-upon goal for the distribution of incomes would place the top 10 percent around two to three times the median. If the US instituted a basic income scheme and narrowed the income distribution in accordance with this goal, the effect would resemble that depicted in Fig. 8.1. Note that the new median income is now higher than the old one.

The question then becomes how to accomplish this narrowing of the distribution in a broadly acceptable way. Policies for redistributing income traditionally have centered around the taxation of income, which can become an ideological football between socialists seeking to tax the wealthy to fund the poor and libertarians seeking to minimize penalties on individual financial success. Other options exist.

[52]Gini coefficients from Wikipedia: https://en.wikipedia.org/wiki/List_of_countries_by_income_equality. Income distributions from World Bank data: http://data.worldbank.org/data-catalog/poverty-and-equity-database.

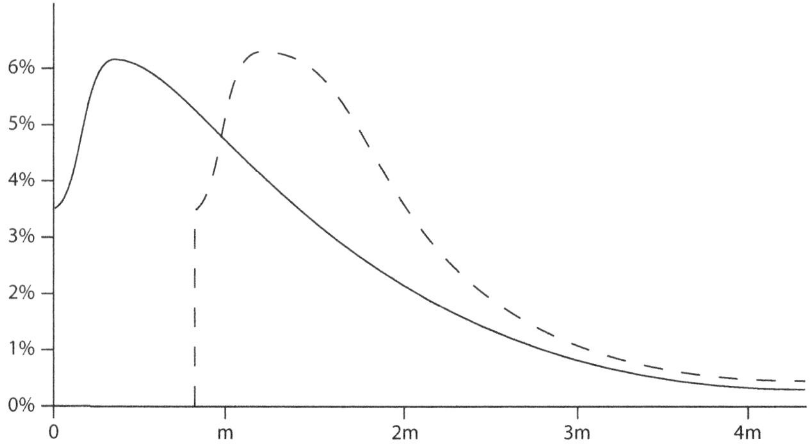

Fig. 8.1 Illustration of a policy of narrowing income distributions and establishing an income floor. (Source: Author's creation, based on income data from the US Census Bureau (Lofquist et al. 2012; Denavas-Walt and Proctor 2014). Note: Solid curve represents a typical current income distribution, scaled according to the median (m). Dotted curve is a hypothetical proposed income distribution in which a basic income scheme establishes an income floor close to the existing median and elective redistribution narrows the income distribution from the top end, such that the 90th percentile falls between two and three times the median)

Dwight Murphey proposed to establish a 'shared market economy' in which everyone received income from corporate stock,[53] thereby sidestepping the redistribution issue. Publicly traded companies would be required to set aside a proportion of shares for government disbursement. It would be as if the government operated a mutual fund without actually trading in the companies' stock. Every citizen upon birth would receive shares in this fund, thereby becoming a minority shareholder in all publicly traded companies. Although potentially an effective way to share wealth and add a degree of public purpose to corporate operations, such a fund would not address the pressing needs many people have for actual income. A related proposal by John Roemer would have the government establish a capital base through taxation to offer to every adult vouchers that could be used to purchase stock through a mutual fund. Profits would be the

[53] Murphey (2009).

stockholder's to keep.[54] As with James Meade's proposal, the success of the scheme would depend upon continually increasing stock market valuations.

Another option to narrow the income distribution is to provide high earners a way to direct some of their income toward a common benefit of some kind. Peter Barnes suggests a Children's Opportunity Trust, which would use the common trust framework to establish a public inheritor of bequeathed wealth for the next generation. The Trust would form a kind of 'temporal commons', linking an end-of-life obligation to pay back to society with "Social Security for the front end of life", as Barnes calls it.[55] Someone passing on their estate at the end of their life could, for instance, will half of it to their heirs and the other half to the Trust. The Trust would pool all such inheritances and use them to create 'individual inheritance accounts' for all children, irrespective of income. A wealthy person's heirs "would still start on third base", writes Barnes, "but others would be in the game".[56]

An advantage of such a scheme is that it would not take money from the treasury, thereby circumventing income taxes. However, this does not preclude also using pre-tax income earned throughout life for redistribution. I call this idea elective redistribution.

People on higher incomes vary considerably in their willingness to support people on lower incomes. Some believe in levying high marginal tax rates on high incomes, whereas others believe they should not have to sink a large portion of their hard-earned wealth into a government 'black hole'. Rather than the government taking a position on this issue, it could give people the option to decide for themselves. For people on incomes above a certain threshold, the option could be provided to direct a portion of pre-tax income to specific purposes, whether a direct transfer of funds or charitable contributions. Those electing to not redistribute income this way would either pay a relatively high marginal rate of income tax to the treasury.

For the sake of illustration, suppose that incomes below the median in the US were taxed at zero percent and incomes above the median were taxed at 10 percent, thereby assuring a basic stream of income tax revenue for the government. This rate structure, by the way, also would be

[54] Roemer (1996), cited in Alperovitz (2005, pp. 44–45).
[55] Barnes (2006, p. 145).
[56] Barnes (2006, p. 111).

consistent with a policy of tax-shifting from income onto material intensity. Incomes above three times the median also would be taxed at 10 percent, and they would be required to allocate an additional 40 percent, or some other relatively high amount, either to income tax or to one or more registered redistributive causes.

A national redistributive program would work a little like existing corporate giving programs in which employees can choose among a menu of registered charities to which to make tax-deductible donations. Under a national redistributive program, however, a household also could have an option to donate to government programs in support of the disadvantaged. These donations would be similar to hypothecated taxes. A third option would be to donate income as cash via the national tax authority. This stream of revenue would be apportioned to lower-income households as unearned income. For recipient households, such income, unlike the income from a basic income scheme, could then be taxable depending upon that household's total income.

Apportionments by income level would have to be established. One possible approach would be to place the donor's gross income into a percentile, then assign the donated amount into a pool of incomes for the mirror percentile at the lower end. All donor households in the 89th percentile of all incomes, for example, would donate to all recipient households in the 11th percentile. In this way, those in the 99th percentile, giving the most income per household, would donate into the very poorest percentile, the first.

The problem with using taxation for redistribution lies with the lack of control taxpayers have over how the money is used. Voters have little or no direct influence over the government budgetary process.[57] Elective redistribution would not be appropriated by the treasury. It would be a direct transfer. By taking ownership over the redistribution of some of their wealth, higher earners would become invested in the outcomes they chose. The scheme also satisfies the principle of strong reciprocity: every higher-income household would know that all others were subject to the same requirement. Over time, the requirement would condition behaviors, which in turn would become norms.

Again, the type of institution that best lends itself to this scheme is a trust. It would not be a planetary trust because it would not protect common wealth; rather, it would be a national redistributive trust. Placing the

[57] See, for instance, Barnes (2014, pp. 86–87).

administration in the hands of a trust would instill confidence among donors that funds would be walled off from government appropriation, and that management of the program would be impartial.

If the scheme could work in the US, it could work in other countries as well. Incomes in many middle-income countries, such as in Latin America, the Middle East and Asia, tend to be more skewed than in the higher-income countries. The ratio of the top 10 percent to the median in these countries ranges from about three up to about six (see Appendix to this chapter). National redistributive programs within these countries would progressively reducing the ratio say, from five, to four, to three. In this way, the very highest earners would be first to undertake such redistribution. Those on incomes below this level would be incorporated into the scheme as the income gap narrowed.

The same approach even could be employed internationally. If I were one of the top 10 percent of earners in the US, I could in principle elect to redistribute some pre-tax earnings to the very poorest of the world. If an elective redistribution program were successfully established domestically then it could be refined to allow a portion of the taxable amount—say, 10 percent of earnings above three times the median—to be allocated to programs improving the lives of low-income people internationally, either through cash disbursements or in other forms. The intent would be to accelerate the elimination of absolute poverty, thereby raising the economic participation rate globally. An international scheme would necessitate an international redistributive trust. In the long run, such a trust would naturally fall under oversight of a global governance council, although in the short term, other candidate institutions are the International Monetary Fund (IMF) or the World Bank.

The program would operate primarily within higher-income countries. As the Appendix to this chapter describes, even the median income in these countries is higher than the top 10 percent in most middle- and lower-income countries. Although international redistribution could be open to middle-income countries, it is doubtful many of those taxpayers would opt for it: they would view it primarily as an obligation of taxpayers within wealthy countries. However, as income disparities among countries narrowed, the program would become open to more participants.

An international elective redistributive program would face a variety of logistical challenges to actually deliver funds to the neediest. Some of the greatest impediments to helping the poorest out of absolute poverty are official corruption, weak oversight and weak law enforcement. It is also the case that those in the greatest poverty have no way to securely store money, nor many markets in which to spend it.

In spite of these logistical difficulties, a precedent of sorts exists. In 2009, Norway entered into a bilateral agreement with Guyana to pay Guyana to refrain from deforestation. Payments are made on the condition that they are spent on programs that actually benefit Guyana, such as infrastructure, education and healthcare. The natural choice of institution for managing the redistribution of wealth to low-income regions on behalf of an international trust is the United Nations. For all its legendary bureaucracy, it is uniquely equipped and experienced to assist in directing resources into recipient communities, having operated programs to do exactly that for decades. It is also respected in many developing countries for its neutrality and lack of political affiliation.

In an international elective redistributive program, member nations of the UN would sign onto the trust, thereby establishing a mechanism by which high-income households could select from a menu of international programs when filing their taxes. Many of these programs already exist, of course: UNICEF, the World Health Organization, Oxfam, Médécins Sans Frontières, to name a few. Some are UN programs; others are independent, yet all have longstanding relationships with the UN. Their assistance is tailored to the needs of particular regions.

As a recipient region began to benefit from disbursements through the trust, the kinds of programs serving it might alter from the basics of food, shelter and medical care to education, employment, infrastructure, banking and judicial systems. Correspondingly, the menu of programs to which a higher-income household could donate would change over time as well.

Member nations wishing to sign on to the program would have to meet certain requirements. They would have to establish a national tax authority able to operate independently from official interference, subject to audit by the trust. Member nations would be required to adjust their income tax structures to enable elective redistribution. They would be required to ensure the reliable distribution of benefits from donors, financial or otherwise, again subject to audit.

The main incentive for a member nation to meet these requirements would be sheer scale of the available wealth distribution to its lower-income citizens. Rather than relying on voluntary donations by concerned citizens in distant countries and on UN membership fees paid by member states, participating programs would be able to avail themselves of an altogether larger funding stream: the pre-tax income from higher-income households worldwide.

A rough estimate of this amount is given in the Appendix to this chapter. Higher-income countries are defined for the purpose of the analysis as those whose median income is higher than the median of the top 10 percent of incomes among all countries in the dataset. This means that someone on an average income in a high-income country, thus defined, earns more than all but the most affluent 10 percent in middle- or low-income countries. Twenty-eight out of 144 countries in the dataset meet this criterion. The analysis assumes that 10 percent of pre-tax incomes in the top 10 percent would be available for disbursal through elective redistribution in place of income tax. This would equate to a higher percentage of incomes above a certain threshold within a progressive taxation structure, which most of these countries practice, although the exact percentage will vary from one country to another. The top 10 percent of incomes is between two and four times the median, with an average around three.

The total disbursable amount is just over a trillion dollars per year, about half of which would come from US incomes. Not only is the US a high-income country but also it is currently by far the largest and its income distribution is somewhat more skewed than most other high-income countries.

Elective redistribution may have ancillary benefits. With income-donation as a pre-tax line item on the paystub of every higher-income individual, the market for charitable aid and assistance would be greatly expanded. Institutions and programs would compete with one another for this pre-tax income. Competition would beget innovation in programs, which would beget improvements in efficacy.

Funds donated internationally would create ties among people across borders. By logging onto their account under the trust, a resident of Beverly Hills, for example, could see the direct effects of their donations upon the lives of people in Malawi. In the grand scheme of things, these connections could even, in certain instances, reduce the likelihood of international tensions or conflict.

An international elective redistribution program would expand the United Nations' role as the world's principal provider of aid and assistance. Coordinating this vast new flow of assistance from the trust would give the UN new relevance. Participating member nations would have a keen stake in its success. Similarly, for development projects, the role of the World Bank would take on new importance, and the IMF for financial flows. These organizations now would handle funds originating not only from contributing nations but also from individual taxpayers via national tax authorities.

Appendix: Elective Redistribution Analysis

Data for the elective redistribution analysis in Chap. 8 come from the World Bank and from the World Income Inequality Database (WIID) administered by the United Nations University World Institute for Development.[58] A common proxy for actual, reported income is gross national income (GNI) per capita. This proxy is used because people's tax returns are not publicly available in most countries. In some less developed countries, taxable income may not, in any case, be a reliable guide for actual income, as many people live beyond the reach of tax authorities or below any taxable threshold.

Since the question of taxing or electively redistributing income applies principally to people in the workforce, this analysis estimates GNI per capita of the labor force. The World Bank publishes estimates of the percentage of each country's population aged 15 to 64, which is a rough approximation of the total potential workforce.[59] For simplicity, unemployment is ignored. GNI per capita of the labor force can then be roughly approximated by dividing GNI per capita by the percentage estimated to be of working age.

The WIID assigns percentages to income deciles, where possible, or to quintiles, or to the bottom and top fifth percentiles. Data on these income distributions come from a variety of sources, including the World Bank. All sources, whether institutional databases or research studies of specific regions, are attributed. Decile data for most countries have most recently been collected since 2000; however, some studies are older. The WIID includes distribution data from all known studies going back in time, which allows for observation of changes in income distributions within a country.

The present analysis estimates average per-capita income by decile by combining the decile data from the WIID with the GNI data from the World Bank. The GNI data are from 2013. Except where the decile data are also from 2013, this means that the analysis mixes data from different

[58] https://www.wider.unu.edu/data.

[59] http://data.worldbank.org/indicator/SP.POP.1564.TO.ZS.

Table 8.1 'High-income' countries for elective redistribution analysis

Economy	*Gross national income (US$ bil.)*	*Labor force (mil.)*	*Median income (US$)*	*Top 10%*	*Ratio*	*Disbursable (US$ mil.)*
Australia	1389	15	75,179	228,539	3.04	34,869
Austria	400	6	63,296	154,019	2.43	8750
Belgium	496	7	62,787	143,999	2.29	10,467
Canada	1670	24	58,223	180,919	3.11	43,250
Cyprus	30	1	31,591	86,190	2.73	699
Denmark	329	4	84,223	200,488	2.38	7205
Finland	252	3	66,597	157,477	2.36	5396
France	2675	41	54,652	169,520	3.1	69,289
Germany	3761	54	61,400	158,877	2.59	86,123
Greece	222	7	27,265	79,575	2.92	5584
Iceland	16	0	69,321	154,465	2.23	330
Ireland	215	3	59,966	188,156	3.14	5624
Israel	286	5	45,549	171,216	3.76	8417
Italy	1975	39	45,343	123,987	2.73	47,797
Japan	4671	78	58,327	104,628	1.79	81,271
Korea, Rep.	1352	36	33,530	95,927	2.86	34,497
Luxembourg	42	0	98,450	238,700	2.42	908
Malta	9	0	28,477	69,682	2.45	197
Netherlands	822	11	67,763	160,373	2.37	17,517
New Zealand	182	3	50,871	157,854	3.1	4670
Norway	477	3	132,201	278,617	2.11	9341
Singapore	282	4	57,727	209,502	3.63	8277
Slovenia	47	1	31,553	66,818	2.12	922
Spain	1330	31	38,027	103,277	2.72	31,778
Sweden	555	6	86,715	181,689	2.1	10,989
Switzerland	681	5	109,937	301,540	2.74	16,343
United Kingdom	2779	41	56,207	176,746	3.14	72,540
United States	17,395	209	65,661	246,154	3.75	514,187
Total					3.08	1,137,238

Source: Author's creation, using data from the World Bank and from the World Income Inequality Database (WIID) administered by the United Nations University World Institute for Development, https://www.wider.unu.edu/data

Notes: High-income countries are defined as having median income greater than the median of the top 10 percent among all listed countries in the WIID database. The table shows the median income, the average income in the highest decile, and the ratio of the highest decile (D10) to the median. The right-most column shows how much income could be disbursable through an elective redistribution scheme if 10 percent of income from the top 10 percent of earners were available for it, instead of income tax years, which could introduce bias. However, the decile data mostly are from neighboring years to 2013 and the income distributions change relatively little from one study to the next. This analysis seeks only a rough outline of the differences in income distributions among countries for the purpose of illustrating the policy argument for elective redistribution.

The full dataset for the analysis is available upon request. It lists 146 countries for which data exist on GNI per capita, population and income distribution by decile, as well as Gini coefficient. It includes the median per-capita income in 2013 US dollars, the income of the top 10 percent, and the ratio of the latter to the former. The lower the ratio, the more even the distribution of income.

For the purpose of the analysis, 'high-income' countries are those whose median per-capita income is higher than the median of the highest deciles (D10) of all listed countries. The median of the highest deciles across all countries is $25,846 (Fiji). Twenty-eight countries have median per-capita incomes higher than this level. They are listed in Table 8.1.

The table shows the ratio of the top 10 percent of incomes (decile 10) to the median. It ranges from just below 2 up to about 3.75. The weighted average of the ratio is just over three. In the example in the table, it is assumed that 10 percent of incomes in the top 10 percent could be disbursable through an elective redistribution scheme in place of income tax. This would be equivalent to a larger percentage of income above a certain threshold within a progressive taxation scheme, which most of these countries practice. The exact percentage of course will vary from one country to another. The total disbursable amount, in millions of US dollars, is shown in the rightmost column.

References

Ackerman, B., & Alstott, A. (1999). *The Stakeholder Society*. Yale University Press.

Alperovitz, G. (2005). *America Beyond Capitalism*. Hoboken, NJ: Wiley.

Barnes, P. (2001). *Who Owns the Sky? Our Common Assets and the Future of Capitalism*. Washington, DC: Island Press.

Barnes, P. (2006). *Capitalism 3.0: A Guide to Reclaiming the Commons*. San Francisco, CA: Berrett-Koehler.

Barnes, P. (2014). *With Liberty and Dividends for All*. San Francisco, CA: Berrett-Koehler.

Barnett, H. J. (1966). Pressures of Growth Upon Environment. In H. Jarrett (Ed.), *Environmental Quality in a Growing Economy* (pp. 15–20). Baltimore, MD: John Hopkins Press.

Beinhocker, E. D. (2006). *The Origin of Wealth: Evolution, Complexity, and the Radical Remaking of Economics.* Boston, MA: Harvard Business School Press.

Boshara, R. (2003). The $6000 Solution. *The Atlantic Monthly, 291*(1), 91–95.

Brown, J. L. (2001). From New Deal to New Opportunity. *The American Prospect, 12*(3), 24–27.

Costanza, R., et al. (2015). *An Introduction to Ecological Economics* (2nd ed.). Boca Raton, FL: CRC Press.

Credit Suisse. (2018). *Global Wealth Databook 2018.* Zürich, Switzerland: Credit Suisse Research Institute.

Daly, H. E. (1991). *Steady-State Economics* (2nd ed.). New York: John Wiley & Sons.

Daly, H. E. (2010). From a Failed-Growth Economy to a Steady-State Economy. *Solutions, 1*: 37–43. (Excerpt from Costanza et al., 2015, pp. 204–209.)

DeNavas-Walt, C., & Proctor, B. D. (2014). Income and Poverty in the United States: 2013. Current Population Reports. Washington, DC: United States Census Bureau.

Dodd, N. (2014). *The Social Life of Money.* Princeton University Press.

Douglas, C. H. (1920). *Economic Democracy.* London: Cecil Palmer.

Douglas, C. H. (1924). *Social Credit.* London: Eyre & Spottiswoode.

Frank, R. H. (1999). *Luxury Fever: Why Money Fails to Satisfy in an Era of Excess.* Simon & Schuster and New York: The Free Press.

Fromm, E. (1976). *To Have or To Be.* New York: Harper & Row.

Hayek, F. A. (1979). *Law, Legislation and Liberty: A New Statement of the Liberal Principles of Justice and Political Economy, Vol. 3: The Political Order of a Free People.* Routledge, London, 1982 reprinting.

Hayward, T. (2005). Thomas Pogge's Global Resources Dividend: A Critique and an Alternative. *Journal of Moral Philosophy, 2*(3), 317–332.

Hughes, C. (2018, April 27). The Wealth of Our Collective Data Should Belong to All of Us. *The Guardian.*

Kelly, M. (2012). *Owning Our Future: The Emerging Ownership Revolution.* San Francisco, CA: Berrett-Koehler.

Kuttner, R. (1998). Rampant Bull. *The American Prospect, 9*(39). https://prospect.org/article/rampant-bull.

LeGrand, J. (1991). *Equity and Choice: An Essay in Economics and Applied Philosophy.* London: HarperCollins Academic.

Lofquist. D. et al. (2012). Households and Families: 2010. *2010 Census Briefs.* Washington, CA: United States Census Bureau.

Meade, J. E. (1993). *Liberty, Equality and Efficiency: Apologia pro Agathotopia Mea.* London: Palgrave Macmillan.

Miller, M. (2003). *The Two Percent Solution: Fixing America's Problems in Ways Liberals and Conservatives Can Love*. New York: Public Affairs.

Mokka, R., & Rantonen, K. (2017). Basic Income and the New Universalism. Finnish Innovation Fund, Sitra Online Article. Retrieved from https://www.sitra.fi/en/articles/basic-income-new-universalism/.

Murphey, D. (2009). *A Shared Market Economy: A Classical Liberal Rethinks the Market System*. Self-published. Retrieved from http://dwightmurphey-collectedwritings.info/SME-TofC-2.htm.

Orrell, D., & Chlupatý, R. (2016). *The Evolution of Money*. Columbia University Press.

Pogge, T. (1998). A Global Resources Dividend. In D. Crocker & T. Linden (Eds.), *Ethics of Consumption. The Good Life, Justice, and Global Stewardship* (pp. 501–536). New York: Rowman and Littlefield.

Roemer, J. (1996). In E. O. Wright (Ed.), *Equal Shares: Making Market Socialism Work*. London and New York: Verso Books.

Schmidheiny, S., & Zorraquín, F. J. (1996). *Financing Change: The Financial Community, Eco-Efficiency and Sustainable Development*. With World Business Council for Sustainable Development. Cambridge, MA: MIT Press.

Srnicek, N., & Williams, A. (2016). *Inventing the Future: Postcapitalism and a World Without Work*. London: Verso Books.

Standing, G. (2014). *A Precariat Charter: From Denizens to Citizens*. London: Bloomsbury Publishing.

Striner, R. (2015). *How America Can Spend Its Way Back to Greatness: A Guide to Monetary Reform*. Santa Barbara, CA: Praeger.

Tobin, J., Pechman, J. A., & Mieszkowski, P. M. (1967). Is a Negative Income Tax Practical? *Yale Law Journal, 77*(1), 1–27.

United Nations Environment Programme. (2009). *Buildings and Climate Change: Summary for Decision-Makers*. Paris: UNEP Sustainable Buildings and Climate Initiative.

Van Parijs, P. (2000). A Basic Income for All. *Boston Review, 25*(5), 4–8.

Van Parijs, P., & Vanderborght, Y. (2017). *Basic Income: A Radical Proposal for a Free Society and a Sane Economy*. Harvard University Press.

Widerquist, K., et al. (Eds.). (2013). *Basic Income: An Anthology of Contemporary Research*. Oxford, UK: Blackwell Publishing.

Wilkinson, R. G., & Pickett, K. (2009). *The Spirit Level: Why More Equal Societies Almost Always Do Better*. London: Allen Lane.

CHAPTER 9

Requirements for Accounting Standards and Practices

In any economy—whether a small, pre-modern one, a materially large, growing one like today's, or a materially large and stable one as might emerge in the future—it is essential to know how an institution is performing in the fulfilment of its functions. The cuneiform tablets of ancient Sumer served exactly this purpose, as did the ledgers later filled by centuries of bookkeepers, until the recent advent of electronic accounting tools. "How are we doing?" would have been a question asked by a chieftain about his community's harvest, by a medieval merchant about his balance of trade, or by a nineteenth-century industrialist about his manufacturing business. It is also a question all households must address from time to time, even if few of them use standard accounting practices to answer it. In a materially significant economy, the question "how are we doing?" becomes broader: the "we" in question becomes the whole planet, comprising both social and natural systems. If the economy is to achieve a measure of material discipline, how does one account for that? This chapter seeks to answer this question, in the process proposing some requirements.

The history of accounting is one characterized by long periods of stasis punctuated by short bursts of evolution. The stasis tells its own story: accounting is a practice benefiting from standardization. While standardization can be maintained within small communities through voluntary means, it usually requires some form of government regulation within large communities, so that everybody knows how to do it the same way.

F. Murison Smith, *A Planetary Economy*,
https://doi.org/10.1007/978-3-030-49296-0_9

As the previous chapter described, regulation makes the market. An investor needs a standard system of accounts to be able to compare businesses and decide how to invest.

Possibly the most important event in the history of accounting was the development of double-entry bookkeeping, in which a single transaction is recorded as two accounting entries, a debit and a credit. Early forms of the system arose independently in Korea and in the early Muslim world starting around 1500 years ago. The earliest preserved double-entry ledgers are Italian from the early thirteenth century. The first codification of the double-entry method was written by Luca Pacioli in his *Summa* of 1494. Giving the trader "without delay information as to his assets and liabilities", the method became widely established, and remains the standard 500 years later.[1]

The regulation and standardization of accounting practices outlines not only what accountants should do but also what they should not do. Within the guidelines of accounting there exists scope, and sometimes the temptation, to 'cook the books'. This is because a set of accounting standards is essentially a kind of taxonomy: it defines various categories of assets and liabilities, income and expenditure; and it states in broad terms the kinds of transactions that go into each.[2] Like any taxonomy, however, it can never be exhaustive, leaving open the possibility for certain transactions to be defined more than one way, thereby allowing nefarious dealings to be concealed.

As a consequence, government regulations on accounting standards and practices have to be adjusted from time to time to close loopholes exploited by renegade practitioners. Recent examples include the Sarbanes-Oxley Act of 2002 in the US, along with similar legislation in a number of other countries, passed in the wake of a series of corporate accounting scandals of the late 1990s and early 2000s.

Accounting for an Economy on a Crowded Planet

By accounting for gains and losses in numerical terms, an individual or organization can measure performance. Although accounting conventionally is done in units of money, it is not limited to this. Measures of material

[1] Jay (2000, p. 38), cited in Beinhocker (2006, p. 276), and Livio (2002, pp. 130–131).

[2] See, for example, the taxonomy of the US Financial Accounting Standards Board at http://www.fasb.org/jsp/FASB/Page/LandingPage&cid=1176164131053.

or energy flows through economic processes also are a form of accounting. A scoreline at the end of a sports game is a form of accounting.

Financial accounting in particular tracks financial performance: the amount of money acquired minus the amount of money paid. Most of the world adheres to accounting standards established by the International Accounting Standards Board. These are known as the International Financial Reporting Standards (IFRS).[3] The prevailing accounting standard in the US has been the GAAP, or Generally Accepted Accounting Principles, administered by two boards, the Financial Accounting Standards Board (FASB) for private entities and the Governmental Accounting Standards Board (GASB) for public ones. Differences between the US and international standards are mostly small.

In and of itself, financial accounting takes no account of material efficiency or any other measure of physical performance, nor does physical accounting consider financial performance. For a materially large economy on a crowded planet, this disjunct could present a risk. What if financial indicators suggested that businesses, or indeed the whole economy, were doing well while the physical indicators warned of impending peril? Which to trust? And how to reconcile the inconsistency? This is the situation in which we find ourselves today.

All accounting standards share a variety of limitations when accounting for natural capacity. In the 1990s, Daniel Rubenstien, then Principal at the Office of the Auditor General in Canada, articulated six such limitations, listed as follows:

1. Accounting has not yet dealt with inherent limits to economic activity.
2. Accounting's definition of an asset is forward-looking but probably not forward-looking enough to accommodate environmental factors.
3. The firm is the traditional accounting entity. However, firms are dependent on resources they do not own in the conventional sense of private property, such as air and water. To properly account for these, the accounting entity may have to be redefined.
4. Accounting is restricted to dealing with legislated social costs, such as environmental fines. It does not account for the full costs of production because society has not legislated that it should.

[3] www.ifrs.org.

5. Traditional accounting records the costs of environmental investments in the costs of production, without recognition of their less tangible benefits. Accruals for remediating or decommissioning an environmental bad are not incorporated, which could penalize rather than encourage environmental responsibility.
6. Accountants see profit as a return on risk. Where does one draw the boundaries of risk?

Making changes to accounting standards is a challenge, not only because a single global standard does not yet exist but also because any changes made to corporate accounting standards also should be made to systems of national accounts.[4]

In recognition of these limitations, attempts have been made to introduce new, separate standards to account for nature, both at the business level and at the national level. These efforts have had limited success. This chapter offers an alternate approach.

Businesses, today being required to report financial performance, are increasingly aware of the risks and opportunities presented by the material coevolution of the economy and nature. That coevolution, in microcosm, is a coevolution of the business itself and nature. The risks and opportunities are not easy to measure or value. In theory, the market could discover these values, if instruments operated to encourage material efficiency. Absent such instruments, businesses are left to evaluate these risks and opportunities as best they can.

The triple bottom line, developed by John Elkington in the early 1990s, was inspired by the maxim 'people, planet and profit'. The idea was anticipated a few years earlier by Freer Spreckley, although never fully developed.[5] The triple bottom line proposed appending to regular financial accounts two additional measures of corporate performance: the environmental and the social. Environmental performance would be measured in terms of material and energy efficiency, waste reduction, and ecological health, among others. Social performance would be measured in terms of ways a business supported fair trade, fair wages and fair hours in its operations, and local communities through services, programs and charitable activities.

[4] Schmidheiny and Zorraquín (1996, p. 142).

[5] Spreckley (1981) and Elkington (1997).

These kinds of activities are all commendable, and many businesses large and small, for-profit and non-profit, engage in some or all of them. They undoubtedly add intangible value to a business: a responsible investor would look kindly upon a company that cares for its natural and social milieu. Yet here lies the rub: the triple bottom line reports mostly intangible value. A business is not required to incorporate intangible value into established accounting practices, nor do such standards exist for it. Although measures of 'environmental performance' can be, and are, tangible, being measured in physical or energetic units, they are not captured in financial accounts, not do those accounts incorporate measures of social performance. Because there is no agreed-upon rate of conversion among the financial, environmental and social measures of performance, the triple bottom line remains a relatively unintegrated accounting system. As Schmidheiny and Zorraquín wrote, "an energy efficiency index without a carbon tax, or a list of contaminated sites without a cleanup cost, means nothing to the markets."[6]

Although environmental performance indicators do not appear on the balance sheet, environmental compliance costs (such as toxic cleanup) are revealed on financial accounts. Taxes or incentives related to carbon emissions, energy consumption or resource throughput, if implemented, also would appear on financial accounts. Mechanisms such as these would bring measures of material intensity into standard accounting structures, thereby aligning environmental performance with financial performance. Materially more efficient companies then would become valued higher than less efficient ones.

Rather than employing the triple bottom line, some approaches have attempted to ascribe a financial value, or range of values, to physical or social factors influencing business prospects: in other words, to monetize them. The intent is to be both forward-looking and backward-looking, projecting ranges of possible future financial performance, adjusting those projections based on actual experience, and reporting actual experience through conventional accounting processes.

A recent report by KPMG summarizes a variety of approaches along these lines, including KPMG's 'True Value' method. KPMG's survey is summarized in the Appendix to this chapter. It reports a plethora of initiatives and methods, reflecting a growing realization of the relationships between financial performance and non-financial factors. Although diverse

[6] Schmidheiny and Zorraquín (1996, pp. 95–96).

in their details, all the approaches share some common themes. One is that they tackle head-on the problem of valuing the non-financial in financial terms. “It is not an exact science”, admits KPMG, and it “cannot fully express ethical aspects of externalities such as human rights or health and safety”. Yet some kind of valuation, however imperfect, is seen as preferable to none. Monetization “is the method that currently offers the most potential to bring considerations of societal value into corporate decision making”.[7] A second common theme among the methods is the creation of three broad categories of value: financial, social and environmental. A third theme, not shared by all approaches but prevalent in many, is the use of standard accounting frameworks to capture non-financial forms of value.

KPMG’s approach is fairly typical of others in that it defines corporate value in terms of earnings balanced against risk. Earnings in turn is a function of revenue, cost and investment. Treated as external to these financial considerations are economic, social and environmental factors, some of which may be beneficial, others which may present risks. An enterprise becomes driven to internalize these factors when influenced by government regulation, or by the market, or by what KPMG calls ‘stakeholder action’, which encompasses pressure from customers, shareholders, suppliers, activists, local communities or the media. Although KPMG does not acknowledge it, nature itself could, in the long run, be the most influential stakeholder in the operations of a business.

Making the heroic assumption that the various ‘externalities’ can effectively be internalized, their financial value then would become reflected on the corporate balance sheet and on a corporation’s risk assessment. Analysis of the financials would prompt a strategic response on behalf of the organization, such as to increase revenue, reduce costs and risks, encourage positive externalities and discourage negative ones. If the response has the desired effect then the financials in subsequent reporting periods should reflect this improvement, provided of course that the monetization of the various externalities sent reliable signals to the accounting process.

The method to account for externalities is straightforward. Earnings, as revenue minus costs, is adjusted for the monetized value of positive and negative economic, social and environmental externalities, resulting in ‘true’ earnings, as KPMG describes it. The method represents a ‘bridge’ from corporate financial earnings to true earnings.

[7] KPMG (2014, p. 44).

Keep the Units Separate

The reason a disjunct between financial earnings and 'true' earnings exists in the first place is because the markets, in which organizations operate, do not send 'true' financial signals to the earnings statement. If they did, then financial earnings would be 'true' earnings by default. It is thus left to corporations to try to correct for this market failure by estimating the financial value of these various external factors. Each organization must separately go through this rather onerous and uncertain exercise. If one considers for a moment how many people-hours are spent across all organizations engaged in these efforts, then it quickly becomes apparent how inefficient this is. It would be much more efficient for all concerned if the market were to discover these values, since it represents at some level the collective intelligence or knowledge of all its participants.

This line of reasoning implies that all that financial accounting really has to do is what it has always done, which is to account for financial inputs. The valuation of non-financial factors should not be conducted within the financial accounting process but should be placed upstream, embedded within the very economic activity in which all organizations participate. It thus becomes incumbent upon policymaking to establish instruments necessary to guide the financial decision-making of organizations toward alignment with non-financial factors, material and social.

What is missing from corporate reporting is not the non-financial expressed in monetary terms but the non-financial expressed in its native units. Such social factors as human rights or human health, being largely qualitative, do not have quantitative units as such. Inevitably, there are limits to what can be accounted for financially.[8] Other social factors, such as income distribution, can be quantified, as the previous chapter discussed. Natural processes, of course, are readily quantifiable, whether in material or energetic terms. One point the various initiatives surveyed by KMPG confirms is that existing accounting methods are readily adaptable

[8] As an aside, this point could support a philosophical argument against public corporations, since public ownership enslaves a corporation to the provision of returns to its shareholders, leaving no room for ethical or other qualitative issues. However, as argued earlier, the growth cycle of a corporation influences the degree to which it may seek profit. If a mature, stable corporation is no longer primarily profit-seeking, then there is really not much value in its remaining publicly traded. By going private, as some large corporations have done, the shackles of shareholder returns can be broken, freeing a corporation to concentrate on delivering broader societal value in addition to financial value.

to physical accounting, allowing financial and material performance to be tracked under a common framework.

It is for these reasons that a material efficiency rating system, mentioned earlier in Chaps. 5 and 8, would become an important component of the transition toward a materially efficient economy. Just as financial rating agencies rely on corporate financial accounts to generate financial ratings, a material efficiency rating system would rely on a parallel set of material accounts. It would be administered by an independent body. Economic instruments to encourage material efficiency could then be tied to the ratings values.

Accounting for Natural Capacity Using Established Standards

Whether following the IFRS or the GAAP, natural capacity can be accounted for using conventional accounting standards without having to create separate 'environmental' or 'social' categories. An integration between the financial and material can be achieved within conventional accounting structures because the core concepts of accounting are unit-independent, which is to say that they can be applied in any units where ownership is clearly defined. By employing the same approach for material as for financial flows, a business's material and financial accounts become integrated in a way that separate evaluations fail to accomplish. The same concepts and processes that are used to translate financial accounts into financial ratings can be used to translate material accounts into material efficiency ratings. Alignment of the material and financial on the corporate balance sheet would represent the very core of economic alignment with natural processes.

Accounting Concepts: Financial and Physical

The Accounting Equation Applied to the Economy and Nature

One of the most basic concepts of accounting is the idea that assets are the sum of liabilities and owners' equity. Put another way, an owner's equity is the value of an asset minus any liabilities against it, such as debt. A house worth, say, \$500,000 on the market, carrying a mortgage of \$300,000, by definition would deliver \$200,000 in equity to the owner. This basic accounting concept rests upon an assumption of clearly defined property rights. An 'asset', by definition, attaches to an owner who, in the example of a private residence, holds title to it. In regions where property rights are poorly defined or poorly enforced, there are unlikely to be many banks willing to lend a mortgage.

A mortgage, as a liability to a property owner, is an asset to a lender. Conversely, a balance in a bank account is an asset to the account holder and a liability to the bank, which must be ready to pay it out at any time.

Although established for the purpose of financial accounting, the concept of assets being the sum of liabilities and owners' equity can apply to any quantity exchanged between two parties. It can apply, for example, to material exchanges between the economy and nature. The economy is owned by human society, in the sense that it is a social construct. Absent humanity, there would be no economy. The economy, then, is an asset to society. Its material intensity has grown over the centuries along with the human population. In a materially small economy having little or no measurable effects upon natural systems, any material liability incurred by economic activity upon natural processes would be negligible. On the other hand, a materially significant economy having high material throughput would find itself borrowing natural capacity. The diminution of natural capacity represents a material debt from the economy to nature. Although a more accurate term might be to say that the economy has 'appropriated' natural capacity, in any case, it is a debt that nature eventually will call in. The modern growth economy is materially exposed to nature through this liability. Therefore, society's equity ownership of the economy, in material terms, is less than 100 percent. Nature has agency.

A planetary economy, having relatively low material intensity, would have a lower exposure to risk from nature. Its material liability would be lower. Society's equity share in the economy would be correspondingly higher. That economy would represent a relatively more conservative, risk-averse position, materially speaking.

The risk associated with eroding natural capacity in the interest of physical growth of the economy is clear. It would be analogous to assuming ever-greater quantities of financial debt for the purpose of growing a financial asset. Such a plan would work only if growth rates continued to outpace the interest on the debt. Yet the material risk taken by a growing economy also is compounded by the absence of a defined material 'interest rate' from nature. This renders the scale of the liability indeterminate. Prudence would dictate it should be minimized. This is essentially the argument for a 'precautionary principle'.

The Balance Sheet

A balance sheet is a snapshot of the general accounting equation at any point in time. It shows the value of assets at the top and liabilities and owners' equity below. The principal categories are illustrated in Table 9.1. The table incorporates an example showing how values might change under policies to encourage corporate material efficiency, such as under tax-shifting from income to material intensity, and with the imposition of a natural capacity use fee. These and other instruments are discussed in chapters to come.

Table 9.1 Principal categories of the balance sheet for a corporate account

Category	*Definition*	*Current environment*	*Tax-shifting + NC use fee*
Assets			
Cash & cash equivalents	Balances in bank accounts and investments maturing within 3 months	$50,000	$50,000
Inventory	Goods kept in stock for sale	$110,000	$200,000
Accounts receivable	Amounts due from customers for goods or services already delivered	$20,000	$50,000
Property, plant & equipment	Assets that cannot readily be converted to cash	$300,000	$320,000
Total assets		$480,000	$620,000
Liabilities			
Accounts payable	Amounts due to suppliers for goods and services already received	$20,000	$50,000
Notes payable	Contractual obligations to lenders	$270,000	$270,000
Total liabilities		$290,000	$320,000
Owners' equity			
Common stock	Amounts invested by owners	$50,000	$50,000
Retained earnings	All net income not distributed to owners as a dividend	$140,000	$250,000
Total owners' equity		$190,000	$300,000
Total liabilities + owners' equity		$480,000	$620,000

Source: Author's creation, from Piper (2013, p. 9), using illustrative data

Note: Rightmost column indicates how values in the various categories might change under policies to encourage material efficiency, such as tax-shifting from income to material intensity and a natural capacity use fee

The effect of tax-shifting and a natural capital use fee would be to increase the value of inventory and of property, plant and equipment. It would also increase the value of accounts payable. It would cost more to make things and consequently they would command higher prices. Following the rule that assets equal liabilities plus owners' equity, the effect would be to increase retained earnings, all other quantities being held constant, as shown in the table.

Retained earnings comprises net income minus any dividends paid out to owners. Assuming no such dividends are paid out, this means that under policies to increase material efficiency, net income will increase. Net income, as the name suggests, is total income (such as from sales or other sources) minus expenses (cost of goods sold, rent, wages and so on). If it costs proportionately more to make things then gross revenue, in *absolute* terms, will increase, and so will net revenue.

In practice, retained earnings usually are reinvested to purchase more inventory for sale or more equipment for production. Under policies to encourage material efficiency, these things would cost more. This reinvestment would appear as capital expenditure in the cashflow account.

Although conventionally the values of all items on a balance sheet are expressed in currency, two of them relate to material assets, known in accounting parlance as 'tangible' assets. They are (1) inventory and (2) property, plant and equipment. Inventory is the market value of the stock of goods available for sale. Property, plant and equipment is the financial value of those physical assets of a business which it uses to operate. Being material in nature, they represent so-called illiquid assets. Cash and accounts receivable are 'liquid' assets, existing only in money form.

Although property, plant and equipment are material assets, corporate accounting conventionally does not also include Earth's natural capacity as a material enabler of corporate activity. One reason for this is because natural capacity is not an 'asset' in an accounting sense, for reasons discussed later. A second, more pragmatic, reason is because the market value of natural capacity cannot readily be estimated *ex ante*. It can only be discovered iteratively by the market itself. The value of property and plant in the market nonetheless will reflect their effects upon natural capacity, at least indirectly, provided that market incentives exist to encourage or discourage certain activities. In an economy where title to property incorporated the natural capacity bounded by it, as suggested in the previous chapter, and where material intensity were rendered expensive, the value of property and plant having low material intensity would be higher than that of property and plant having high material intensity, all else being equal.[9] Assuming no other differences in liabilities, this value would translate into relatively higher owners' equity. We see such an effect, for example, in the difference between comparable properties having solar power and those not having it. In today's economy, that difference is small relative to the value of most properties but nonetheless significant.[10]

Cash Flow and Material Flows

Whereas a balance sheet is a snapshot of financial accounts, statements of income and expenses, and of cash flows, are longitudinal. They show amounts carried from one accounting period to another.

The income statement shows accruals, which are obligations to pay. Income is an obligation of others to pay a person or an institution. Expenses are the obligations of a person or institution to pay others. When I receive my utility statement, I incur an expense to the utility, even if I have not actually paid it yet. This expense represents an accrual to the utility company.

Once an expense is paid, an accrual turns into cash flow. A key difference between an income statement and a cash flow statement is in the way they record events. If a time lag exists between the incurring of an expense and its payment, an income statement will record the expense in the period in which it was incurred, whereas a cashflow statement will record it in the period when it was paid. If I set up online automatic payment for my utility bill then the time lag goes almost to zero, the expense and payment occurring almost simultaneously.

Another key difference between an income statement and a cashflow statement is that the cashflow statement records certain transactions not captured in the income statement. Certain kinds of transactions are considered neither revenue nor expenses. Examples are loans and capital expenditures. A loan is an increase of an asset (cash) and of a liability (notes payable). It will appear on a cashflow statement because it is a cash inflow.[11] Similarly, a capital expenditure is an increase of an asset (property, plant or equipment) and of a liability (accounts payable). It is a cash outflow.

[9] Conventionally, the value of an asset is recorded as its historical cost. In the case of an asset held for a long time, such as real estate, this method may underestimate the actual market value of an asset at any time. Although this convention may introduce a bias in valuation, it is nonetheless consistent (Piper 2013, p. 70).

[10] Hoen et al. (2015).

[11] Piper (2013, pp. 28–29).

The way cash flows are accounted for is similar to the counting of material flows. Given that any economic entity, whether a business, a public entity or a household, uses material flows from nature and expels material flows back to it, there exists a relation between the material balance of Earth and the material balance of an economic entity. The material balance of an economic entity is converted, however imperfectly, into financial form, which manifests itself as a cashflow statement. Items on the cashflow statement relating to the production of goods and services produce economic outputs that, in turn, have material effects upon nature. Table 9.2 illustrates the concept.

Table 9.2 is another way of expressing the inputs and outputs of the economic system model in *Economics of a Crowded Planet.* However, it reveals an additional level of detail, which is how economic inputs and outputs are translated into cash flows. As the column headers show, those inputs and outputs can be expressed in material terms or in money terms,

Table 9.2 Linkage of material flows (in tonnes, *t*) through Earth's natural systems and cash flows ($) through a generic economic entity

Natural output (t)	*Economic input* (t *or* $)	*Cash flow statement* ($)	*Economic output* (t *or* $)	*Natural input* (t)
				Solar flux
Ecological production	Supply chain	Cash paid to suppliers for capital expenditures	Property, plant & equipment	Effluent
Stored material		Cash paid to suppliers for operating expenses		
		Cash receipts from customers	Products	Effluent
		Cash paid for wages		
		Cash paid for taxes and fees		
		Dividends paid to shareholders		
		Cash received from issuing new shares		
		Net change in cash	Net change in economic production	Net change in natural capacity

Source: Author's creation

Note: Cashflow categories are adapted from Piper (2013, pp. 30–31)

whereas natural inputs and outputs can be expressed only in material (or energetic) terms, and cash flows by definition are expressed in currency.

Conventionally in accounting, capital expenditures are grouped under investment, whereas operating expenses are grouped with cash receipts from customers, wages, taxes and the like under cash flows from operations. In Table 9.2, capital expenditures and operating expenses are grouped to show their connection to the physical supply chain originating from material inputs from nature.

Both the conventional cashflow statement and a statement of the material balance of nature in the presence of a large economy show similar things, albeit in different units. The cashflow statement identifies categories of flows from distinct kinds of activity, such as from operations or investment. It sums them to show a net change in cash from one period to the next. If that value is positive then the entity in question made money during that period, whereas if it is negative, the entity lost money.

Similarly, a material flow statement for nature would show solar flux as a form of material receipt. In the presence of a large and growing economy, the amount of material 'paid' by nature to the economy would be significant and increasing. The net change in natural capacity, as a function of material flows to the economy and effluent flows from it, would be negative, representing a divestment from natural capacity. In the presence of an economy increasing in material efficiency, however, lower material payments to the economy and lower receipts of effluents from it would incrementally turn the divestment from natural capacity into a re-investment in it. Net change in natural capacity would become positive.

What Table 9.2 omits is the recirculation of material within the economy from economic outputs to economic inputs, thereby reducing the dependence upon material inputs from nature. Instead of appropriating most of its material inputs from nature, the economy would appropriate some of the solar flux to drive its own material recirculation. Daily solar flux is abundant relative to humanity's total daily energy consumption.[12]

The key to effecting this change lies in the financial signals that the economy provides to every entity preparing a cashflow statement. If, for

[12] The International Energy Agency (IEA 2016) estimates total human energy consumption in 2014 at around 109,000 terawatt-hours (TWh), a terawatt-hour being one billion kilowatt-hours (kWh). This equates to around 299 TWh per day. Based on an average solar irradiance to Earth of about 6 kWh per square meter per day, total daily insolation is about 2.7 million TWh per day, or about 10,000 times daily human consumption.

example, economic instruments shift tax expenses from wages onto the supply chain, or require the payment of natural capacity use fees based on an enterprise's material intensity, then the enterprise will have an incentive to reduce its supply chain expenditures and to make its products easier to recirculate at the end of their useful lives; in other words, to become materially more efficient. It is the signals the economy sends to every reporting business that would effect a conversion from physical units to financial ones. Financial reporting then can run in parallel with material reporting, using common templates, thereby allowing an evaluation of the degree to which policies for material efficiency may influence financial decision-making.

Before describing how that would work, two important additional accounting concepts must be introduced: depreciation and double-entry bookkeeping.

Depreciation, Financial and Physical

Accounting conventionally depreciates the cost of fixed assets, such as equipment, over a period corresponding to an asset's useful life. Similarly, the value of intangible assets, such as patents, copyrights and trademarks, is amortized over an asset's life. For a tangible asset, the cost may be depreciated over a few years, or anywhere up to twenty or thirty for factories or power plants. Methods for estimating depreciation vary: it may be linear, deducting a fixed amount from the book value of an asset each year, or proportional, deducting a percentage. Depreciation also can be based on the recommended number of hours in an asset's service life.[13]

At the end of its useful life, a fixed asset typically will have a residual value of some kind. Businesses often estimate useful life as the period before which significant maintenance costs will be incurred. A business will seek to replace an asset close to the end of its useful life through sale into a secondary market of some kind, reclaiming the residual value. A buyer paying a residual value for an asset would expect to incur maintenance costs on it. Car rental companies, for example, typically hold onto vehicles for three to five years, or up to a certain mileage, and then sell them to private parties.

The canonical assumptions of financial depreciation rest on a seldom-articulated thermodynamic basis, which is that things wear out over time. A car rental company can depreciate a new car over three to five years because it knows that, at the end of the car's useful life, it can purchase a

[13] Piper (2013, Chaps. 12 and 13).

new one. Hypothetically, if the automobile manufacturing sector were to shrink, as happened during the Second World War, a car rental company might be prompted to depreciate its assets over a longer period. Cuba is an example of a country whose automobile fleet has been depreciated since the 1950s over an indefinite period.

A capital asset is not autopoietic, or self-regenerating. It must either be regenerated exogenously, such as through maintenance by people, or replaced. In this sense, it is fundamentally different from productive assets employing the autopoiesis of living things, such as farmland. That production is driven by sunlight, an exogenous input of free energy. Thermodynamically, this is a process of organization, which represents a reduction in entropy.

Growth is synonymous with *appreciation*. Having life-cycles of their own, living things also die, which represents a loss of organization, an increase in entropy, or a process of thermodynamic *depreciation*.[14] Life on Earth is a thermodynamic balance of appreciation and depreciation. The maintenance or replacement of a capital asset therefore requires an exogenous input of free energy. Depending on the kind of asset, part of this energy could be delivered by human labor; part of it may come from embodied solar energy, in the form of ecological resources or solar electricity; or part of it could come from stored solar energy, in the form of mineral fuels. When a car rental company purchases a new car, it pays a price that reflects the cost of producing the car. However, that cost does not necessarily reflect the full, thermodynamic, cost upon natural capacity because the drawdown on the mineral base is not usually accounted for, nor are atmospheric emissions and other effluent streams related to its production and distribution.

In a planetary economy, much of the material input to a new capital asset would come from material recirculation within the economy, relatively little from nature. The input of free energy similarly would come from sources not diminishing natural capacity, to the greatest possible extent. Incentives for material efficiency would result in the development of capital assets that last longer and have higher residual values, being designed to allow the greatest possible amount of material to be reclaimed.

Even in a materially stable economy, though, financial depreciation schedules still will be orders of magnitude shorter than most natural cycles.

[14] "The natural tendency [of a structure] is depreciation. Its loss of potential energy and increase in entropy is an expression of the second law [of thermodynamics]" (Odum 1994, p. 313).

If the present economy were suddenly to vanish, the regeneration—or appreciation—of natural capacity back to pre-human levels would take place over many thousands of years. The mineral base already drawn down by humanity would take millions of years to regenerate.

It is not the responsibility of the accounting profession to resolve this disjunct. Instead, it is the responsibility of policy to create conditions favorable to the regeneration of natural capacity in the presence of a materially large economy. Financial depreciation, in turn, will reflect these new incentive régimes by extending depreciation schedules and recording higher residual values.

Double-Entry Method

If assets equal the sum of liabilities and owners' equity then any transaction must affect both sides of the equation, otherwise it would not balance. A business purchasing a piece of equipment using available cash would record the transaction as a transfer from a cash account to an equipment account. These accounts are both asset accounts on the balance sheet, as Table 9.1 showed. The cash account would decrease by a certain amount and the equipment account would increase by the same amount. One is a debit, the other a credit. The transaction conventionally is recorded as a 'journal entry' on a general ledger, showing a debit entry to cash and a credit entry to equipment. This is the essence of the double-entry method established by those medieval Italian merchants.

Although journal entries conventionally are made in currency, they could equally be made in material terms. Obviously, it does not make sense to log the purchase of a piece of equipment in material terms, since it is a financial transaction. However, any material interchange between an economic entity and nature could be so recorded. The felling of a tree for firewood increases the asset side of a household's balance sheet, yet it also increases the liability side, in terms of a corresponding loss of natural capacity. If another tree were planted in its place, then this would represent an investment in future natural capacity to replace the capacity lost. Because the tree would take a few decades to reach maturity, there would remain a residual liability in natural capacity representing the shortfall in natural capacity during the new tree's growth. This residual liability is analogous to accumulated interest on a loan. All these quantities would be recorded in material units rather than financially.

The double-entry method applies not only to balance-sheet accounting but also to the income statement. A journal entry to a revenue account is recorded as a credit, and a journal entry to an expense account is recorded as a debit. Where in financial terms a business seeks to maximize its net revenue, which is gross revenue after expenses, there does not currently exist an overarching incentive for businesses to become materially efficient, even if some businesses report financial benefits from doing so. If such an incentive existed then, in material terms, it would result in a tendency to minimize net material intensity. The 'revenue' part of the income statement would be the total material intensity of a business over a specific period: that is, the sum of resource inflows and effluents. The 'expense' part of the statement would be the amount of material intensity offset by measures to improve natural capacity. Clearly, the terms 'revenue' and 'expense' are poorly suited to this concept, yet if we replace them with 'material intensity' and 'natural capacity replenishment', then the double-entry method can be applied to this longitudinal component of accounting as well as to the balance sheet. The felling of a tree would be recorded in kilograms of material intensity, and the planting of a seedling to replace it as kilograms of natural capacity replenishment.

Tracking Financial and Material Performance in Parallel

In this way, the tools of accounting can be used to run the financial and material records of performance in parallel. Policies and instruments encouraging material efficiency, or the regeneration of natural capacity more generally, would necessitate a parallel accounting of the material intensity of an operation, alongside its financial accounting. This would be the only way to know how an operation were performing in material terms relative to a prime directive of material discipline.

The parallel approach outlined here compares financial and material performance under a common structure, allowing an organization to see how economic instruments to encourage material efficiency would affect both material and financial performance. It would allow a policymaker to evaluate the efficacy of the instruments in question.

A numerical example employing a GAAP-like template serves to illustrate the point. In Table 9.3, a simple income statement of a hypothetical operation, such as a farm, is run over ten years. Net income is assumed for simplicity to equal operating income, meaning that non-operating income and expenditures cancel each other out. Over ten years, policies are phased in to encourage material efficiency. These increase the cost of goods sold and of operating expenses, costs which are passed on to the customer

Table 9.3 Illustrative example of parallel financial and material accounting using income statement structure

Year-on-year rates of change		*2%*	*2.5%*	*3%*	*3.5%*	*4%*	*3.5%*	*3%*	*2.5%*	*2%*
Income statement ($000s)	**Y1**	**Y2**	**Y3**	**Y4**	**Y5**	**Y6**	**Y7**	**Y8**	**Y9**	**Y10**
Sales	$5000	$5100	$5228	$5384	$5573	$5796	$5999	$6178	$6333	$6460
Cost of goods sold	$1500	$1530	$1568	$1615	$1672	$1739	$1800	$1854	$1900	$1938
Gross profit	$3500	$3570	$3659	$3769	$3901	$4057	$4199	$4325	$4433	$4522
Total operating expenses	$1500	$1530	$1568	$1615	$1672	$1739	$1800	$1854	$1900	$1938
Operating income	$2000	$2040	$2091	$2154	$2229	$2318	$2399	$2471	$2533	$2584
Net income	$2000	$2040	$2091	$2154	$2229	$2318	$2399	$2471	$2533	$2584
Material statement (tonnes)	**Y1**	**Y2**	**Y3**	**Y4**	**Y5**	**Y6**	**Y7**	**Y8**	**Y9**	**Y10**
Material throughput										
Total material inputs	10,000	9800	9555	9268	8944	8586	8286	8037	7836	7679
Outputs of production	6000	6000	6000	6000	6000	6000	6000	6000	6000	6000
By-products	4000	3800	3555	3268	2944	2586	2286	2037	1836	1679
Of which, recirculated:										
Inputs to operations	2000	2040	2091	2154	2229	2318	2399	2471	2533	2584
Outputs from operations	1500	1530	1568	1615	1672	1739	1800	1854	1900	1938
Material intensity, kg	10,500	10,030	9451	8768	7987	7115	6372	5749	5239	4837
Natural capacity regenerated, kg	500	510	523	538	557	580	600	618	633	646
Net material intensity, kg	10,000	9520	8928	8229	7430	6536	5773	5131	4606	4191

Source: Author's creation

through sales. The top row in the table shows the year-on-year rates of change in these values brought about by the policies.

Below the income statement is a material statement, organized in a roughly similar fashion to the income statement. The intent is to arrive at a net material intensity, which would be a reportable statistic, used for regulation, taxation or incentives to encourage material efficiency. In year 1, total material throughput is 10,000 tonnes. Of this, 6000 tonnes comprises the outputs of production, which are sold, and the remaining 4000 tonnes are by-product. Of this by-product, 1500 tonnes is recirculated, to be stored and re-used or repurposed. Out of the 10,000 tonnes input to production, 2000 tonnes comes from within the enterprise itself, as stored material. The difference between the amount input to production and the amount recirculated from production means that either the store is being drawn down or it is being replenished from another source. If the enterprise is a farm, then some of its inputs to production will come from nature.

Any material not recirculated represents the enterprise's material intensity. The example says nothing about how the external inputs to operations would be evaluated to derive the relative proportions of resources extracted from nature compared to material recirculated within the economy. That is a policy question having to do with labeling. For the purpose of accounting, it is not necessary to distinguish between inputs from nature and inputs from suppliers because ultimately all material inputs not recirculated within the economy come from nature anyway. When economic instruments are applied to these quantities, their effects will ramify up the supply chain. For instance, if the farm uses a hundred tonnes of artificial fertilizer in a year then any instruments applied to the manufacture of fertilizer might render it more expensive. The farm might switch to buying compost if it is thus rendered relatively cheaper; however, the input of compost still would form part of the farm's material footprint. If the farm substituted compost grown on-site for external supplies of it then this would increase its recirculation and decrease its material footprint. Instruments encouraging the reduction of material intensity naturally would encourage material recirculation, although all enterprises ultimately would be limited in how much they could feasibly recirculate. Even so, the more individual enterprises that were recirculating material, the more the economy as a whole would do so.

In the example in Table 9.3, material intensity is calculated as the sum of the total material inputs and by-products minus the total material recirculated. Natural capacity regeneration also is measured, and this is subtracted from the material intensity to produce the net material intensity. In the example of a farm, natural capacity regeneration could, for instance, entail the ecological remediation of a riparian habitat on the property or a tree-planting program. Regeneration of natural capacity is the material analogue of operating expenses, as it represents material provided back to natural systems. This material can be measured in physical units, such as in kilograms, with varying degrees of confidence, depending upon the specific ecological remediations performed.

Over time, the enterprise in Table 9.3 reduces its material input by the year-on-year percentages shown in the top of the table. It also increases its material recirculation and natural capacity regeneration by these same percentages. For simplicity, the example assumes that the enterprise produces the same amount of product in all years. The overall effect is to redirect a substantial amount of material into recirculation, thereby reducing the net material intensity.

Through labeling and measurement at each stage in the supply chain, it should be possible to trace the percentage of any product through the economy derived from recirculated resources. That percentage would reflect the material efficiency not merely of an end product but also of all the upstream operations leading to it. This kind of tracking is not yet mandated. However, in a planetary economy, it would be well established, and would form the basis of some kind of material efficiency rating system, to which economic instruments for material efficiency could be attached. Today's standard accounting structures provide a framework to accomplish this, as the example above illustrates.

Systems of National and Global Accounts

Having discussed ways a business might answer the question, 'how are we doing?' within a planetary economy, it is natural then to ask the question, how is the economy itself doing? During the twentieth century, the focus was at the level of national accounts rather than on the performance of the global economy, the relevance of the economy's scale not yet having been fully appreciated. In the twenty-first century, it becomes necessary to scale up. Fortunately, the same practices and standards used for national accounting apply quite readily at the global level, albeit with some important conceptual differences.

National Economic Accounting

As economics textbooks explain, it would be an error to create an accounting of a whole region by simply summing the accounts of all the reporting entities within it. This is because an account of a whole region measures inflows to the whole region and final outflows from it. Many reporting entities within a region are intermediate in nature, passing goods and services down the supply chain, in return for payments flowing upstream. Summing all institutional accounts would count many quanta of value more than once.

A common and commonsense way to eliminate intermediate values from an aggregate account is to calculate the value added at each step.[15] The sum of the values added equals the aggregate value of a final good or service. For instance, the aluminum supplied to an automaker has a certain market value per tonne. This value would be subtracted from the market value of a finished automobile to calculate the value added through the assembly of a vehicle.

The same kind of principle applies to the measurement of any system, using any units. It can be used to measure the material flowing through the economy. The field of environmentally extended input-output analysis does just this. The principle applies also to the measurement of material and energy flows through ecosystems.

Standard System of National Accounts and of Environmental-Economic Accounting

Although simple in principle, the practice of national accounting is complex, as any review of the standard System of National Accounts (SNA) will reveal. The System of National Accounts is a statistical framework developed by the Inter-Secretariat Working Group on National Accounts, which comprises the United Nations, the European Commission, the Organisation for Economic Co-operation and Development, the International Monetary Fund and the World Bank. More than 600 pages long, the SNA provides, in its own words, "a comprehensive, consistent and flexible set of macroeconomic accounts for policymaking, analysis and research".[16] First published in the 1950s, the SNA has been through five editions as of 2008.[17]

[15] Colander (2010, p. 529).
[16] ISWGNA (2008, p. iii).
[17] ISWGNA (2008, p. xlvii).

Details on the SNA and on its sister standard, the System of Environmental-Economic Accounting (SEEA) are provided in the Appendix to this chapter. The SNA defines a variety of financial and non-financial asset classes and liabilities, the non-financial ones including 'produced' assets—from economic activity—and 'non-produced' assets, which include legal assets, such as contracts, as well as natural resources. The SNA defines an 'asset boundary' delineating 'economic' natural resources from 'non-economic' ones. Economic natural assets are those over which effective ownership rights can be exercised, thereby bringing them under control of one or more economic entities.[18]

Where the SNA values assets in money terms, the SEEA incorporates a physical accounting in parallel, using a common structure. In other words, it does for national accounts what the structure outlined above does for corporate accounts. It also expands the asset boundary for physical accounting to incorporate non-economic natural assets.

In the past, one of the criticisms leveled against systems of national accounts by some of the economists of nature was their conflation of natural resource depletion with economic income. Resources removed from nature were incorporated into conventional economic accounting as income rather than as a reduction in capital. Herman Daly traced the error back to A.C. Pigou, who had mistakenly rejected Irving Fisher's definition of the national dividend, which kept stocks separate from the services they provided. Fisher correctly stated that the service provided by a capital stock is an addition to capital, and therefore counts as income. Pigou rejected this distinction because it went against "the ordinary use of language",[19] a rejection which surely must be one of the laziest arguments ever given against a scientific method. As long as we do not fall into inconsistencies, Pigou argued, "the choice of definitions is a matter, not of principle, but of convenience".[20] He could hardly have been more wrong. Generations of economists and accountants fell prey to exactly the kind of inconsistencies against which Fisher had warned. The effects of this error upon the exploitation of Earth's natural systems could hardly be overstated.

Kenneth Boulding acknowledged the error as far back as 1949, advising that the objective of economic policy should be "the maintenance of

[18] ISWGNA (2008, pp. 212–214).
[19] Daly (1991, p. 203).
[20] Pigou (1920, p. 35).

capital stock with as little consumption or production as possible".[21] Wants, agreed Daly, are not satisfied by the depreciation of capital stocks. The conventional counter-argument in favor of treating natural capacity as an income stream in national accounting was based on a presumption of substitutability. Human technology, the argument goes, can substitute for natural capacity. The position has no basis in empirical observation but is simply a theoretical convenience.[22] Theoretical niceties aside, common sense tells us that the atmosphere cannot be substituted by human technology. It is, like all other natural capacity, a complement to human society.

The current editions of the SNA and the SEEA acknowledge the distinction between stocks and income, staying closer to Fisher's original definitions. This allows the depletion of natural capacity (the term used here) or environmental assets (the SEEA's term) to be treated as a diminution of capital stock, with its attendant effects upon future income streams.

For all its improvements over prior systems, however, the language of the SNA remains unashamedly anthropocentric, reflecting the deep influence of conventional economics. The economy is assumed as the primary system of analysis, with nature categorized as a line-item input rather than as the economy's container and fundamental enabler. Nature is defined as a 'non-produced asset', presumably because all that matters is economic production. The SNA represents inside-looking-out thinking codified into a system of value produced, vetted and endorsed by five of the world's most powerful supranational agencies, and employed as the standard by most national governments around the world. In particular, its distinction between 'land' and 'non-cultivated biological resources' is, to the ecologist, preposterous: a meaningless distinction which could only have been invented for the convenience of anthropocentric conceptions of property.

One of the strengths of the SEEA, in contrast, is its ability to show financial and physical performance side by side. With a few minor adjustments, the SEEA's supply and use tables, summarized in the Appendix, could in fact be adapted for corporate accounting. Yet like the SNA on whose foundation it is built, the SEEA also is fundamentally anthropocentric, omitting from consideration any 'environmental asset' not having current or expected future economic use. This omission begs the question, what is an asset? It begs another, broader question, which is how far can the use of conventional accounting terminology be extended to

[21] Boulding (1949, p. 79).
[22] Daly (1996, p. 77).

encompass nature before the terminology fails? These are not questions purely of semantics. They have practical implications for the way the performance of a planetary economy would be measured.

What Is an Asset, and What Is a 'Natural Asset'?

In the 1980s and early 1990s, the ecological economists extended Fisher's definition of 'capital' to include nature. This was an error, a transplanting of terminology from one system to another without sufficient critical evaluation of its applicability. As Herman Daly acknowledged, Fisher was not originally referring to nature in his distinction between capital and income, only to artificial capital, the capital of the economy. The concept, the ecological economists believed, was also applicable to nature.[23] Since the 1980s, they have used the term 'natural capital' to refer to the 'stocks' of nature that yield flows of benefits to the economy, a terminology which has made its way into the SEEA.

The terminological extension seems innocuous enough. After all, one can measure the amount of biomass available for inputs from ecosystems to the economy, or the amount of nonrenewable natural resources available as economic inputs. Salah El Serafy, for instance, proposed investing receipts from the liquidation of nonrenewable resources into renewable 'natural capital'.[24] In 1992, Daly and Cobb developed an alternate system of national accounts, not only extending Fisher's original definition of capital to include nature but also transplanting some other financial terms onto natural resources. For instance, 'offtake', such as from a renewable resource, was defined as equivalent to a financial dividend. 'Divestment' is a reduction in a renewable stock or capital, the stock becoming 'depreciated' and less able to support future production. 'Income' is a return from production from a stock. 'Net income', then, is gross income minus depreciation.

From these terms, it was clear that what natural resource biologists and bioeconometricians called 'yield' from harvested resources is analytically the same as income. Maximum sustainable yield therefore is the maximum production from a stock that does not reduce its size. The parallels with financial stocks appeared self-evident.

[23] Daly (1991, p. 203).
[24] El Serafy (1989).

Daly and Cobb's Index of Sustainable Economic Welfare (ISEW) sought to capture the value of the physical growth of the economy by incorporating the consumption and depletion of 'natural capital'. The index also incorporated measures of income inequality and foreign debt burden.[25] The values of the ISEW that Daly and Cobb derived did not paint a rosy picture for the growth economy of the 1990s. Yet it never supplanted GDP as the standard index of economic performance. Daly may have been correct that the ISEW makes no more, and probably fewer, arbitrary judgements than standard GDP accounting.[26] However, its reliance on combining quantities measured in different, potentially incommensurable units, rendered it less useful.

The terminology of 'environmental assets' organized into stocks of 'natural capital' is problematic for global economic accounting for two reasons. One reason relates to the term 'asset' and the other to the term 'capital'.

The SNA's definition of an 'asset' rests upon well-defined property ownership, which it defines as 'legal ownership'. It also defines a different form of ownership, which is economic ownership. "The economic owner of … goods and services, natural resources, [and] financial assets and liabilities", writes the SNA Working Group, "is the institutional unit entitled to claim the benefits associated with the use of [such entities] in the course of an economic activity by virtue of accepting the associated risks." The key phrase in this definition is 'entitled to claim'. In the absence of legal title, such entitlement is assumed to arise through eminent domain. "When government claims legal ownership of an entity on behalf of the community at large", the Working Group continues, "the benefits also accrue to the government on behalf of the community at large. Thus, government is both the legal and economic owner of these entities."[27] Hence, if a region were invaded and occupied by a foreign power, any eminent domain thus acquired would deliver to the invaders control over access to the resources contained therein. The colonization of the Americas by Europeans is an example of a transfer of economic ownership: one made, of course, against the will of the indigenous peoples.

At the level of an economic actor—an individual, a household or an organization—ownership means legal ownership, at least where property

[25] Cobb and Cobb (1994) and Daly and Cobb (1994).

[26] Daly (1996).

[27] ISWGNA (2008, p. 41).

rights are well-defined and enforced. For most economic units, legal ownership and economic ownership overlap. There exist markets for the transfer of ownership—in both of its meanings—from one party to another. The situation is different at the national level because nations cannot be traded. I cannot just go and buy Belize, for instance. There is no title, as such, to Belize, nor is there a market for it. As a consequence, the concept of an 'asset' for national accounting has to be constructed, which is what the SNA does. The SNA defines an asset as "a store of value representing a benefit or series of benefits accruing to the economic owner by holding or using the entity over time".[28] As the standard for national accounting, the SNA therefore says that an asset can only be something that is owned, in the economic sense of having control over access to it. This is why the SNA places certain natural systems outside the asset boundary.

If the definition of an asset is linked to ownership—legal or economic—then an asset can only be an artifice. For nature to be considered an 'asset' requires the presence of one or more people to derive benefit from it *and* to control access to it. The presence of human beings on Earth would bring forth an asset where before there existed only nature. By this logic, it follows that nature, in the absence of people, is not an asset. Even though the concept of an asset may be a human construct, it is clear that this conclusion is fallacious. An orangutan reaching for a fruit in a tree takes hold of the fruit, at which point the orang has control over it, and the fruit becomes an asset to the ape. A dominant male elephant seal controls a harem of females: they are an asset to him. Biologists generally use the term 'resources' to refer to those elements of an organism's surroundings which have not yet become 'assets' in this sense. However, the distinction here is between elements of nature that may be assets to *other* elements of nature—measured in physical or energetic units—and elements of nature that may be assets to the human economy, measured in money.

Well-meaning though recent attempts may have been to place a range of dollar values on 'natural assets', in the hope of incorporating them into economic accounting, they have only tried to value the unvaluable. This brings us to the second problem: the use of the term 'capital'.

By extending the Fisherian definition of 'capital' to include nature, the ecological economists made an implicit assumption that nature has the property of economic capital. It does not. A stock of economic capital is

[28] ISWGNA (2008, p. 39).

inert. Nature, on the other hand, is autopoietic: it lives. Although a farm may harness the power of nature to make food, its status as a capital asset does not suggest any living force. A capital asset is a human construct used to describe and value artifacts or real property. It is not the *capital* asset of a farm that is autopoietic but the portion of nature bounded by its property line.

Some elements of that natural production are useful, such as the cabbages or onions growing on the farm. Others, such as weeds, are not. The management of weeds and pests incurs upon the farm a cost of production. This is because a farmer delineates between 'useful' biological production, which has economic value, and 'useless' biological production, which has no economic value. There is no agricultural market for weeds; if there were, they would not be called weeds. Yet nature produces them anyway.

The production of useful agricultural products, such as cabbages and onions, represents a stream of economic income. It may therefore be tempting to apply the Fisherian definition of an asset to that which produces the cabbages and onions, namely biological production. That application is a fallacy, not only because an asset must, by definition, be owned, as noted above, but also because nature itself does not have the property of an economic stock. Howard Odum's energy circuit language, developed to model ecological processes, contained an element known as a store or a tank which, when combined with elements used to model interactions among variables, allowed Odum to simulate a wide variety of dynamics related to the accumulation and release of resources. Some had the property of capacitance, where resources accumulate up to a point at which the store discharges; others had more complex dynamics, exhibiting changes in behavior based on certain thresholds. Odum's modeling of ecological dynamics did not treat storage in an accounting sense, as a linear adjustment to a stock, but in a dynamic sense, reflecting observed behaviors of ecological communities.[29]

It is telling that conceptions of nature as a 'stock' do not adapt well to the characterization of flows of effluents from the economy. No stock of capital, as such, is depleted; rather, effluents from the economy produce a loading effect upon natural systems affecting their dynamics. The SEEA hints at this reality where it refers to the "environment's capacity" to accommodate this or that economic activity.

[29] Odum (1994).

Replacing the static concept of 'capital' with the dynamic one of 'capacity' allows both the input and output sides of the relationship between the economy and nature to be considered on the same terms. It also forces us to recognize that changes to nature resulting from the material intensity of the economy will not be like the linear adjustments to capital stock in accounting. They will most likely be non-linear, unpredictable, and non-repeating.

Ironically, the ecological economists were guilty of inside-looking-out thinking. They attempted to shoe-horn nature into the artificial constructs of financial accounting, where it does not fit. Although accounting methods can be used to track material flows, accounting concepts do not apply well to the totality of nature. The ecological economists were guilty of the same mistake as Alfred Pigou, confusing a common-parlance meaning of a term with a technical meaning.

Rather than trying to value 'natural assets', it may be preferable to abandon the idea of nature as an asset in an accounting sense. Even in common parlance, nature is far more than an asset. It is the system that fundamentally enables the economy. It creates the conditions for the economy to come into existence in the first place, and to persist. The term 'natural asset' is a misnomer.

Methodologically, it is preferable to keep the monetary measures separate from the physical ones. The SNA and the SEEA do this, although they do not keep the terminology separate. It is appropriate to consider both economic and natural processes in terms of stores and flows, but it must be understood that the behaviors of those stores and flows may be very different in money-based systems compared to physical systems. All the more reason to consider them distinctly. It is not so much the structure of the SNA and the SEEA that is at fault: the structure is clear and functional. What should be critically examined is some of the terminology, and the implicit assumptions and ontologies that underlie it.

GDP, for all its limitations, is at least a clean measure of economic performance, in that it expresses the currency value of economic activity. The policy decision to be made is to subsume it within a measure of the *physical* expansion of the economy. Economic performance in physical terms will be a more critical measure of human welfare *in the long run* because the long-term stability of the economy will depend upon it. Once the economy achieves a high rate of physical performance, the focus could legitimately turn to GDP as a measure of its financial performance. Also, by keeping financial and material measures separate, the material performance of the whole economy need only be tracked using one unit of measure: tonnes per unit time.

Global Economic Accounting

At the level of a geopolitical unit, such as a nation, as at the level of an enterprise, it is clear that a system of accounts on a crowded planet should measure two quantities in parallel: flows of money and flows of materials. It should not make any kind of *ex ante* determination of the equivalence between money and materials, as that equivalence is not discoverable to a sufficient degree of confidence to be reliable. Accounting is principally linear; nature is not.

The ultimate measure of the performance of a planetary economy, in both financial and material terms, is a global measure. This is the only measure that really matters, if the economy is to achieve a degree of material alignment with nature while delivering prosperity to its members. An enterprise or a nation may be able to implement measures enabling it to do particularly well in both material terms and financial terms, yet if the whole economy performed poorly by these measures, their efforts would be for naught.

A system of global economic accounts would serve as a reference point for national and corporate performance. It would allow the identification of nations or other geopolitical units that underperform relative to the whole world, thereby prioritizing those needing special attention. It would allow us to identify at a global level how reductions in material intensity affect natural capacity over time, as indirectly measured through such indicators as climate, sea level, acidity, aridity, biodiversity and nutrient cycles. A standard, therefore, is needed, as well as a performance-tracking mechanism, one freely available worldwide.

The standard in question would exist to measure both financial and material performance, separately but in parallel. If global economic accounting were to establish this standard, then it would percolate into both geopolitical accounting and corporate accounting. A parallel measure of financial and material performance thus would encompass all economic activity, not only that part of it measured financially. This standard would seem an absolute necessity to the guiding of economic activity toward long-term stability and material efficiency.

In the SNA and the SEEA, a template exists for national accounting. At the corporate level, a similar template could readily be established. Indeed, the absence of a standard template for corporate material accounting renders a global standard all the more necessary. This section suggests a structure and some factors to consider.

In the SEEA, an opening stock of produced assets and an opening stock of environmental assets each are adjusted by monetary and physical supply and use to produce a closing stock of produced and environmental assets. The earlier discussion on corporate accounting suggested a structure to avoid the terminological error of defining environmental 'assets' to the economy in terms of static stocks. It simply displayed material income and expenditures alongside financial ones, as a way to show the potential effects of incentives for material efficiency on financial performance.

A structure for global accounting, then, would bear some resemblance to the SEEA table in the Appendix (Table 9.8), which links the supply and use tables to the asset accounts; except it would not have a 'Rest of the World' column, nor an 'Environmental assets' column. The SEEA table also mixes physical and monetary terms in the 'Produced assets' column, where they should be kept separate.

Table 9.4 reorganizes the SEEA table so that monetary supply and use is placed within the context of material flows, allowing items in units of money and in physical units to be viewed in common columns. In this way, it is structurally similar to Table 9.3, linking material flows with cash flows in corporate accounting. The concept of 'environmental assets' is gone. In this structure, there are only produced, or economic, assets. The structure allows an accounting of material efficiency by identifying recirculated material as a category of economic supply. For any given volume of input to the economy, if recirculated material is increased then effluents should correspondingly decrease, thereby raising material efficiency. It is hypothesized that, over time, increases in material efficiency of the economy would have a beneficial effect upon natural capacity.

Beneath the material accounting is a financial one, occupying only the columns relating only to economic activity, by definition. There are no monetary imports or exports, of course, because the table represents a system of global accounting.

The economic categories in Table 9.4 can be disaggregated into major economic sectors, such as industries, households and government, as the SNA and SEEA do. Then, under economic use, one would record intermediate and final consumption. The purpose of this organization, though,

Table 9.4 Structure for a system of global material and financial accounting, side by side

Material accounting (tonnes)							
Outputs from nature	Produced (economic) assets	Economic supply		Economic use	Produced (economic) assets		Inputs to nature
Ecological production Stored material	Opening stock	Recirculated material Extracted material	Output	Gross capital formation	Other changes in volume of stock	Closing stock	Solar flux Effluent
Financial accounting (currency)							
	Produced (economic) assets	Economic supply		Economic use	Produced (economic) assets		
	Opening stock	Output		Gross capital	Revaluations	Closing stock	

Source: Author's creation

Note: 'Nature' in the column headers denotes earth's natural systems is to show visually how the financial accounting is a subsystem of the physical accounting, reflecting the subsumption of the economy's endogenous flows of money within the physical flows of material around, within and through it.

The existence of a structural identity for accounting across scales of organization, from the global down to the corporate, or even the household, represents a self-similarity allowing the performance of economic subsystems to be compared: a form of hierarchical tractability that could be extremely useful. The boundary of financial accounting would encompass all transactions where money or other monetizable value is exchanged between parties within the economy. Financial accounting would not attempt to value anything beyond these boundaries, instead restricting itself to the conventions that have served it well throughout history.

The onus for material performance would fall upon the establishment of policies and instruments for material discipline. A parallel accounting of material flows and money flows would enable an evaluation of the material performance of these mechanisms, as well as their effects upon financial performance. The task for policymaking within a growth economy then would be to improve material performance as rapidly as possible. A parallel material-financial accounting structure would allow policymakers, corporate managers and indeed private individuals to evaluate performance by both of these measures. No tenuous assumptions need be made about how material performance translates into financial performance.

By keeping the units separate in material and financial accounting, the material and financial performance of enterprises, nations and the whole economy can be rendered transparent and intelligible, thereby allowing the design, evaluation and selection of economic controls best suited to different phases of a transition from a growth economy toward a materially efficient, stable one.

Appendix: Accounting Systems

Approaches to Valuing Externalities in Financial Accounting

Table 9.5 Approaches for valuing externalities in financial accounting

Reporting standards		
Name	*Institution*	*Description & link*
Integrated Reporting	International Integrated Reporting Council (IIRC)	New approach to corporate reporting which communicates full range of factors affecting ability of organizations to create value over time. www.theiirc.org
Natural Capital Accounting workstream	EU Business and Biodiversity Platform	Decision-making framework and set of principles to help companies determine what form of natural capital accounting to adopt, and to identify best practices and tools. http://ec.europa.eu/environment/biodiversity/business/ workstreams/ Workstream1-Natural-Capital-Accounting/index_en.html
Sustainability Accounting Standards	Sustainability Accounting Standards Board (SASB)	Helps publicly listed companies in more than 80 industries disclose material sustainability factors in compliance with US Securities and Exchange Commission (SEC) requirements. www.sasb.org
Sustainability Measurement and Reporting System (SMRS)	The Sustainability Consortium	Standardized framework for communicating sustainability-related information throughout product value chain. Aims to help companies design better products. www.sustainabilityconsortium.org
The Prince's Accounting for Sustainability Project (A4S)	HRH The Prince of Wales	Systems, tools and guidance to enable accounting and finance community to integrate measures of environmental health, social wellbeing and economic performance into financial decision-making, accounting and Reporting. www.accountingforsustainability.org
Measurement initiatives		
Name	Institution	Description & link
Environmental Profit & Loss (EP&L) Statement	BSO/Origin (Eckart Wintzen), 1990; Puma (Jochen Zeitz), 2010	Cradle-to-grave evaluation of environmental effects along supply chain of a business. Uses input-output models. http://about.puma.com/ puma-completes-first-environmental-profit-and-loss-accountwhich-values-impacts-at-e-145-million

Economics of Ecosystems and Biodiversity (TEEB)	UNEP, Germany, UK	Method to bring values of biodiversity and ecosystem services into decision-making at all levels. www.teebweb.org
Natural Capital Protocol	Natural Capital Coalition (formerly TEEB for Business)	Harmonized framework for valuing natural capital in investor decision-making. www.naturalcapitalcoalition.org
B Impact Assessment	B Impact Assessment	Standards, benchmarks and tools enabling companies to assess, compare and improve social and environmental effects. http://bimpactassessment.net
KPMG True Value	KPMG	Three-step approach enabling companies to (i) assess their 'true' earnings including externalities, (ii) understand future earnings at risk and (iii) develop business cases that create both corporate and societal value. www.kpmg.com/newvision
Redefining Value	World Business Council for Sustainable Development (WBCSD)	Program helping WBCSD member firms to standardize tools to measure and manage effects upon society and Environment. www.wbcsd.org
Shared Value	Shared Value Initiative	Management strategy focused on creating business value by identifying and addressing social problems. www.sharedvalue.org
Social Return on Investment (SROI)	SROI Network	Framework based on generally accepted accounting principles used to help manage and understand an organization's social, economic and environmental outcomes. www.thesroinetwork.org
Total Impact Measurement & Management (TIMM)	PricewaterhouseCoopers	New language to assist companies in understanding overall effects of their activities. www.pwc.com/totalimpact
Integrated Profit & Loss	True Price	Framework for quantifying and valuing economic, environmental and social effects of an organization's operations. http://trueprice.org

Source: Author's creation, adapted from KPMG (2014, pp. 36–37)

The SNA and SEEA

The commonly used terms 'gross domestic product' (GDP) and 'gross national product' (GNP) derive from statistics generated by applying the SNA to economic data. GDP is the gross product within a geographical boundary, whereas GNP is the gross product of a nationality—the citizens and businesses of an economy—no matter where they are located.[30] Where the textbooks define GDP simply as the sum of consumption, investment, government spending and net exports within a geographical boundary, the SNA provides three distinct ways of computing it.[31]

The SNA's basic architecture is very similar to the structure of corporate accounts. It identifies various accounts for production, income and capital, all of which feed into a final financial account. From the final financial account, a balance sheet may be produced. The SNA also goes into considerable detail on standards and methods for measuring economic activity in various sectors of the economy: labor, corporate activity, government activity, non-profits, households and so-called informal aspects of the economy, including 'unincorporated enterprises'. National economies in the SNA include a 'sector' known as the 'rest of the world', which is all economic activity outside the national boundary.

A nation's balance sheet organizes the various assets and liabilities within the various sectors, along with the total, as illustrated in Table 9.6.

The SNA divides stocks into non-financial assets, financial assets and financial liabilities. Assets and liabilities share common sub-categories, as Table 9.6 illustrates. Being purely financial, the SNA includes no such category as 'non-financial liabilities'.

Non-financial assets are further sub-categorized into produced and non-produced assets, 'produced' meaning physical artifacts, and 'non-produced' being a catch-all for everything else not otherwise classified as a physical artifact or a financial asset. "These three types of assets have little

[30] Colander (2010, p. 582).

[31] These are as follows: (1) "the sum of gross value added of all resident producer units plus that part (possibly the total) of taxes on products, less subsidies on products, that is not included in the valuation of output" or (2) "the sum of the final uses of goods and services (all uses except intermediate consumption) measured at purchasers' prices, less the value of imports of goods and services" or (3) "the sum of primary incomes distributed by resident producer units" (ISWGNA 2008, p. 34). GDP represents one of several categories of 'balancing item', which is calculated as the difference between assets and liabilities. Others include 'value added', 'operating surplus', 'disposable income', 'saving', 'net lending or borrowing' and 'current external balance'. ISWGNA (2008, p. 49).

Table 9.6 Categories used in the balance sheet in the System of National Accounts

Stocks	Non-financial corporations	Financial corporations	General government	Households	Non-profits serving households	Total economy	Rest of the world	Total
Non-financial assets								
Produced non-financial assets								
Fixed assets								
Inventories								
Valuables								
Non-produced non-financial assets								
Natural resources								
Contracts, leases and licenses								
Goodwill and marketing assets								
Financial assets/liabilities								
Monetary gold and special drawing rights								
Currency and deposits								
Debt securities								
Loans								
Equity and investment fund shares/units								
Insurance, pension and standardized guarantee schemes								
Financial derivatives and employee stock options								
Other accounts receivable/payable								

Source: Author's creation, adapted from ISWGNA (2008, Table 13.1, p. 258)

Note: Shaded cells indicate non-applicability

in common", admits the Working Group, "except that they are all non-produced and non-financial."[32] Natural resources falling under this category differ from agricultural products in that they are not produced economically. Further, as the SNA explains, "not all environmental resources qualify as economic assets".[33] Some lie outside the so-called asset boundary defined by the SNA. These include "naturally occurring resources [for which] it is not feasible to establish ownership [such as] the air or the oceans". Other natural resources outside the asset boundary include as-yet-unknown ones and those which are "so remote or inaccessible that, in practice, they are not under the effective control of any [economic] units". Natural assets, according to the SNA, must not only be

[32] ISWGNA (2008, p. 212).
[33] ISWGNA (2008, p. 212).

owned but also be capable of bringing economic benefits to their owners in order to be accounted for. Biota located where "no institutional unit is able to exercise effective ownership rights over them … fall outside the asset boundary". However, "fish stocks in the high seas which are subject to international agreement on how much may be caught by individual countries may be counted as falling within the asset boundary".[34]

The SNA goes on to classify types of natural resources, organizing them into 'land', in a quasi-Ricardian sense, including soils but excluding surface biota; 'mineral and energy resources', which are economically exploitable resources within Earth's crust; and 'non-cultivated biological resources', which are product-yielding biota (presumably surface biota) whose ownership is defined but which are not cultivated. Think of forests from which people harvest fruits or game, for example.[35]

Valuation of these resources, the SNA admits, is not generally available from market prices. Yet they still may provide economic value not included in the value of the associated land. The SNA recommends valuing them by the present value of future returns expected from them.[36]

If a disaggregation of the natural resource accounting categories is required, the SNA suggests following the guidance of its companion System of Environmental and Economic Accounting. The SEEA is developed by the same group of supranational agencies as the SNA, under the auspices of a United Nations Committee of Experts in Environmental-Economic Accounting.

The SEEA consists of a 'Central Framework' which aims to be both a "conceptual framework for understanding the interactions between the economy and the environment" and an international standard for reporting statistics on these interactions. As an accounting system, it is designed to be compatible with the SNA, as well as with international industrial and product classification systems, and with frameworks for developing environmental statistics.[37]

The SEEA is accompanied and complemented by two additional publications. One is *SEEA Experimental Ecosystem Accounting*, which synthesizes current knowledge about the application of an accounting approach to the measurement of ecosystem stocks and flows. The intent is to

[34] ISWGNA (2008, p. 213).
[35] ISWGNA (2008, p. 214).
[36] ISWGNA (2008, p. 265).
[37] UNCEEA (2014, pp. vii–ix).

provide a measurement framework for ecological assets consistent with the SEEA. The other is *SEEA Applications and Extensions*, which describes ways the SEEA may be used for monitoring, policy analysis and the development of indices.[38]

The main difference between the SNA and the SEEA is that the SEEA considers physical flows. It uses the same economic 'asset boundary' as the SNA, and employs the same accounting concepts of stocks and flows to provide a framework for physical accounts. "Thus, the scope in physical terms is not limited to those items with economic value", writes the Committee of Experts. "It is recommended that those environmental assets that have no economic value be clearly distinguished."[39] Unlike the SNA, the SEEA separates 'soils' from 'land' in its treatment of natural resources.

The structure of the SEEA allows physical measures to be considered alongside monetary ones. It has, in its authors' words, "a capacity to organize physical and monetary data that have common scope, definitions and classifications into combined presentations".[40] In this sense, it functions similarly to the parallel reporting mechanism outlined earlier for corporate accounts. It employs the same accounting convention of stocks and flows as the SNA, considering 'the environment' holistically as a single stock of all living and non-living systems. The environment is the source of all natural inputs to the economy, these being flows. Outputs from the economy to nature, which are also flows, are termed 'residuals'.[41]

The SEEA defines two categories of natural 'assets'. The first conforms to the asset boundary of the SNA. It is "individual components of the environment that provide materials and space to all economic activities", in other words, those resources which are directly used. This class of assets is the focus of the SEEA Central Framework. The second class of assets is ecological in nature, comprising the ecological interactions and processes which deliver so-called ecosystem services to the economy. Ecosystem services derive from life's autopoiesis—tendency for biomass to regenerate—thereby providing material inputs for the economy; as well as from the capacity for ecosystems to maintain and regulate climate and water

[38] UNCEEA (2014, p. viii). One of many programs that supports the SEEA in its ecological valuation is the Wealth Accounting and Valuation of Ecosystem Services (WAVES) program of the World Bank.

[39] UNCEEA (2014, p. 7).

[40] UNCEEA (2014, p. 8).

[41] UNCEEA (2014, pp. 12–13).

systems, and metabolize effluents from the economy. Another 'service' is human enjoyment of nature. Ecosystem services fall not under the Central Framework of the SEEA but under its Experimental Ecosystem Accounting. Although the SEEA does not say so explicitly, the implied reason is because this class of assets does not readily fit into the standard accounting framework employed by the SEEA. It is included in the experimental accounting framework because the SEEA recognizes the importance of "analyzing the extent to which economic activity may reduce an ecosystem's capacity to generate ecosystem services".[42]

Both the SNA and the SEEA employ what they call 'supply and use tables' to account for flows of value, monetary or physical, through a national economy. The flows of value over a given period allow an accounting of the changes to assets and liabilities over that period. Table 9.7 illustrates the form of the monetary supply and use table, as employed in both the SNA and the SEEA, and the physical supply and use table, as employed only in the SEEA.

Total supply in money terms is the sum of industrial output within a national economy and imports from the rest of the world. Total use is the sum of intermediate consumption by industry (the use of products and services to make other products and services), final consumption by households and government, gross capital formation and exports to the rest of the world. Value added is the balancing item.

The physical table takes the same form as the monetary table, although is a little more complex, incorporating flows from nature and back to it. As a consequence, the supply section has three line-items: natural inputs, economic products and residuals. The use section has the same three line-items. The use of natural inputs is the extraction of flows from the natural environment. The use of products is the same as in the monetary table, although here measured in physical units. The use of residuals represents the recirculation of material back into the economy or the flow of effluent back into the environment.

Table 9.8 shows how the flow measurements in the supply and use tables update the asset accounts.

Tacked on to the right of the monetary and physical supply and use tables are the monetary and physical asset accounts. On the monetary side, the opening stock of economic capital is adjusted by gross capital

[42] UNCEEA (2014, pp. 13–14). It should be noted that the SEEA does not use the term 'autopoiesis': that is my phrasing, borrowed from Maturana and Varela.

Table 9.7 Supply and use tables employed in the System of National Accounts (SNA) and the System of Environmental-Economic Accounting (SEEA)

Monetary supply and use	Industries	Households	Government	Accumulation	Rest of world	Total
Supply table						
Products	Output				Imports	Total
Use table						
Products	Intermediate consumption	Household final consumption expenditure	Government final consumption expenditure	Gross capital formation (including changes in inventories)	Exports	Total use
	Value added					
Physical supply and use, employed in the SEEA alone	**Industries**	**Households**	**Accumulation**	**Rest of world**	**Environment**	**Total**
Supply table						
Natural inputs					Flows from environment	Total supply of natural inputs
Products	Output			Imports		Total supply of products
Residuals	Generated by industry	Generated by final household consumption	From scrapping and demolition of produced assets			Total supply of residuals
Use table						
Natural inputs	Extraction of natural inputs					Total use of natural inputs
Products	Intermediate consumption	Household final consumption	Gross capital formation	Exports		Total use of products
Residuals	Collection and treatment of waste and other residuals		Accumulation of waste in controlled landfill sites		Residual flows direct to environment	Total use of residuals

Source: Author's creation, adapted from Tables 2.1 and 2.2 in the UNCEEA (2014, pp. 17–18)

Note: Empty dark gray cells are null by definition

produced over the period of measure, other changes in the volume of assets, and revaluations, to give the closing stock over the period of measure. On the physical side, the opening stock of economic capital is adjusted by gross capital formation, residuals from physical supply and residuals from use, as well as other changes in the volume of physical assets, to give the closing volume of physical stock at the end of the period.

Table 9.8 Connections between supply and use tables and asset accounts in the SEEA

						Asset accounts (physical and monetary terms)	
		Industries	Households	Accumulation	Rest of world	Environment	Total
						Opening stock	
Monetary supply and use table	Product supply	Output			Imports		
	Product use	Intermediate consumption	Household final consumption expenditures	Government final consumption expenditures	Exports	Gross capital	
Physical supply and use table	Natural inputs — supply						Extracted natural resources
	Natural inputs — use	Inputs of natural resources					
	Product — supply	Output			Imports		
	Product — use	Intermediate consumption	Household final consumption		Exports	Gross capital formation	
	Residuals — supply	Residuals generated by industry	Residuals generated by final household consumption		Residuals received from rest of world	Residuals from scrapping of produced assets; emissions from landfills	
	Residuals — use	Collection and treatment of waste and other residuals			Residuals sent to rest of world	Accumulation of waste in controlled landfills	Residuals flowing to environment
						Other changes in volume of assets (e.g., natural growth, discoveries, catastrophic losses)	
						Revaluations	
						Closing stock	

Source: Author's creation, adapted from Table 2.4 in the UNCEEA (2014, p. 21)

Note: Empty dark gray cells are null by definition

The opening stock of environmental assets is adjusted only physical flows, by definition. This adjustment comprises extracted natural resources, residuals flowing from the economy as effluent, and other changes in the volume of environmental assets, such as through natural processes, giving the closing stock of environmental assets.

The SEEA places residuals flowing from the economy to the environment as environmental 'assets' because they affect the asset balance. Essentially, they are negative assets that "may affect the capacity of environmental assets to deliver benefits". Similarly, notes the SEEA, other changes in the volume of assets may be reflected in the changing capacity of environmental assets.[43]

References

Beinhocker, E. D. (2006). *The Origin of Wealth: Evolution, Complexity, and the Radical Remaking of Economics*. Boston, MA: Harvard Business School Press.

Boulding, K. E. (1949). Income or Welfare? *Review of Economic Studies, 17*, 77–86.

Cobb, C. W., & Cobb, J. B. (1994). *The Green National Product: A Proposed Index of Sustainable Economic Welfare*. Lanham, MD: University Press of America.

Colander, D. C. (2010). *Economics* (8th ed.). New York: McGraw-Hill.

Daly, H. E. (1991). *Steady-State Economics* (2nd ed.). New York: John Wiley & Sons.

Daly, H. E. (1996). *Beyond Growth: The Economics of Sustainable Development*. Boston, MA: Beacon Press.

Daly, H. E., & Cobb, J. B. (1994 [1989]). *For the Common Good*. Boston, MA: Beacon Press.

El Serafy, S. (1989). The Proper Calculation of Income-Depletable Natural Resources. In W. J. Ahmad, S. El Serafy, & E. Lutz (Eds.), *Environmental Accounting for Sustainable Development* (pp. 10–18). Washington, DC: World Bank.

Elkington, J. (1997). *Cannibals with Forks: The Triple Bottom Line of 21st Century Business*. North Mankato, MN: Capstone Publishing.

Hoen, B., et al. (2015). *Selling into the Sun: Price Premium Analysis of a Multi-state Dataset of Solar Homes*. Lawrence Berkeley National Laboratory Report for U.S. Department of Energy, Berkeley, CA.

IEA. (2016). *Key World Energy Statistics*. Report of the International Energy Agency, Paris.

ISWGNA. (2008). *System of National Accounts, 2008*. Inter-Secretariat Working Group on National Accounts, United Nations, New York.

Jay, P. (2000). *Road to Riches or The Wealth of Man*. London: Weidenfeld and Nicholson.

KPMG. (2014). *A New Vision of Value: Connecting Corporate and Societal Value Creation*. Report of KPMG International Cooperative, Switzerland.

Livio, M. (2002). *The Golden Ratio: The Story of Phi, the World's Most Astonishing Number*. New York: Broadway Books.

Odum, H. T. (1994). *Ecological and General Systems: An Introduction to Systems Ecology* (Revised ed.). University Press of Colorado.

Pigou, A. C. (1920). *The Economics of Welfare* (4th ed., 1932). Palgrave Macmillan, UK (reprinted 2013).

[43] UNCEEA (2014, p. 21).

Piper, M. (2013). *Accounting Made Simple*. SimpleSubjects, LLC. Retrieved from www.obliviousinvestor.com.
Schmidheiny, S., & Zorraquín, F. J. (1996). *Financing Change: The Financial Community, Eco-Efficiency and Sustainable Development*. With World Business Council for Sustainable Development. Cambridge, MA: MIT Press.
Spreckley, F. (1981). *Social Audit – A Management Tool for Co-operative Working*. Leeds, UK: Beechwood College.
UNCEEA. (2014). *System of Environmental-Economic Accounting 2012—Central Framework*. United Nations, European Union, FAO, IMF, OECD, World Bank.

CHAPTER 10

Money and Finance in a Planetary Economy

Up to this point, the discussion has focused mostly on the economy's material relation with nature. However, another critical flow is involved in the economy's relationship with nature: money. The economy's prudent financial management will be central to its long-term alignment with nature. This chapter discusses the kinds of institutions, policy, accounting and instruments that may be required.

Money is the medium through which most of the economy operates. It is the blood supply of the economic corpus, enabling its cells and tissues to enter into exchanges with one another. The financial system is the sector of the economy that processes money. The way the financial sector operates today, as earlier chapters have indicated, is not altogether in line with a prime directive of material discipline. An economy operating in alignment with natural processes would handle its money quite differently: principally, to reduce market volatility, thereby increasing long-term confidence in the financial sector and stabilizing the economy as a whole. Ancillary benefits would be seen in the distribution of incomes and in welfare more generally. Financial reforms, and changes to the way we think about money, which also broaden the economy's overall prosperity, would help to bring the economy into alignment with nature.

The discussion of money and finance in this chapter serves as a precursor to the discussion of economic controls to follow. Two issues stand out. First, the value of money presently does not reflect the state of natural capacity. Second, the financial system is unstable, prone to endogenous

F. Murison Smith, *A Planetary Economy*,
https://doi.org/10.1007/978-3-030-49296-0_10

volatility, resulting in economic myopia. A review of our perceptions about money, its nature and functions, will provide valuable context for these questions.

Brief History of Money

Agriculture and civilization prompted the development of numerical systems for counting. These systems were an essential development to allow people to track quantities and values of agricultural product, as well as quantities of other products needed to support agriculture: the plough, beasts of burden, tools for working the land, tools for building structures, to name a few. Significantly, the earliest samples of writing from 3200 BC in Uruk are accounts of deposit banking, exchange transactions and lending. By the second millennium BC, the ancient Babylonians had established official banking laws, along with sophisticated accounting and governance systems. Money, as a medium of exchange and valuation, was well established by this point. The government of ancient Mesopotamia used silver 'shekels' as a system of account. The shekels did not actually circulate but instead were stored in temples. The government levied taxes in shekels, which were paid in durable goods or agricultural produce at fixed valuation rates.[1]

By ancient Greek times, coins were in circulation. Early forms of banking and arbitrage arose. Roman coins were first minted in the temple of Juno Moneta, the protector of funds. She gave her name to the word 'money'. The Forum became Rome's main financial hub, offering depository and lending functions, and financial exchange.[2] Because different regions used different monies, the trading of money as a commodity in its own right emerged. Merchants would evaluate market conditions in different places and move their gold or silver accordingly, converting it into coinage of the desired currency for maximum return. During the days of the wooden sailing ship and the beast of burden, this process was slow. Today, financiers and currency traders move funds around the world electronically, almost instantaneously.[3]

Paper currency emerged in China in the ninth century AD. In the late thirteenth century, Marco Polo gave an extensive report of Kublai Khan's

[1] Orrell and Chlupatý (2016, pp. 15–20).
[2] Orrell and Chlupatý (2016, p. 25).
[3] Striner (2015, p. 5).

paper currency, which later became accepted as legal tender as far west as the Baltic Sea. Meanwhile, in Italy, paper money emerged by way of receipts goldsmiths provided to record customers' deposits of gold coin in their strongboxes. These receipts became tokens of a promise to pay, and thus an implicit loan agreement with the goldsmith. The bench across which these transactions took place was the '*banco*'. By becoming a bank loan, money in its modern sense was born, though its ties to commodities did not altogether disappear. Even today, a currency crisis may trigger investors to return to such commodities as gold.

The emergence of paper money as a promise to pay also triggered the emergence of the practice of lending beyond the total value of the reserve. Goldsmiths assumed, reasonably, that not all depositors would want to retrieve all their coins at once. This is fractional reserve banking, a practice which, if leveraged many times over, such as by speculation, can lead to financial instability.[4] Some commentators have argued that fractional reserve banking itself formed the very basis of the capitalist economic system.[5] When a loan is created by bookkeeping, money—specifically, bank money—is fed into the financial system. When the loan is repaid, that money is destroyed. The money supply, as Orrell and Chuplatý describe, is thus "like a bathtub full of water with both the tap on and the plug out".[6] Because money's value derives from its scarcity, as Chap. 1 discussed, the extension of credit in amounts exceeding the money supply acts as a catalyst for economic growth.

Indeed, beginning late in the seventeenth century, many national governments in Europe created central banks, in response to numerous runs on private banks where creditors lost confidence in the banks' ability to repay. The central banks, the Bank of England being an early example, held coin or gold as the lenders of last resort. They had legal power to set reserve requirements for the commercial banks. These requirements specified the amount and form of legal tender the commercial banks must hold relative to their obligations in paper money. The central banks formally recognized paper money as legal tender, in addition to coin, and centralized the issuance of bank notes.[7] "It is a well-established fact", wrote Alexander Hamilton in 1790, in encouraging the US Congress to create a

[4] Litaer (2001, pp. 303–305).

[5] Striner (2015, p. 25), quoting Geoffrey Ingham.

[6] Orrell and Chlupatý (2016, p. 111).

[7] Huber and Robertson (2000, p. 21).

Bank of the US, "that banks in good credit can circulate a far greater sum than the actual quantum of their capital in gold and silver."[8] This being the case, argued Georg Knapp 130 years later, "the soul of the currency is not in the material of the pieces but in the legal ordinances that regulate their use".[9]

Money's evolution from measuring real commodities to the exchange of currency as a commodity in its own right, to the leveraging of bank reserves, reached an additional stage during the latter half of the twentieth century, which is a global one. Money also has become increasingly ethereal, taking the form of data in online accounts, moved through electronic transactions. Not only might a regional banking system exhibit instability due to speculation or a variety of other factors but also this instability may propagate throughout the entire global financial system. Hypothetically, if every account holder in the world were to try to withdraw all their money in the form of cash from the world's commercial banks on the same day, the banks, having leveraged their assets (your assets to be precise), would not have sufficient funds cover the demand. The banks would fold or be closed by their national governments. The central banks themselves might have insufficient funds to repay all creditors, at which point they too would go bankrupt. The global economy would quickly come to a juddering halt. Creditors would have no further recourse, except, perhaps, at the ballot box or on the streets.

What Is Money and What Does It Do?

While some social scientists argue *ad nauseam* about what money is, the rest of us do not pay much attention to the form it takes, provided we have enough to spend. What we call 'money' exists in many forms. Currency issued by the government is 'sovereign' money, or 'fiat' money. It may take the form of hard cash ('payment' money) or of credit issued by a central bank. Not all currencies are government-issued. Electronic currencies, for instance, are not, although this state of affairs could change during the twenty-first century. Non-cash money lent by banks often is referred to as 'bank' or 'finance' money, arising not through the creation

[8] Striner (2015, p. 8), quoting Hamilton's Report to the House of Representatives, First Congress, Third Session, "The Second Report on the Further Provisions Necessary for Establishing Public Credit", December 14, 1790.

[9] Mellor (2016, p. 140), quoting Knapp (1924).

of legal tender but through creative accounting. Instruments treating financial assets, liabilities or currency as commodities in their own right—derivatives and collateralized securities, for instance—are considered forms of 'private' money, arising out of private transactions. Not all forms of money can be used to buy goods and services. Bank money and private money cannot: they have to be converted into payment money. If I hold $1 million in credit default swaps, for example, I cannot spend them directly; instead, I have to liquidate them into legal tender. One of the roles of government, Nigel Dodd points out, is to regulate the conversion between finance money and payment money.[10]

Underlying these distinctions are several properties of money. Originally it was, and still is, a unit of accounting. As Orrell and Chuplatý describe, the practices of numeracy, logic and rationalism incrementally spread through human cultures as the use of money expanded.[11] All monies have a number that signals a value. By being interchangeable with goods and services, money effectively stores the value of the goods and services it is used to purchase. One can imagine money, in abstract, as a store of the values of all assets that could be purchased with it. The act of purchase itself is a form of exchange. Money, then, conventionally is described as simultaneously a unit of accounting, a store of value and a medium of exchange. Yet because money can be lent and borrowed, it also functions as a form of debt. In this sense, to many social scientists, money is a social relation: an opportunity to cancel a debt, or a medium through which a debt may be paid.[12] Money's increasing detachment from its roots in precious metals has led some to view it as pure circulation, devoid of content.[13]

Yet it is not completely devoid of content. It contains information. Some authors, such as Kenneth Boulding, Nicholas Georgescu-Roegen and Howard Odum, argued that the information contained within money fundamentally is thermodynamic. This argument was made in an attempt to describe social processes within their natural context in common units of free energy. While energy is a principal currency of natural dynamics, it is only one of numerous currencies of human social dynamics, money being another. A thermodynamic description of money never took hold

[10] Dodd (2014, Chap. 6).

[11] Orrell and Chlupatý (2016, Chap. 3).

[12] See Dodd (2014, Chap. 3), in particular his discussion of Alfred Mitchell-Innes' theory of money (pp. 102–106).

[13] Dodd (2014, pp. 194–195).

because it was asking money to do something for which it was never designed. The information contained within money as it flows through the economy relates primarily to endogenous values—that is, social ones—money itself being a social construct. It is a measure and a repository of value agreed between two or more parties to a transaction. The value in question will be agreed among parties irrespective of an item's thermodynamic value.

Viviana Zelizer describes money flows as 'circuits of commerce' that "describe the micro foundations of economic life and are distinct from economic formations such as markets, networks or firms". Money thus 'enacts' social structure; it is created through relational work.[14]

Similarly, Richard Striner describes money as "a form of financial energy" transmitted from one party to another, "in many ways equivalent to physical energy, except that this energy is created by the human mind. Yet it brings to life real goods and services, transforming our physical world."[15] Along the same lines, Orrell and Chlupatý emphasize the difference between money objects and money flow. Money objects, like physical quanta, are "transferable entities, created by a trusted (or simply obeyed) authority, that have the special property of a defined monetary value". That value is expressed as a number, which allows interchange among goods and services within the economy. Money "is a means to attach exact, timeless, Pythagorean numbers to the fuzzy and transient concept of real-world value". Money flows through the transaction of real-world value mediated by money's dimensionless value.[16] The process of purchasing is similar to the transformation of potential energy into kinetic energy. "The potential energy", Striner writes, "is the *power* to pay; the kinetic energy is *payment*" (emphasis his).[17]

As a whole, then, money is a circulation of purchasing power within the economy, that power itself being measured and accounted for in money's numerical units. As Orrell and Chlupatý note, what we call 'price' in economics is possible only with the existence of money because market prices ultimately are attained from the use of money objects. Market prices thus become an emergent property of the system of using money objects.[18]

[14] Dodd (2014, pp. 292, 293), citing Zelizer (2004) and Collins (2000).

[15] Striner (2015, p. 23) (emphasis his).

[16] Orrell and Chlupatý (2016, pp. 45–47).

[17] Striner (2015, p. 69).

[18] Orrell and Chlupatý (2016, p. 46).

Money was never designed originally to reflect the value of natural capacity, for two reasons: one, because the value of natural capacity is not directly measurable; and two, because in a materially insignificant economy, money never needed to measure it in the first place. But what if money could *indirectly* measure natural capacity?

Money and Natural Capacity

Through all its phases of evolution, money's value has been maintained not only through scarcity but also through its relation to some agreed-upon underlying value. In early times, the minting of coin in gold or silver tied money's value to that of those precious metals. As the use of banknotes spread, the agreed-upon underlying value remained physical money, coined in gold or silver. A trustworthy banknote could be presented at a bank and be converted into physical coin. With the advent of checking and demand deposits, a checking account, or any portion thereof, could be liquidated into cash in the form of banknotes. The value of the banknote became a function of the trustworthiness of the government's central bank to pay out in coin if so demanded. But because such demands became increasingly rare during the nineteenth century, the notes themselves became, for all intents and purposes, the underlying value. Occasional exceptions, such as during the hyperinflation of 1920s Germany, served as a reminder why money fundamentally was not auto-valuing. Yet the abandonment of the gold standard by the Bank of England in 1931 set the stage for a new measurement of underlying value. National currencies now were measured against one another. A market for gold remains to this day; however, the dollar value of gold fluctuates with global market dynamics rather than being set by the monetary policy of sovereign governments.

The Bretton Woods Accord of 1944 attempted to formalize relationships among the currencies of the allied nations by tying the value of the US dollar to gold and maintaining exchange rates of the other currencies relative to the dollar at near-constant values. The International Monetary Fund was established to bridge temporary imbalances of payments. The International Bank for Reconstruction and Development—now part of the World Bank—was established to provide for postwar investment. The stability of the Bretton Woods system lasted until 1971, when the Nixon administration decoupled the value of gold from the dollar. Arbitrage in the trading of gold on the open markets, coupled with high public spending and inflation in the US had caused the dollar to become overvalued.

The Nixon administration bowed to the inevitable conclusion that exchange rates could not be held constant.

With currencies now floating against one another, the underlying value of any one of them essentially became the perceived trustworthiness of their government's monetary policy compared to that of other governments. There was no singular underlying value, only an expectation of relatively responsible monetary policy, one that avoided inflation while ensuring enough credit remained in circulation to maintain a steady rate of economic growth. The uncoupling of the dollar from gold took place a few years before the financial markets were revolutionized by information technologies. Whether Nixon foresaw this revolution is moot: his motivation had been to restore America's balance of payments.[19] Yet the conversion of currencies into bits of data led, eventually, to the emergence of electronic currencies, such as Bitcoin. These currencies function as money, their creators claim, because their supply is limited and they are costly to create, requiring considerable computing power. The underlying value of electronic currencies is the investment in computing architectures required to generate each unit. The electronic currency community refers to this process, appropriately enough, as 'mining'.[20] Other forms of money have other underlying values. Local exchange trading schemes (LETS) and related community-based exchange currencies are underlain by the work performed by one person for another and the good faith upon which the work is undertaken.

The evolution of money is a story of its progressive divorcement from the physical world. It has reached a stage early in the twenty-first century where the 'underlying value'—if one can call it that—against which money is measured is a form of collective trust in monetary policy and a collective trust in a network of computers. Speculatively, but not beyond possibility, an exogenous electromagnetic shock, such as a massive solar flare, could erase the planet's data networks and leave people scrambling for assets of physical value. This catastrophe scenario raises the question whether there might exist an underlying physical store of value for any and all forms of money. For millennia, it was gold. Gold is a product of geological processes, which are in turn a subset of all natural processes. The underlying value for all human monies, fundamentally, has to be natural capacity. It enables people to invest their energies into providing services to one

[19] Striner (2015, p. 58).
[20] Dodd (2014, pp. 362–372).

another; it enables people to develop physical technologies for the provision of goods and services; and it provides sources of energy to power all this activity.

Rather than the value of money remaining untethered, as in the present day, it could become regrounded if in some way it reflected the state of natural capacity. This relationship would not protect against solar flares so much as to help to guide the economy toward alignment with natural processes. As Fig. 1.11 depicted, credit is extended based on an *expectation* of growth. If the value of money were a function of natural capacity, then the extension of credit would be based on an expectation of improved natural capacity in the future. If we take material efficiency as a proxy for natural capacity, then the extension of credit would be based on an expectation of increased material efficiency rather than of economic growth in material terms. Economic growth in financial terms thus would become a function of increasing material efficiency.

This is the kind of thinking underlying Richard Douthwaite's idea of energy-backed currency units. The idea is premised on the assumption that the only institution capable of taking thermodynamics into account within the economy is government. In Douthwaite's conception, a finite supply of energy-backed currency units would be issued to governments by a transnational authority, such as the IMF, on a per-capita basis. The coupons would be used by those governments to purchase emissions rights and would be distributed to citizens and corporations. (The proposal assumes that energy production creates emissions.) This form of money would not be financial money but 'ecological' money, to use Douthwaite's term, governed not through the supply and demand of credit but of energy. Economic expansion would occur through increasing energy efficiency—wringing more economic production out of each unit of energy credit—or by switching to low- or zero-emission sources of energy.[21]

The appeal of this system is its inherent theoretical simplicity. Energy-backed currency units would qualify as money by virtue of being in demand and representing some underlying value: in this case, reduced atmospheric emissions and greater energy efficiency. However, as a kind of self-imposed thermodynamic austerity on the part of participating governments, it would, in practice, be open to free-ridership on the part of

[21] Douthwaite (2000, Chap. 4).

non-participating governments. Either everybody would have to be in the scheme or no-one could be.

The top-down nature of the scheme is another weakness. Aside from the free-rider problem, there remains a problem of economic risk, in that it may not be possible to anticipate all the possible economic effects of such a scheme in advance. To mitigate these risks, the scheme would have to be phased in.

If we look at the idea through the hierarchical structure of a global governance council then it could have a role to play in measuring natural capacity, although it might be one of numerous schemes operating at different levels of organization. The scheme also would have to be broadened to encompass not only a nation's emission rights but also their material intensity more generally. The scheme would work only if substantially all national jurisdictions were already members of a global governance council and they all signed on to the scheme. Material intensity coupons would be issued to national governments according to national populations. Trading in the coupons among member nations would deliver much-needed revenue to small, poor, low-intensity countries, thereby helping them out of poverty. As overall material efficiency increased, the availability of coupons would decline.

By the time such a scheme as this were feasible, it is possible that nation-states as we know them might have changed almost beyond recognition, as Chap. 5 argued. A more likely scenario is a global governance council providing a structure within which iterative, bottom-up solutions could be tried. These would include policies and instruments for material efficiency implemented initially among smaller jurisdictions, later adopted by larger ones, during which time membership of the global governance council itself continued to expand. As the prices of goods and services increasingly reflected the economy's overall material discipline, the preconditions for a form of global thermodynamic money might emerge. However, it could not feasibly be imposed from above.

Governance of Money and Finance for Long-Term Stability and Prosperity

Material discipline on the part of the economy can be accomplished only if the real assets of the economy—that is, the physical ones—are not parasitized by extractive ownership. The emergence of market fundamentalism as a form of economic policy in the 1970s enabled the growth of a vast 'financial' economy speculating on the assets of the real economy. The value of trades in this universe of speculation is multiples of actual gross domestic product. The mortgage crisis of the mid-2000s emerged in part because of what some have termed 'casino finance' in mortgage-based securities, and in part due to a lack of government oversight on the banks. Homeowners were allowed to make risky purchases, and hedge funds engaged in speculation on credit default swaps and other exotic mortgage-based investment vehicles tied to bundles of these mortgages. "Everyone in the chain was under the same pressure to produce", explained John Katovich. "Not only the mortgage broker but also the CEO of the bank, the bank itself, the lawyers who designed the collateralized debt obligation structures, the ratings firms, and so on." When loans started to go bad, speculation increased. Houses were left empty and the real economy went into a deep, extended recession.[22]

The role of government regulation is to structure the rules of the game to safeguard not only the productivity of the real economy but also, on a crowded planet, the natural capacity on which it depends. "If somebody buys bad meat from a careless slaughterhouse", writes Marjorie Kelly, "we would not lay the blame on [the buyer]. Yet tainted mortgages remained legal [even after the recession] and were in fact encouraged by the design of extractive ownership."[23]

Numerous ideas have been proposed over many decades for reforms to monetary policy and finance, driven by a variety of motivations. Three main questions recur: (1) who should have control over the money supply, (2) whether and how the financial system should be stabilized and (3) ways that financial reform could distribute incomes more evenly.

All three have a bearing, albeit indirect, on the economy's long-term ability to become aligned with nature and to become broadly prosperous and participatory. The question of who should have control over the money supply takes us back to the more fundamental question of the nature of money, and the ways its various forms may shape life in a future economy. These questions become the context for the proposals examined below.

[22] Kelly (2012, Chap. 2).
[23] Kelly (2012, p. 98).

Governing the Money Commons

Whereas the material commons encompasses natural capacity and society's use of it, a large-scale, long-term question, the money commons exerts a much greater influence over people's day-to-day decision-making. Because money-flow, unlike material flow, is entirely endogenous to the economy, it is an artifact.

Like the supply of material resources, the supply of money is limited. The propensity for economic actors to hoard it limits its supply further. At the same time, however, the process of lending—the relationship facet of money—leads to its 'self-expansion', using Kojin Karatani's term,[24] to a point where debt obligations may far outweigh the total currency in circulation—the artifact side of money—leading to periodic financial and economic instability.

Some writers, such as Adam Levitin, see the problem as the enablement of money's self-expansion through lending by commercial banks, a conflict of interest engendered by the commingling of depository and lending functions. Others point the finger at national governments, who cannot be relied upon to manage the currency well. Pierre-Joseph Proudhon was an early proponent of the 'disintermediation' of banking—that is, of establishing banking institutions by the people for the people. One hundred and fifty years after Proudhon, Satoshi Nakamoto, the pseudonymous founder of Bitcoin, made a similar argument in favor of electronic currencies.[25]

Yet it was Edwin Riegel, writing in 1976, who came perhaps closest to the idea of monetary governance imagined here. "When the people of the world have a common monetary language, completely freed from every government, it will so facilitate and stabilize exchange that peace and prosperity will ensue even without world government." The nation-state, Riegel argued, would no longer be able to dilute its money. A common monetary language would enable a 'union of peoples' rather than any kind of political union.[26] Although Riegel's claim that this future would be free of strife invites skepticism, his notion of a common monetary *language* is promising. Riegel was not advocating a single global currency but rather a shared understanding about what money is and what it is for, how it

[24] Karatani (2003), cited in Dodd (2014, p. 82).

[25] Dodd (2014, Chap. 8). Australian information technology consultant Craig Wright claims to be Nakamoto.

[26] Riegel (1976, p. 28), quoted in Dodd (2014, p. 360).

should be used and how it should not be used. No single nation would dictate the norms of money but rather such norms would be enshrined within a supranational institution of some kind.

Because money comes in so many different forms—sovereign currency, scrips, LETS, electronic currencies, to name a few—the governance of money and, by extension, of finance also might take different forms within a planetary economy. Because certain aspects of banking and money-creation as currently practiced are deleterious to the economic goals of material discipline and widespread prosperity, some kind of organizing structure would be needed, an institution or collection of institutions allowing governance to succeed at different levels of social organization. Just as society's relationship to the natural commons on Earth needs a form of governance that can be simultaneously global, regional and local, so the money commons also needs a form of governance that can accommodate all scales. The dominance of fiat money may wane over time, particularly given technology's increasing ability to connect everyone electronically. 'Money' likely will become a collective noun for different types of money operating at a variety of organizational and geographical scales. For the system of money to enable, rather than to inhibit, the transition toward long-term stability and wide prosperity, it must be set up with these long-term goals in mind.

Existing Proposals

Existing proposals for monetary governance cluster around two somewhat overlapping themes: one is state control over the issuance of money and the other is the restriction of banks' ability to create bank money.

Legal tender created by government typically enters general circulation through commercial banks. Central banks and commercial banks thus are connected within the money cycle, although, as Striner explains, cash issued by the government essentially is an operational representation of an underlying force of credit. Credit is expanded or contracted by the central bank through its manipulation of the interbank interest rate. Commercial banks then extend their own credit through their lending operations. Meanwhile, government operations themselves must be funded. This is accomplished within today's system by drawing revenue from the monetary pool through taxation or by borrowing.[27] Government borrowing

[27] Striner (2015, p. 85).

essentially amounts to the state borrowing its own money at interest. If debt dries up or becomes unsustainable, so does the money supply. "Debt", argues Mary Mellor, "cannot be the basis of money supply because a debt-based system is always subject to crisis."[28]

If or when commercial banks lend irresponsibly or in excess, the entire economy can be affected. Ever since the emergence of fractional reserve banking during the Renaissance, a tension has existed between bank money and fiat money. Banks, writes Mellor, "are not creating loans in 'Lloyds florins' or 'HSBC guineas'. They are creating bank accounts that represent the public currency." There is nothing inherently preventing them from creating loans in their own currencies. It is simply convenient for them to do so in units of national currency. The problem for the public, however, is that "this new debt-based public currency has to be 'backed' by public monetary authorities, usually the central bank".[29]

Calls to rein in the commercial banks began as early as the eighteenth century. Some commentators, such as Thomas Jefferson, proposed that "the issuing power of money should be taken from the banks, and restored to the people to whom it belongs". Abraham Lincoln expressed a similar sentiment.[30] Karl Marx recognized that credit money "creates the illusion that capital is self-expanding", leading to the formation of what he called 'fictitious capital'.[31]

Financial and economic stability, it was argued, could be possible only if the commercial banks met certain reserve requirements imposed by the central banks. Yet the cash basis of banking could be circumvented altogether if commercial banks acquired an ability to transfer financial assets between accounts directly in non-cash form. That ability arrived with the advent of the first electromagnetic means of communication, the telegraph and the telephone—hence, the term to 'wire' money—and later with electronic banking. These technologies enabled the growth of so-called demand deposits or sight deposits, which are deposits that can be withdrawn without delay or penalty. Checking accounts, for instance, hold demand deposits. The telegraph and the telephone, by the way, also enabled stock traders to follow stock prices on exchanges almost in real time, the prices being printed on ticker tape.

[28] Mellor (2016, p. 41).
[29] Mellor (2016, p. 41).
[30] Huber and Robertson (2000, p. 22).
[31] Dodd (2014, p. 55).

In the wake of the stock market crash of 1929, Irving Fisher formulated a plan, based on the ideas of Frederick Soddy, for what he called '100 percent banking'. The only institutions who lend money of their own creation, Fisher noted, are banks; everyone else lends 'pre-existing money'. The government, he argued, should take away this ability.[32] Fisher's plan later became known as the Chicago Plan, having been developed and elaborated by other Chicago economists in the 1940s, including Henry Simons, Frank Knight and Milton Friedman.[33] The plan proposed for commercial banks to hold a cash reserve of 100 percent of all demand and savings deposits. In this way, its proponents argued, "deposits would become again the true and safe cash deposits they were thought to have formerly been".[34] Yet, as Joseph Huber and James Robertson point out, the Chicago economists misunderstood the actual nature of money within a telegraphically connected financial system. It was no longer cash in someone's pocket but information on a bank statement. Fractional reserve banking, they write, "is now opaque, inherently unsafe and unstable, almost impossible to control, and too expensive". Because commercial banks are able to lend non-cash money against their non-cash deposits, the system of money and finance has become perceived increasingly as unaccountable. In fact, financial regulations—or the lack thereof—allow deposits into checking accounts to create funds for lending by the banks. A 2014 study by the McKinsey Global Institute estimated the total debt in the world at around $199 trillion,[35] which is about two-and-a-half times the world's total GDP that year, three times the capitalization of the world's stock markets and 25 times the value of the world's gold reserves.[36] Assuming, perhaps anachronistically, that gold represents the ultimate monetary reserve, this means that the aggregate reserve ratio of the world's financial system is about 25.

In accounting terms, the balance in a checking or savings account is treated as a liability on the bank's accounts. The liability is question is not actually the funds themselves but is instead the deposit account, as a kind of contract. A deposit account represents a promise by the bank to pay on demand. Meanwhile, the actual funds become legally the property of the

[32] Fisher (1936).
[33] Dodd (2014, p. xii).
[34] Huber and Robertson (2000, p. 23).
[35] Dobbs et al. (2015).
[36] Orrell and Chlupatý (2016, p. 119).

bank. As such, they can be used by the bank for lending. This means that if you deposit $1 million in cash into your bank account, those bills are no longer technically yours. You cannot just go in and retrieve them the way you could retrieve gold bars or a diamond necklace stored in a lockbox. This is the distinction between real assets, such as gold, coin, precious stones, land or structures, and financial assets, functioning as promises to pay. Nonetheless, wrote Hyman Minsky, "the public uses bank deposits as money, and banks use central bank deposits as money".[37] While cash in hand is legal tender—fiat money, issued by the government—demand (or sight) deposits at commercial banks are not. To the banks, money is a commodity.[38] Yet, through lending, banks are still able to increase the money supply. They lend by placing demand deposits into the checking accounts of their borrowers. Those demand deposits can be used as legal tender to make purchases. Somehow, the process of lending appears to have created 'money', in some shape or form, as if from nowhere.

Actual legal tender issued by the government thus "sits beneath the debt pyramid that makes up the monetary and financial system", according to Dodd.[39] Ultimately, of course, this debt pyramid resolves into legal tender. If a single sequence of debts is called in, then someone at the end will walk away with cash in hand. However, if debts were called in en masse, as happened in the subprime mortgage crisis of the late 2000s, then many creditors in the pyramid would lose money, and it could collapse.

Within today's globalized financial system, the banks' regulatory freedom "[allows] political sovereignty [to become] subordinate to monetary sovereignty", according to Takayuki Nakahara.[40] The complexity of existing financial regulations, writes Levitin, has rendered the politics of financial regulation as "an insider game that inherently favors market participants". Only market participants know where the loopholes are. "Because market participants have unequal upside benefits and downside risks due to privatized gains and socialized losses", Levitin notes, "they effectively hold a call option on economic growth."[41]

[37] Dodd (2014, p. 117), citing Minsky (1986).

[38] This concept of money, argues Papadopoulos (2009), is 'neoclassical', in the sense of diminishing or ignoring the significance of regulatory authority over the money supply.

[39] Dodd (2014, p. 108).

[40] Nakahara (2013).

[41] Levitin (2016, p. 442).

Peter Warburton once likened fractional reserve banking to counterfeiting, albeit completely legal.[42] This legal creation of non-cash money is used to fuel rampant speculation in other non-cash vehicles, such as in financial derivatives. The practice appears tantamount to legal counterfeiting used for legal gambling. "By generating a volatile international environment through over-banking", writes Dodd, "[finance has become] dangerous to *society* insofar as it is a threat to its *money*" (italics his).[43]

This threat is double-edged because government—and ultimately the people—are on the hook as the funders of last resort. In the wake of the 2007–2008 financial crisis, governments poured trillions of dollars of fiat money into the commercial banking sector to protect against a run on commercial banks. "Why did the banking sector trigger such largesse", asks Mellor, "when the poor and vulnerable and the planet did not?"[44] Part of the reason, she argues, is because commercial banking has become "parasitical on the public capacity to create and guarantee the public currency".[45] After the bailout forced governments into deficit, their debt was passed onto the public in the form of austerity measures which ate into government services. There is no 'backing' for fiat money, whether publicly or privately created, other than the provisioning capacity of the people who live within a monetary community.[46] In other words, if everybody stopped working and went on strike, the banks would fail and the government would be unable to bail them out.

Bailing out the financial sector puts a squeeze on already-constrained government budgets, which are funded partly through government borrowing. This is borrowing of the very same public currency that the central bank poured into the system. In many countries, therefore, infrastructure continues to deteriorate, government services continue to shrink, and raising income taxes remains politically unpopular. While this system may counteract a government's impulse to print money, the interest on government borrowing renders it an expensive means of raising funds.[47] These are reasons why some reformers have proposed ways for government to take back control over its money, arguing that, in a democratic system, it is the people's money.

[42] Warburton (1999), cited in Dodd (2014, p. 199).
[43] Dodd (2014, p. 122).
[44] Mellor (2016, p. 2).
[45] Mellor (2016, p. 5).
[46] Mellor (2016, p. 160).
[47] Orrell and Chlupatý (2016, p. 111).

Three proposals are reviewed. The first, pure reserve banking, aims to replace fractional reserve banking to restore stability to the financial system, thereby removing from government the obligation to support the financial sector. The second, pure money, is a mechanism for government to fund public needs through the direct issuance of money, without placing any specific constraints upon fractional reserve banking. The third, sovereign money creation, combines elements of both.

Pure Reserve Banking

Adam Levitin describes the concept of pure reserve banking in a 2016 paper in the University of Chicago Law Review. Addressing the risk of financial instability brought about by fractional reserve banking, he proposes not to constrain it through reserve requirements but to eliminate it altogether through institutional reform. Under pure reserve banking, depository banking would be separated from lending. The essence of the idea can be traced back to the Chicago Plan. Yet in Levitin's conception, commercial banks would be limited to providing only depository and payment functions. Lending would become a function of the capital markets, provided by money brokers. "Historically", writes Levitin, "a separation between deposits and lending was not possible, but it is now feasible with today's deep and efficient capital markets."[48]

A customer typically would maintain two types of account: a depository account with a bank and a brokerage account with an investment broker. Because the two institutions would be entirely separate, banks would fund their operations not through returns on lending, as is common today, but solely from fees to provide their depository and payment services. Levitin argues that they would exercise greater market discipline in the provision of these services. Safekeeping, after all, is meant to be a risk-free function. Using deposits to fund loans, as is common in today's system, "inevitably poses risk to deposits, thereby undermining the safekeeping function".[49] A customer's brokerage account, on the other hand, would become their conduit for investment in the financial markets as a means to earn a return. Funds in brokerage accounts would form the basis of lending. As Levitin points out, credit cards and other forms of debt already can be linked to brokerage accounts. Banks, on the other hand, no longer would offer them.

[48] Levitin (2016, p. 357).
[49] Levitin (2016, p. 357).

Customers seeking loans would work with money brokers, who either would invest their capital directly in the markets or work with dedicated investment brokers. A money broker would qualify a customer for a loan based on a set of underwriting guidelines. These practices, Levitin points out, are already common.[50] However, the institutional separation of the depository and lending functions would not place the deposits of bank account holders at risk if loans went bad. The only parties at risk would be the money brokers. Knowing there would be no government bail-out, they would be inherently more risk-averse in their practices than banks are today. Pure reserve banking thus would, in Levitin's words, "protect the money supply from the markets and the markets from the money supply",[51] a situation in stark contrast to that following the Lehman Brothers' failure in 2008.[52]

In pure reserve banking a deposit in a checking account would not be a loan to a bank but, legally speaking, a bailment. "The bank", Levitin writes, "would have fiduciary obligations to the depositor to account for the funds that were deposited." The main benefit of depositing funds with a bank then would be not so much the depository function itself but the payment function: the ability to move funds from one account or institution to another.[53]

Not only would pure reserve banking help to stabilize the money supply but also it would create a safe base of capital. "Deposits could be deployed by depositors from their deposit accounts to their brokerage accounts to recapitalize those firms that are temporarily undervalued because of a collapse in the lending markets", Levitin writes. The separation of deposits from lending not only would create greater market stability *ex ante*, such as by reducing the risk of market bubbles, but also would assist in stabilizing the financial markets following a crash.[54]

[50] Levitin (2016, pp. 423–424).

[51] Levitin (2016, p. 357).

[52] Levitin (2016, p. 426). Mellor (2016, p. 4) points out that Lehman was allowed to fail whereas commercial banks were bailed out because, as an investment bank, Lehman had no automatic public guarantee. It did not take customer deposits with the obligation to refund them, only investor money, which was subject to risk. Lehman failed, however, because its interests were so intertwined with commercial banking—specifically borrowing and speculation—that its collapse threatened to bring down the whole system.

[53] Levitin (2016, p. 426).

[54] Levitin (2016, p. 437).

this way, government-created money would flow into the existing monetary pool already subject to the monetary policies of the Federal Reserve. However, it would be created "over the head of the entire banking system", to use John Commons' words, specifically for purposes spelled out in Congressional legislation.[61] The people's elected representatives, Striner explains, would create new money to be spent—not lent—into use for specific purposes.

"The legitimate public needs of the United States are enormous", Striner points out. They encompass such relatively uncontroversial sectors as infrastructure, public safety and national security.[62] If Congress decides that the public infrastructure, for instance, needs improvement or expansion then the government would not need to borrow money to make this happen. It would simply create its own money and pay it into the accounts of the public employees or government vendors tasked with the work. If additional public needs at the state or local levels were identified, then funds could be transferred to those entities for the same purpose. In this way, government at any level would have access to additional funds from outside the regular budgetary process. The process would amount to a government-funded expansion of public budgets. Public budgets, in turn, would shrink, because relatively more public expenditure would be funded directly, thereby reducing public borrowing and narrowing the public deficit.

The effect, naturally, would be stimulative. National, regional and local governments would have more funds to do necessary work in the public good. It also would be inflationary, as Striner acknowledges. Because the inflationary effects would be unpredictable, a phased introduction would guard against over-stimulation, with smaller and progressively larger amounts being injected into the system by the legislature as the program took effect. The central bank would retain control over its levers on the money supply to counter any inflationary effects.

Sovereign Money Creation

Similar to Striner's pure money idea is a collection of proposals for government to control the money supply and at the same time constrain or prevent fractional reserve banking. The original idea, called 'seigniorage reform' or sovereign money creation (SMC), was proposed by Joseph

[61] Commons (1931), quoted in Striner (2015, p. 44).
[62] Striner (2015, pp. 59–63).

Huber and James Robertson in 2000, although a similar proposal made by the American Monetary Institute predated it by a few years. SMC subsequently influenced the Positive Money movement in the UK and the *Vollgeld* movement in Germany.[63]

SMC reform comes in two parts, as Robertson explains:

1. "Central banks should create the amount of new non-cash money (as well as cash) they decide is needed to increase the money supply, by crediting it to their governments as public revenue. Governments should then put it into circulation by spending it.
2. It should become infeasible and be made illegal for anyone else to create new money denominated in an official currency. Commercial banks will thus be excluded from creating new credit as they do now, and be limited to credit-broking as financial intermediaries."[64]

Under seigniorage reform, demand or sight deposits would acquire the status of legal tender, existing on a level playing field with bullion, coin or bank notes, a status which today they do not have. As such, they would fall under the control of the central banks. This is not to say that the government would control your private checking account. Rather, the central bank would deposit new funds into government accounts for the administration of the day to spend according to its policies and plans. However, because demand accounts would be legal tender, the commercial banks could no longer include such accounts on their books. They would be recognized formally for "what they now actually are", according to Huber and Robertson, which is "accounts containing non-cash money [belonging] to customers, just as customers' wallets and purses contain cash money that belongs to them". Therefore, "current accounts will cease to be accounts belonging to the banks. They will be containers of money belonging exclusively to bank customers."[65]

Because money will enter circulation as debt-free seigniorage, deposit accounts would become simply assets to their holders rather than the property of a bank and a promise to pay out. "By detaching current

[63] Two helpful online resources track developments in the field and provide references to recent work: Sovereign Money (https://www.sovereignmoney.eu/books/) and Positive Money (http://positivemoney.org/publications/).

[64] Huber and Robertson (2000, pp. 1–2). Similar reform has been proposed by the American Monetary Institute: www.monetary.org.

[65] Huber and Robertson (2000, pp. 23–24).

accounts from the banks' balance sheet", wrote Huber and Robertson, "the problem of how to prevent banks from creating non-cash money will be solved. Banks need not be forbidden to create sight deposits. They will no longer be able to."[66]

They would still be able to lend money, but only from their own cash deposits, so-called operational funds, held in their current accounts with the central bank. Customers' deposit accounts would represent funds under management by a bank, in a similar sense to funds under management by a broker. Managed funds are, by definition, not the property of a broker. Thus, there would exist a clear distinction between customers' funds and the banks' own capital, a distinction that does not exist today. By the convention of double-entry accounting, Huber and Robertson explained, "payments into current accounts will always have to be matched by payments out of other current accounts, or paid in as cash … except when a central bank is creating new money as public revenue". No more fractional reserve banking. Any other creation of legal tender would constitute counterfeiting or forgery, being punishable as such.[67]

In this way, commercial banks would no longer be in the business of creating new money. That ability would now reside solely with a central bank, operating presumably in the public interest rather than in the interests of private shareholders. In today's system, explains Mary Mellor, new money created by a central bank is debt-free at the point of creation.[68] The problem is that the central bank places it into circulation through commercial banks, who convert it into debt, both private and public. This debt creates an expansionary effect in the economy, as the interest on it must be repaid from somewhere—such as the origination of more debt—an effect which, if not offset by improvements in material efficiency, would place ever greater demands upon natural systems. Additionally, investment decisions using public money traditionally have been considered private matters because they access private credit, offered through commercial banks. Yet if public money is being used for investment, asks Mellor, surely the public should be consulted about it? A financial system under public oversight, she argues, would never have been allowed to use such large amounts

[66] Huber and Robertson (2000, p. 24).
[67] Huber and Robertson (2000, p. 3).
[68] Mellor (2016, p. 4).

of bank-created money for speculation, as happened in the lead-up to the 2008 financial crisis.[69]

New non-cash money created by central banks could be used in a variety of ways, Huber and Robertson point out, such as paying down debts or reducing taxation. It would be up to the government of the day to decide how to invest the funds, just as it would be the responsibility of the central bank to strike a balance between economic stimulus and guarding against inflation.

If commercial banks were no longer able to issue debt through fractional reserve banking, how would people or institutions obtain debt? A commercial bank's own money would exist either as cash in the till or as non-cash money in its operational account with the central bank. Loans would be made to customers from these sources, most of which, in turn, would have been borrowed from bank customers and from other banks. Funds will be transferred into a loan-brokering bank's account at the central bank, and thence into the current account of the borrowing customer. "The stock of circulating money", Huber and Robertson explained, "will thus remain unchanged, except for any additions created by the central bank and spent into circulation as public expenditure."[70]

One additional institutional modification would be the creation of a firewall between the money-issuing function of a central bank and its other banking functions. Such separation would render the creation of money completely transparent. The issuing activities of central banks would create new money, while their banking activities would not. The issuing department would regularly publish how much new money had been issued and to whom. The Treasury would have no influence over the amount of new money issued, while the issuing department of the central bank would have no influence over how the money were used. An alternate recipient of newly created money could be the banking department of the central bank, in the form of a non-interest bearing loan. Such funds could be used to loan to commercial banks, for example.

Huber and Robertson's ideas sparked a flurry of economic research and policy development during the 2000s. Andrew Jackson and Ben Dyson built upon Huber and Robertson's foundation to develop a more detailed justification for the idea and a more detailed description of the mechanism by which new money would be issued, spent and circulated within the

[69] Mellor (2016, p. 52).

[70] Huber and Robertson (2000, p. 25).

economy.[71] Sovereign money creation might not dampen lending by commercial banks so much as reduce the frequency of bad loans. Having more disposable cash, households would be relatively more inclined to pay down existing loans, thereby reducing default rates and encouraging banks to make new loans through the new mechanism.

Huber and Robertson identified several benefits of sovereign money creation. In particular, private and public borrowers would no longer pay interest to private banks for new money created by those banks. The stratospheric earnings of the 0.001 percent would no longer be possible, although normal banking revenue would remain unaffected. In today's system, the creation of demand deposits by commercial banks allows them to "cream off a special profit", according to Huber and Robertson. "They lend the money to their customers at the full rate of interest, without having to pay any interest on it themselves." This special profit "represents a sort of private tax—a privileged subsidy in favor of the banks against everyone else".[72] This tax has negative effects upon the wider economy. It restricts general savings, raises prices, increases welfare spending, taxation and public borrowing, reduces private and public investment, and diminishes opportunities for employment. Under SMC, the authors argue, private and public budgets would no longer have to pay for the special banking profits arising from the creation of debt-money by commercial banks. In addition, governments would acquire seigniorage revenue from the creation of new money. These benefits would go to people and businesses, in the form of more money in their accounts with which to add value to the real economy.[73] A 2014 article in the *Financial Times* estimated that under sovereign money creation, the UK could run a fiscal deficit of about 4 percent of GDP without having to borrow or levy taxes.[74] Under a transition to SMC, sales of luxury private yachts would diminish, while sales of products and services supporting broader prosperity would rise.

Commercial banks would no longer be able to shape the way the economy develops, argued Huber and Robertson:

[71] Jackson and Dyson (2013, figure 10, p. 22).
[72] Huber and Robertson (2000, pp. 31–32).
[73] Huber and Robertson (2000, pp. 39–40).
[74] Wolf (2014).

> A debt-free money base, a less indebted government, a better-balanced government budget, a lowered tax burden, a better moneyed civil society—all these will contribute to a higher level of net income and a larger capital base for both businesses and private households. This will help to make them less dependent on subsidies and allowances and external capital, and better able to provide for themselves and one another.[75]

Because the monetary system would become simplified it also would become more stable, losing some of the positive feedback loops that tend to destabilize the current financial system and, by extension, the wider economy.[76]

Another benefit would lie in the economy's relationship with nature. As Douthwaite observed, the present money supply system would collapse if it were denied continuous expansion.[77] This expansion becomes increasingly untenable in an ever-more crowded world, particularly while it remains tied to material intensity. Seigniorage reform would enable a controlled approach toward a low-growth state, as population growth leveled off and material efficiency increased. The issuance of new money would not be the poorly controlled bonanza it is today but a macroeconomic policy decision guided by a directive of financial and economic stability.[78]

References

Collins, R. (2000). Situational Stratification: A Micro-Macro Theory of Inequality. *Sociological Theory, 18*, 17–43.

Commons, J. R. (1931). *Institutional Economics: Its Place in Political Economy.* New York: Macmillan.

Dobbs, R., et al. (2015). *Debt and (Not Much) Deleveraging.* McKinsey Global Institute. Retrieved from www.mckinsey.com/insights/economic_studies/debt_and_not_much_deleveraging.

Dodd, N. (2014). *The Social Life of Money.* Princeton University Press.

Douthwaite, R. (2000). *The Ecology of Money.* Schumacher Briefings, Book 4, UIT Cambridge Ltd., UK.

Fisher, I. (1936). 100% Money and the Public Debt. *Economic Forum*, Spring Number, April–June: 406–420.

[75] Huber and Robertson (2000, p. 40).

[76] Orrell and Chlupatý (2016, p. 181).

[77] Douthwaite (2000, p. 27).

[78] Huber and Robertson (2000, p. 42).

Huber, J., & Robertson, J. (2000). *Creating New Money: A Monetary Reform for the Information Age*. Report of the New Economics Foundation, London.
Jackson, A., & Dyson, B. (2013). *Sovereign Money*. London: Positive Money.
Karatani, K. (2003). *Transcritique: On Kant and Marx*. Cambridge, MA: MIT Press.
Kelly, M. (2012). *Owning Our Future: The Emerging Ownership Revolution*. San Francisco, CA: Berrett-Koehler.
Levitin, A. J. (2016). Safe Banking: Finance and Democracy. *University of Chicago Law Review, 83*(1), 357–455.
Litaer, B. (2001). *The Future of Money: Creating New Wealth, Work and a Wiser World*. London: Century.
Mellor, M. (2016). *Debt or Democracy: Public Money for Sustainability and Social Justice*. London: Pluto Press.
Nakahara, T. (2013). *Régulation* Approach to Financial Crisis: Potential Economic Policies Under Structural Changes of Institution. *World Review of Political Economy, 4*(4), 484–507.
Orrell, D., & Chlupatý, R. (2016). *The Evolution of Money*. Columbia University Press.
Papadopoulos, G. (2009). Between Rules and Power: Money as an Institution Sanctioned by Political Authority. *Journal of Economic Issues, 43*(4), 951–969.
Riegel, E. C. (1976). *The New Approach to Freedom*. Los Angeles: Heather Foundation.
Striner, R. (2015). *How America Can Spend Its Way Back to Greatness: A Guide to Monetary Reform*. Santa Barbara, CA: Praeger.
Warburton, P. (1999). *Debt and Delusion: Central Bank Follies that Threaten Economic Disaster*. London: Allen Lane.
Wolf, M. (2014). Strip Private Banks of Their Power to Create Money. *Financial Times*, April 24.
Zelizer, V. A. (2004). Circuits of Commerce. In J. C. Alexander, G. T. Marx, & C. Williams (Eds.), *Self, Social Structure and Beliefs: Explorations in Sociology* (pp. 122–144). Berkeley: University of California Press.

CHAPTER 11

Monetary and Financial Requirements

As the foregoing and many other writings have acknowledged, the global financial system is a network of promises to pay. Rarely acknowledged, though, is that it is not only a network of promises to pay within the economic system but also a collection of promises to pay nature, albeit 'later' at some undefined time. The present system of money and finance, writes James Robertson "imposes a perverse calculus of values, compelling or encouraging almost everyone in the world to compete against one another for a greater share of planetary resources and, in doing so, to turn planetary resources into waste". Meanwhile, the so-called experts in charge of money and finance manage the system to serve their own interests.[1] Nature could, of course, call in its debts at any time. Part of the thrust of this book is to advocate for the back-payment of debts to nature, as well as for the forward-payment of expected future obligations. This chapter proposes some requirements for a financial system to encourage that.

Proposals Compared

As the three proposals outlined in Chap. 10 emphasize, the current global financial system effectively has become a triumph of commodity money over fiat money. Each proposal stressed the significant economic and social costs of this state of affairs. But does that mean that commodity money

[1] www.jamesrobertson.com.

F. Murison Smith, *A Planetary Economy*,
https://doi.org/10.1007/978-3-030-49296-0_11

should be restricted, and does it mean that governments alone should be allowed to issue money?

Before addressing this question, several strengths and weaknesses of the proposals, some which have been discussed in the literature, deserve mention. The proposals also share certain synergies and differences, summarized in Table 11.1.

Criticisms of government money creation express concern that it could be inflationary and it might erode the independence of the central bank. The difference between the pure money proposal and sovereign money creation regarding the money-creating institution raises the question whether the legislature or the central bank would the preferable choice. Proposals to separate deposits from lending, in turn, run the risk of being *de*flationary, by tightening access to debt. They also could result in capital flight. Further, they might leave loopholes for speculation in other forms of money not denominated in legal tender.

As Striner, Huber and Robertson all point out, the risk of inflation from government money creation would be slim, provided that the central bank retained control over monetary policy. That control would be exercised by manipulating interest rates or altering the amount of legal tender in circulation. Mellor notes that any inflationary effects also could be dampened through taxation, simply returning money "that has done its work in creating public benefit" to the government. Any amount not taxed would circulate within the economy, representing a net gain to the economy as a whole.[2] Huber and Robertson prefer direct control rather than indirect control over the circulation of money. Unlike quantitative easing through

Table 11.1 Similarities and differences between pure reserve banking, pure money and sovereign money creation

Proposal	*Status of sight deposits*	*Bank lending*	*Government money creation*
Pure reserve banking	Bailment in legal tender	Prevented	N/A
Pure money	Bank money	Permitted	By legislature
Sovereign money creation	Legal tender	Restricted to banks' cash deposits	By central bank

Source: Author's creation

[2] Mellor (2016, p. 74).

the lowering of interest rates, it would not risk a ballooning of debt.[3] Because, under sovereign money creation, the government itself would have no say in the amount of new money created, there would be no risk of inflation from ill-conceived government spending programs, only through an error of macroeconomic judgement on the part of the central bank. Such an error could quickly be rectified by reducing the amount of new money issued.

But could the central bank remain independent? In Striner's pure money proposal, it would, because the money-issuing authority would be the legislature. Under sovereign money creation, however, additional arrangements might be necessary to ensure that the monetary authority based its decisions on what would be good for the economy as a whole, not over how the money were spent. As Mark Joób describes, "a central aim of the sovereign money concept … is to restore democratic control over the monetary system."[4] As a consequence, the members of the monetary authority should be democratically elected in some way: perhaps not directly by the people, argues Joób, but vetted and voted in by the legislature, similar to how members of the judiciary are chosen. Once installed in their positions they would act, presumably, with the same independence as the judiciary; however, they probably would serve fixed terms, like the leaders of the world's main central banks.

There are advantages and disadvantages to the central bank or the legislature as the money-issuing authority. The pure money and sovereign money proposals agree that commercial bankers, in Striner's words, "should not be the ones to decide when and how we, the people, should fund our necessities".[5] Yet he departs from the sovereign money community over how the money should be spent. The pure money proposal does not do away with the budgetary process but instead adds a second, distinct channel for funding government programs, which is direct legislation. Programs that the legislature deemed sufficiently important not to become subject to budgetary whims could be established over the head of budget-making, and be funded directly by new money created by the legislature itself. Striner makes a strong argument that budgetary policy should not be the impetus for money creation. It is too far removed from public

[3] Huber and Robertson (2000), Chaps. 2 and 4; Jackson and Dyson (2013), Part 1.
[4] Joób (2014, p. 7).
[5] Striner (2015, p. 83).

influence: the everyday needs of the public should determine it, he states, citing a program for full employment as an example of such a need.[6]

Yet money creation could fall prey to legislative whims just as it could fall prey to budgetary ones: not only in terms of the type of programs funded through new money creation but also in terms of the volume of new money put into circulation. If the legislature pumped too much new money into the system, the central bank would be left scrambling to mop some of it up. On balance, the monetary system would be more stable and more accountable if the central bank had authority over the *amount* of new money placed into circulation—this being a macroeconomic question not trusted to the legislature—and the legislature had authority over how to spend it, whether through special appropriations or through the regular budgetary process. As Chap. 16 will discuss, perhaps the most effective system would be a combination of the two, in which the government requested funds from the central bank for its operations, and the central bank then assessed the request according to macroeconomic criteria prior to issuing funds.

The membership of the monetary authority could go beyond Joób's proposal to give the legislature authority over it, instead giving the people a direct say. After all, writes Mary Mellor, "the public needs to recognize that the public currency is theirs, because the only thing that 'backs' it is the people themselves".[7] Possible mechanisms for this public representation remains an open question: it would be a valuable avenue of inquiry, one in keeping with the general institutional requirement for the governance of a planetary economy to become more participatory.

On the separation of deposits and lending, a first concern is whether this separation could be deflationary. Since sight deposits could no longer be converted by banks into bank money for lending, consumer debt might be harder to obtain. However, under sovereign money creation, any incipient deflationary tendency would be counteracted by an incipient inflationary tendency from the government's issue of new money, which ultimately would find its way into the bank accounts of consumers.

Another deflationary effect could come from capital flight. Capital flight is framed as a kind of first-mover penalty. Any country that separated bank deposits from lending would risk losing its commercial banking

[6] Striner (2015, p. 83), citing William Greider: The people's money, *Policy Winners*, May 2014 newsletter, pp. 2–3.

[7] Mellor (2016, p. 193).

sector to other countries who did not. Banks and other market participants might determine that they could make more money elsewhere. Yet, as Levitin describes, the loss of fractional reserve banking would be counteracted by an influx of funds into capital-market lending. Lending overall would decline somewhat but only by the volume of capital currently poured into risky, highly leveraged loans. Money-making through arbitrage on shadow-banking instruments, on the other hand, would no longer be possible in a country where deposits and lending were separated. Capital would flee for markets where that was allowed, for certain.

Does this mean that a country such as the UK would lose its sizable commercial banking sector if it separated deposits from lending? Not at all. British firms still would need investment. Under sovereign money creation, banks would still be able to invest operational funds in the capital markets to support capital-market lending. And jurisdictions not implementing the separation would experience higher regulatory costs over their banking sector. Once one government separated deposits from lending, the argument goes, other governments would have an incentive to do so as well. If the majority of OECD countries were to adopt it then there would be few remaining desirable markets for capital to flee to.

Sovereign money creation, with its attendant separation of deposits and lending, could become established initially within small economies from which the risk of capital flight was correspondingly small. In a 2015 report assessing the prospects of sovereign money creation for Iceland—a small economy with its own currency, a small homegrown commercial banking sector, and significant commercial investment from outside—Frosti Sigurjonsson argued that the country has little to lose, since its commercial banking sector is limited to three large banks serving more than 90 percent of the customer base. Under sovereign money creation, they would lose their ability to create money. Yet this may not necessarily place them at a competitive disadvantage. Their knowledge of the Icelandic market is unmatched by foreign banks, and yet "large foreign banks with superior economies of scale … already have a competitive edge on Icelandic banks, when lending to large Icelandic export companies".[8]

All three proposals discussed above focus on principally on fiat money and bank money. This focus may be too narrow. In practice, the schemes may risk being gamed by parallel types of currency and private monies not subject to sovereign law. As Sigurjonsson points out, "the motivation for

[8] Sigurjonsson (2015, p. 93).

creating money substitutes is strong in any monetary system, as the issuer is rewarded with seigniorage." Not only electronic currencies but also novel financial instruments could thwart the twin aims of controlling the issuance of money and reining in speculation. According to Nigel Dodd, the proportion of circulating money falling under state control is only about 15 percent.[9] Some authors, such as Joób, suggest that bank-created securities, such as derivatives, eventually could take over the function of money within the financial system.[10] Investment banks already are exploring ways to do business in electronic currencies. As the lender of last resort, a central bank might still come under pressure to alleviate a banking crisis in unofficial currencies by bailing out commercial banks in official currency: that is, by converting the former to the latter. As Levitin writes,

> Simply regulating depositaries while tolerating the continued existence of the shadow-banking market would make little sense, because the benefits of Pure Reserve Banking would be lost through arbitrage with the shadow-banking market. A clear cleavage between Deposits and Lending requires not only maintenance of 100% reserves but also the abolition of Deposit substitutes that enable regulatory arbitrage. [11]

Both Levitin and Joób advocate for the abolition of what Joób calls 'near monies' (a.k.a. private money), which would impair a central bank's control over the money supply. However, as Sigurjonsson correctly points out, government control over money creation should extend only to money that is "by law acceptable as payment in commerce and for settlement of debts and taxes".[12] Other modes of exchange, such as community scrips or LETS, would be exempt.

Reclaiming public seigniorage is an admirable goal because it gives the commons control over the scale of the common liability. In this sense, it reflects a convergence rather than a conflict of interests within the financial system. Sovereign money creation in particular appears to offer several benefits to a planetary economy. Whether or not funds thus created were used by government for programs to stimulate improvements in material efficiency, sovereign money still would appear to benefit the economy in

[9] Dodd (2014, p. 213).
[10] Joób (2014).
[11] Levitin (2016, p. 417).
[12] Sigurjonsson (2015, p. 92).

other ways supportive of economic stability, most notably in financial stabilization and improving income equality.

There is a synergy also with pure reserve banking. Rather than making it "infeasible and illegal" for banks to create new money denominated in official currency, legislation simply would require financial institutions to choose between depository and lending functions. The key to success, however, would be for *all* depository and lending functions to be separated, no matter which form of money were used for them. Government then would control the issuance of legal tender, which may or may not be convertible into any number of other forms of currency, but which would continue to be used by government to fund its programs and to collect taxes. Meanwhile, the separation of deposits and lending in all monies would ensure that government-issued money would not be converted into other forms of money for unregulated debt-creation or speculation.

Private monies and novel financial instruments would not be subject to this separation as it would not apply to them. A derivative, for instance, is a contract between two parties. Although valued in money terms, it cannot be deposited like money, only converted into it. With banks unable to lend sight deposits, there would be no social risk to the banks from overheated speculation in securities. Speculators in the shadow-banking markets would be on their own.

Even though the twenty first century could witness a proliferation of monies, it is also possible that a progressively global integration of sovereign monies eventually could lead to a consolidation of fiat currencies, particularly if the creation of sovereign money eventually fell under the oversight of a transnational institution of some kind. Some private monies, such as electronic currencies, appear to have certain properties that render them forward-compatible with this outcome: they are globally available, their individual tokens are unique, and they invest heavily in data security.

The sovereign money community and other banking reformers have touched upon the importance of reclaiming democratic control over money and finance. "Reclaiming public seigniorage in money creation", writes Mellor, "is essential if democracy is to replace debt as the foundation of the public currency."[13] Although money has both a sovereign heritage and a commercial heritage, she notes, modern banking has combined the two. It is not 'bank' money, however: "it is a liability on the people."[14]

[13] Mellor (2016, p. 156).

[14] Mellor (2016, p. 157).

To Randall Wray, "the real debate should be over the proper role of government, how it should use the monetary system to achieve public purpose".[15] Levitin, quoting Christine Desan, points back to the origins of money as a constitutional project:

> Control over the monetary system is a fundamental part of the sovereign condition. When that control is outsourced … there is an inherent loss of self-determination by the polity. Restoring political control over the monetary supply is a critical part of the democratic project.[16]

Yet the focus of these writers has remained singularly national. Banking, after all, is carried out in national currency and taxes are collected in it. When we look into the distant future, two questions arise. One relates to the nature of money in a materially large economy coevolving with nature, and the other relates to the kinds of institutions governing money. The two questions are intertwined. If the emergence of a materially stable, widely prosperous economy necessitates a form of hierarchical governance, as outlined in Chap. 5, then part of this governance must concern the flows of money within the economy. Some universal rules should be in place to ensure financial stability, income stability and substantially full social participation in material discipline. The separation of deposits from lending might be one such rule. It would prevent fractional reserve banking while still permitting financial speculation.

If global governance emerged as a result of a progressive diffusion or dilution of the nation-state over the coming decades, then the implications for seigniorage would be profound. In this future world, seigniorage could be established at any social level. We see this already with the creation of electronic currencies and with the establishment of local exchange systems. Unregulated seigniorage could become just as destabilizing as unlimited fractional reserve banking. A system of global governance would be able to formalize institutions and requirements for seigniorage so as to ensure all currencies in circulation, whether local, regional or global, conformed to certain minimum standards. This would be the future 'shared monetary language' Edwin Riegel referred to.

[15] Wray (2011, p. 17).

[16] Levitin (2016, pp. 446–447).

Norms

If recent financial instabilities and the evolution of monies are any guide then, over the coming decades, society's perception of 'money' and our attitudes toward it may become much more pluralistic and distributed. "An open, multifarious and flexible monetary system", writes Dodd, "describes the state of the monetary landscape in which we are increasingly finding ourselves, where alternatives to the dominant state fiat money system are proliferating."[17] Yet underlying these trends there remains the normative challenge, discussed in Chap. 3, of moving beyond modernity's rampant acquisitiveness. Money is a mediator of both positive investment and unchecked greed. Society evolves along a cycle of norms, institutions, policies, instruments and behavior which, for a prime directive of material discipline, would have to be steered toward positive investment and away from individual greed. The way we perceive money will inform the way we govern it, and the way it is governed will, in turn, influence society's norms about it, as it has already done for centuries.

In a planetary economy, money would be much less a thing to acquire in and of itself because a prevailing norm of material discipline would influence people's attitudes and decision-making. It would go hand in hand with another norm supporting basic living standards for all, in part to ensure full participation in material discipline. Personal enrichment still would be perceived as acceptable, and even encouraged, although it would be tempered by a recognition of one's role within the wider social and planetary systems. Further, the modern impulse to sacrifice one's time for earnings would, in this future world, have been replaced with an impetus to devote relatively more time to non-monetary enrichment, basic income needs having been met. Within this context, what would be considered 'money' could then take many different forms, some of which cannot be anticipated today.

Institutions and Technologies

Governance of money for a planetary economy would have to be flexible, simple, accommodating and comprehensive.

The twentieth century and the early years of the twenty-first have witnessed a tug-of-war between the nation-state and the financial world over

[17] Dodd (2014, p. 387).

money. Money symbolizes power, in that it communicates the identity of the entity creating it. During the era of the nation-state, it has represented the state's ability to regulate people's ability to purchase. Yet, as a medium of exchange, money itself can become—and has become—commoditized and financialized. In the financial world, money is a symbol, and a mediator, of relationships. As the financial system has globalized, so those relationships also have globalized. In so doing, they have undermined the power of the nation-state. More than this, the use of public currencies by private finance is democratically unaccountable. The public has no say in how much new money in legal tender the banks create, nor for what it is used.[18] Absent any supranational governance, the global financial markets run the risk of operating too freely and creeping into a free-market anarchy. There has to be governance of some kind, it has to be global, and it has to represent the common interest.[19]

Michael Hardt and Antonio Negri predicted that the nation-state, as the controller of fiat money, would become supplanted over time by a global network of capital flows, from which an institution, or institutions, may emerge to govern them. If the exploitative capitalism of the past were to give way to a more regenerative system, then the wealth produced by capital flows would come under the control of the 'multitude', to use their term, for the benefit of the all. This is not to say that the global collective would control the world's money flows in order to suppress capitalist practices but rather that capitalism would support the global collective. Hardt and Negri imagine the wealth generated by capitalist activity being spread out rather than concentrated, the power of the markets being used to help lift everybody out of poverty. "Finance", they write, "grasps the common in its broadest social form and, through abstraction, expresses it as value that can be exchanged."[20] However, finance would require an institutional structure to encapsulate this collective will and to guard against the concentration of wealth. It would have to support the creation of sovereign monies at least while sovereign states continued to exist.[21] Its governance

[18] Mellor (2016, p. 68).

[19] Dodd (2014, p. 246), citing Hardt and Negri (2009).

[20] Hardt and Negri (2009, p. 158).

[21] Dodd (2014, p. 261) describes one of the flaws of the current Eurozone as preventing member states from directly monetizing. They have no power to do so. Instead, under the 1992 Maastricht Treaty, this power was devolved to the European Central Bank. There thus arises a disjunct between monetary control, which is supranational, and financial control, which resides partly at the national and supranational levels. This disjunct has created both

would have to enable the participation of the many to ensure long-term social stability; in fact, it would have to ensure representation across all hierarchical levels of society, from individuals to nations. This is what Costas Panayotakis and others call economic democracy.[22]

A Global Monetary Authority

The Austrian economist Friedrich Hayek advocated for the state to relinquish its control over money, allowing multiple private currencies to coexist. A profusion of currencies could emerge in the economy of the future, both within and across sovereign jurisdictions. Yet, counter to Hayek, the track-record of traders within the financial markets does not inspire confidence in a self-regulating, self-policing private system. In the late 2000s, for instance, the London Interbank Offered Rate, which acts as a baseline for other interest rates within the financial sector, was found to be manipulated by a cartel of bankers.[23]

The monetary system of the future does not need to be all-private or all-public, nor is it likely to be. The history of money and finance is littered with examples both of subterfuge by private financial institutions and of serial default by sovereign states. Hayek argued, contrary to the prevailing economic orthodoxy, that the economy is such a complex system that its management should not be entrusted to an institution as ponderous and siloed as government; the market itself contains all the necessary intelligence to self-regulate. Although Hayek was correct about the complexity of the economy, his dismissal of government regulation was based on an assumption that government would adopt the very same linear, mechanical thinking of which his economic colleagues were guilty. It is not *whether* the monetary and financial systems should be regulated in the public interest but *how*.

Given the likely proliferation of local and virtual private currencies alongside sovereign ones in the decades to come, a form of governance is needed that will accommodate all of them, require them to meet common standards, and enable them to compete against one another, finding their

economic and political strains within the European Union, particularly between the stronger and weaker economies. Striner argues that the architects of the Euro might have had it backward. It is not monetary unification that is needed but regional and global banking controls.

[22] Panayotakis (2011).

[23] Orrell and Chlupatý (2016, p. 149), citing Hayek (1978).

niches within the global monetary ecosystem. This governance should embody a common set of monetary norms, establishing policies for monetary governance. It would be structurally and functionally hierarchical, in the same way the global governance council outlined earlier would be hierarchical, thereby allowing all levels of society to be represented. In this way, monies operating at local, regional, national and global levels all would have a place to register, to meet certain common standards, and to interact with one another. Local, regional, national and global institutions established for seigniorage over these monies also would have a means of representation in governing the world's money.

Whether this institution actually would be part of a global governance council, adjunct to it, or entirely separate, is an open question. Separation might be preferable for long-term stability because it would resemble the current separation of political governance and central banking that, by and large, has successfully protected against the hijacking of monetary policy by government whim. Yet by sharing a degree of hierarchical homology with the global governance council, such an institution also would be able to liaise effectively with the council on issues at a variety of levels.

The institution would not be a bank because it would not issue its own money. It would be a 'global monetary authority' or some similarly named institution. Like a global governance council, it would be organized and administered by its members. These would be public jurisdictions, private monetary seignieurs[24] who passed a set of commonly agreed-upon requirements, and members of the public, in a quasi-sortitional arrangement providing much-needed public scrutiny. Organized this way, the authority would provide a framework for regulation of the monetary ecosystem and global finance in the interests of no-one in particular and of everyone.

This institutional structure has a precedent of sorts: the Federal Reserve model. The US Federal Reserve is not, despite its name, an arm of the US government. It is an independent non-profit corporation, syndicated into twelve regional reserve banks owned by a consortium of commercial banks.[25] Similarly, the Bank of England is a quasi-governmental non-profit organization independent of the UK government. Although monetary regulation has a well-established history of independence from sovereign government, it also renders a central bank independent from, and

[24] Institutions engaged in seigniorage. Not to be confused with 'seigneurs', as in feudal lords (*Oxford English Dictionary*), although both words have the same root.

[25] Orrell and Chlupatý (2016, p. 105).

therefore unaccountable to, democratic governance. As government loses control of its money, argues Mellor, the public, in turn, loses faith in political democracy.[26]

Imagine extending the idea of an independent central bank to include any registered entity, public or private, that meets certain seigniorage requirements. Imagine the money-regulating authority passing from commercial banks to member public jurisdictions and monetary seignieurs. Imagine the regulating institution becoming global in scope rather than national, accommodating a degree of representation from the public through sortition. This would be a global monetary authority.

Like a global governance council, a global monetary authority would not have direct jurisdiction over any of its member governments, be they national, regional or local. Instead, members of the authority would vote on policy requirements and monetary controls to be enacted within participating jurisdictions. For instance, rather than the reserve requirements of commercial banks being regulated by national central banks through the adjustment of interbank interest rates, which might vary from one country to another, now the world's commercial banks—many of which are multinationals anyway—would be held to reserve requirements established at the global level. Bank regulation would become simpler and more effective. Alternately, the authority might establish requirements and guidelines for pure reserve banking policies to be enacted into law by member legislatures. The tradeoff for a member government in being part of the authority would be in relinquishing a degree of sovereign independence in return for becoming a stakeholder in global financial policymaking.

Another facet of financial governance relates to payment networks. Like the proliferation of monies within the economy, a proliferation of payment networks also is under way. While banking remains the dominant payment network early in the twenty-first century, a variety of others have emerged in an attempt to disrupt it, including online payment platforms, social networking platforms and mobile telecommunications providers. Users typically pay small charges of a few percent per transaction for the convenience of using these networks.[27] A global monetary authority would support multiple payment networks by establishing, maintaining and policing common standards for making payments.

[26] Mellor (2016, pp. 7, 8).
[27] Dodd (2014, pp. 378–380).

A global monetary authority charged with a prime directive of material discipline and an economic directive of widespread prosperity would regulate the world's monies and its financial system to ensure a return primarily to the commons, only secondarily to private shareholders. The global financial system is a common asset: as such, it should serve the common interest. A global monetary authority would provide an institutional structure through which the common interest in the global financial system could be served. The intent would not be to undermine commercial finance but instead to guard against its tendency to excess. The institution effectively would encourage generative capitalism while discouraging exploitative capitalism.

A Public Financial Rating Agency

The establishment of common standards for money and payments implies also common standards for financial ratings. These would ensure that any two parties to a transaction anywhere in the world would have clear information about the creditworthiness of the other. A financial rating agency governed under a global monetary authority, whose members would be the issuers of the world's monies, would also ensure that the rating process itself were conducted in the common interest. Today, this is not the case. In the US, for example, most financial credit ratings are provided by three private companies—Standard & Poor's, Moody's and Fitch Group. They are not regulated, although a degree of oversight is provided by an industry group, the Securities Industry and Financial Markets Association. It is well known that these companies are not sufficiently independent from the companies they are asked to rate. The capture of the ratings agencies by financial firms came to public attention during the subprime mortgage crisis of the 2000s, having been long established by then.

Placing financial ratings under the common governance of a global monetary authority would ensure that projects or enterprises seeking a financial rating would no longer appear more creditworthy on paper than they actually are. Coupled with the establishment of a material efficiency rating agency under a global governance council, this institutional revision would be made in the interest of the stability of the global financial system and of the economy more broadly.

I Am My Own Bank

The next few decades might see not only profound institutional changes but also profound technological ones. The emergence of electronic currencies and the ability of individuals and organizations to move money globally in electronic form may, in the long run, render the bank as we know it obsolete. These technological and institutional evolutions go hand in hand.

Whereas banks in the historical past were physical stores of coin, today they are electronic accounting platforms. This being the case, why should a bank have control over the electronic accounting of your money, thereby being able to engage in fractional reserve banking, effectively creating new money for themselves out of your money? As Richard Striner remarks, "who among us would not enjoy the privilege of creating money through accounting tricks that are legal?" The reason they are legal is because of the influence of the banking lobby on legislators.[28]

Striner did not pursue the implication of this question, whether it might be possible for anyone to leverage their private assets to engage in fractional reserve lending. If all a bank offered were electronic accounting then, in principle, anyone else could create a platform for this purpose outside the traditional banking system. All they would need is an ability to receive and pay out funds electronically. As an account holder, not only I would receive income into my account and use it to pay expenses, but I also would have an ability to lend funds to others based on some multiple of my reserves. The success of such agent-to-agent lending would be predicated on the same relationship that supports the rest of the financial system: trust between lender and borrower. I would, in effect, become my own bank.

The implications are profound. Commercial banks could be replaced by electronic banking at the agent level, complemented by institutional financial brokerage. The system would enable the issuance of microloans too small for traditional banks. It also would provide a common platform for peer-to-peer funding, such as crowdfunding. It would become a platform for community-based finance, whether the community in question were geographically located, virtual, or some mixture of the two.

Unlike today's electronic banking, such an online platform would accommodate any number of money types, from national currencies to

[28] Striner (2015, p. 51).

electronic currencies, reward points and local exchange currencies, to name a few. The account-holder would be able to consolidate all their monies into one place. Exchange rates among the various monies would enable convertibility.

An institutional separation of deposits from lending might not be needed in this scenario because the aggregate financial risk would be spread among millions of actors rather than consolidated into a small number of large market players. In essence, money lending would revert to its ancient form, practiced by individuals rather than by large financial institutions, now enabled by global electronic communications de-locating the lending activity. At the same time, though, if fractional reserve lending proved destabilizing, too complex to regulate, or prone to consolidation into large groups, then the pure reserve banking model would have to be applied.

This scenario illustrates how fundamentally the global financial system could change over the coming decades, enabled by global communication networks and characterized by a profusion of monies. Properly regulated, peer-to-peer financial relationships hold great promise for grassroots efforts to align economic activity with natural processes, unconstrained as they would be from the policies and agenda of financial multinationals. Many ideas would fail—and would not be bailed out, of course—but a few would succeed spectacularly, and would spread through the system. Novel approaches to material discipline would emerge, hand in hand with novel approaches to financing, the overall rate of innovation within the economy increasing significantly as a result.

Policies and Legislation

A global monetary authority would provide the requisite institutional structure for two critical regulatory functions of the global economy: monetary governance and financial regulation.

Monetary Governance

Monetary governance is undertaken at least in part by the manipulation of interest rates. This is the role of today's central banks, operating at the national level. A global monetary authority would be able to assume this role over all member states: membership would be contingent upon accepting this authority. Given a sufficient number of economically large member states, the monetary governance exercised by the authority effectively would encompass the whole, interconnected, global economy.

Lest this proposition sound like a top-down imposition of austerity measures along the lines of those imposed by the European Union on Greece during the 2010s—measures which supposedly were to suppress inflation, even though the risk of inflation was low—the distinction between an institutional arrangement and a policy should be clarified. A global monetary authority's ability to adjust the money supply would be independent of whatever monetary policy the authority might exercise. While one member state might disagree with other member states about the monetary policy for its currency, the emergence of a global monetary authority in the first place would be predicated on a realization that the global money system is composed of an increasingly diverse array of monies, many of them non-sovereign. The authority would formalize co-ordination among member states on the supply of their sovereign currencies; yet, at the same time, it would enable joint regulation of the world's private monies in conjunction with private seignieurs. In effect, it would regulate the supply of many monies at the same time, each according to its needs. If a national currency ran into trouble for any reason, then the collective action of the global monetary authority would be available to stabilize it. Meanwhile, consumers and businesses would be able to transact in a variety of other currencies, as needed. In this way, one can see the benefits of their using a technology platform integrating all their forms of money.

Manipulating the money supply through interest rates is only one form of monetary governance. Another major form is through legislation. Acting through its members as a forum for establishing policy directions, a global monetary authority would establish policies for governments to enact into legislation. Policies naturally would be tailored to the level of government in question. Policies governing local currencies, for instance, would differ from those governing national currencies, and they would be designed with local assemblies in mind. Local governments likely also would have more leeway than national governments to experiment with their own monetary programs, independent of whatever the global monetary authority may recommend. This leeway would be reflected in the policymaking of the authority itself, framed, perhaps, as 'recommendations' rather than as 'requirements'. National governments, on the other hand, being in charge of currencies in wider circulation, would need to more closely co-ordinate their monetary policies to ensure consistency and stability; hence, the need for policy 'requirements' rather than recommendations.

A global monetary authority would establish requirements for membership of private monies and for their regulation. One way to prevent the establishment of 'cowboy' monies, as one might call them, operating outside the regulation of the authority, would be for member states to enact legislation requiring private seignieurs to register with the global monetary authority. While the range of potential private monies would not be restricted—to the contrary, innovation in this area would be encouraged—certain minimum standards would ensure quality, accountability, traceability, security and, above all, confidence. Like the policymaking for public monies, the maintenance of standards and the regulation of private monies would involve all stakeholders within the institution: government entities, non-governmental members and the public.

As Orrell and Chlupatý describe, private monies vary considerably in whether they can be lent. Some, such as scrips, LETS, rewards points and air miles, by their design cannot be lent. Electronic currencies, on the other hand, in principle can be lent, even though, as of the 2010s, a market for debt in these monies has not yet been established. One of the requirements for membership of a global monetary authority would be a prohibition on fractional reserve banking on the part of the issuing authority. This would prevent banks from inventing their own currencies specifically for the purpose of multiplying it through loan origination, to earn returns on the interest.[29]

The consolidation of monetary governance into a global monetary authority would enable a systemic, and systematic, consideration of policies and legislation in support of material discipline and widespread prosperity. Viewed within the context of this institutional structure and an increasingly diverse global monetary ecosystem, the concept of sovereign money creation would become much broader than merely pertaining to sovereign currency. What Huber and Robertson imagined as government control over the creation of national currency in effect would become a form of regulated money creation in which any issuing authority, operating under the oversight of a global monetary authority, would make its currency available to government entities needing funds for public programs. The flow of monies could occur at the local level, for example,

[29] Already, commercial banks are developing their own currencies, which presumably they could use for fractional reserve lending. See Orrell and Chlupatý (2016, pp. 193, 210).

from a local issuer to a local government, or at regional, national or global levels. The issuer of an electronic currency with global circulation, for instance, could make an offering of funds available to any jurisdiction in need of them. A wide variety of different monies, then, would be available at any one time. Conversion rates among them would be market-driven rather than imposed by the authority itself, in the same way as currencies fluctuate relative to one another today. The value of any form of currency, after all, is a function of the market's confidence in its issuer. Common standards established by a global monetary authority for all member currencies would at least ensure a minimum level of confidence.

Within this context, legislation enabling sovereign money creation by national governments might not be as all-important as when one looks only at national currency, as the SMC community does. Membership of a global monetary authority by any government entity would entitle it to monies issued by any member seignieur, public or private. Your local town, for instance, could pay for needed road works using rewards points. Alternately, it could create a public bank to issue a local form of money, to be used in the public interest within its jurisdiction. It would need to enact legislation enabling it to do this, of course, and the legislation would be a precondition for membership of a global monetary authority. A national government, on the other hand, already having a currency in circulation, would enact legislation enabling sovereign money creation along the lines described above. SMC would supersede the expensive and inefficient government borrowing process, enabling the central bank to issue new money directly.

Access to funds or an ability to create them, in whatever form, would create a significant incentive for government entities to join the authority. Prospective private money-issuers, by the same token, would have an incentive to join as well. Requirements for financial probity, and the penalties for violating it, would be correspondingly stringent. Monies issued by local public banks and by private issuers alike would be subject to common controls by the authority, such as to guard against inflation. A global monetary authority would become not only a global regulator for money issuance but also, simultaneously, a global clearing-house for it.

Perhaps the clearest pathway toward this future state might lie in the creation of local or regional public banks. These banks could deal not only in national currency but also in private monies or in their own money. Money creates markets, not the other way around. Money created this way would avoid the kind of financial instability associated with highly leveraged lending. A balance would need to be established between creating enough new money to form a market for programs serving local or regional needs and avoiding price inflation. To the extent that local and regional governments forming public banks would begin to enter into exchange relationships with one another, and to establish common standards and best practices for monetary control, here one would find the seed of an institution having some of the properties of a common monetary authority.

Nationalizing the Currency

Until a shared monetary language emerged, it would fall to individual national governments to reclaim their public currencies. Sovereign money creation in the US, for instance, would effectively be 'nationalizing' the dollar. Legislation or, if necessary, a constitutional amendment, would declare public ownership and management over the public currency. No longer would the US dollar be treated primarily as a private currency, for private benefit, yet backed by the people. It still would circulate within the private money system but that system would not be able to create dollars. The same legislation would replace the Federal Reserve with a National (or Public) Monetary Authority (NMA), which would be owned not by consortia of commercial banks but by the public, in whose name it would operate. It would not be part of government, although legislation would establish funding for its administration as a guaranteed appropriation from government revenues, much like the judiciary today. The NMA could operate a trust-type of structure, although it would have to be adapted to its special functions. Tasked with managing the money supply for monetary and financial stability, the NMA would have the power to issue new money, to reclaim money from circulation as needed, and to manage the flows of dollars into and out of the country. The latter power would be important as a safeguard against international currency manipulation. If necessary, the NMA would buy or sell dollars on the international markets as a way to adjust the domestic money supply.

Having nationalized the dollar, the next step would be to establish a mechanism for sovereign money creation and reclamation. The money system would have to satisfy four principal criteria:

1. An ability to meet the budgetary needs of government, subject to macroeconomic conditions, such as limiting inflation;

2. Mechanisms for reclaiming money in circulation, such as through government taxation or, if needed, directly, via 'quantitative tightening';
3. The elimination of government debt and the reduction of the government deficit to a low, manageable level;
4. The ability to provide an adequate flow of money into the commercial finance sector to support private lending.

Figure 11.1 illustrates the flows within this money system. Similar to Jackson and Dyson's depiction of sovereign money creation, this one shows flows of money among the NMA, the government, the commercial finance sector, businesses and households. It differs, however, in the ability of the NMA to tighten the money flow directly for macroeconomic reasons, either by limiting the government budget or by reclaiming funds from the commercial banks. The same mechanism applied in reverse—that is, supplying more money to the banks—is quantitative easing, an action the Federal Reserve has authority to take today. Quantitative adjustment would allow the NMA to adjust the amount of money available for private lending without affecting the government budget. It would represent a separate, parallel path of economic stimulation or dampening.

The money flows would follow the sequence illustrated in Fig. 11.2, which shows each of the sectors in Fig. 11.1 placed into a swim lane.

In any given year, the government would submit its budget to the NMA for evaluation against established macroeconomic criteria. If the government's budget request were so high as to potentially create a risk of inflation, then the NMA would recommend reducing it. However, by having the option of quantitative tightening by removing money from the

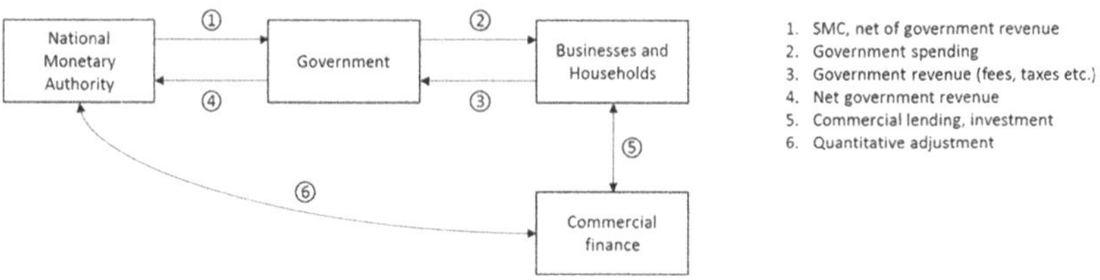

Fig. 11.1 Illustration of a US national money system established through legislation to nationalize the dollar, separate commercial deposits from lending, and replace the Federal Reserve with a National Monetary Authority. Source: Author's creation

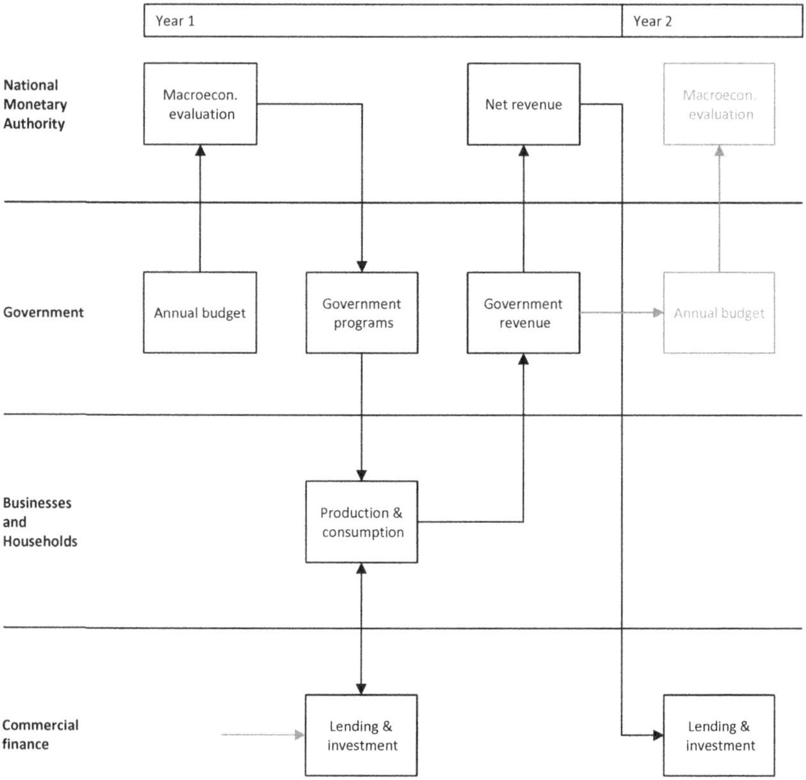

Fig. 11.2 Sequence of flows of money within a US sovereign money system. Source: Author's creation

commercial finance sector, the NMA would be able to exercise a degree of flexibility if the government's budgetary need were critical. Ultimately, the NMA would have to have the last word on the budget, in the interest of long-term economic stability.

The NMA would issue new money to the government to spend. This spending would place funds into the hands of businesses through government contracting and into households through the payment of dividends, such as basic income. This new money then would circulate within the economy through other business and consumer expenditure. That expenditure would incur taxes, such as a material intensity tax and income tax, or fees, such as a natural capacity fee. These funds would flow back to government. Government revenue less than its expenditure in any given year would create a budget deficit for that year, whereas revenue greater than expenditure would create a surplus. Any surplus would represent net revenue flowing back to the NMA. This net revenue would allow the NMA to return money into circulation through the commercial finance sector the following year, subject to macroeconomic criteria. In this way, the supply of commercial credit would be maintained, while at the same time preventing a ballooning of private debt.

Under this system, it would be acceptable for the government to maintain a small year-on-year deficit, provided it did not vary much, and the government debt were paid off. Assuming that legislation nationalizing the dollar were passed within an environment of significant government debt, as exists early in the twenty-first century, an extra quantity of new money initially would be created by the NMA to help pay down that debt, perhaps over a period of ten years or so. This in and of itself would infuse the economy with trillions of new dollars, principally through the finance sector. The infusion would soften the blow of eliminating fractional reserve banking. A projection of the government budget under these assumptions is provided in Chap. 16.

That these changes represent a fundamental structural adjustment to the national money system hardly needs emphasizing. It could take at least twenty years for its effects to fully propagate, and for the system to settle into a new, more stable rhythm. The early years might feel like a financial crisis, not least because of uncertainty the changes would generate within the financial markets. Monetary and financial regulators may have to operate for several years in crisis mode, making quick macroeconomic adjustments as issues presented themselves. It is entirely probable that trillions of dollars of value could be wiped off the US financial markets in the early going. Out there in the real economy, however, businesses would benefit from increased government investment. Long-overdue improvements to public infrastructure would be made; and households all at once would find themselves financially on a much more stable footing.

Of course, the market effects of these changes would be lessened if legislation in the US were coordinated with similar legislation in other OECD countries. Hypothetically, although perhaps not realistically, if all the OECD countries, home to the world's major stock markets, passed similar legislation around the same time then there would be few if any places for capital flight. Even if only a handful of major economic powers instituted these reforms then they would set in motion changes that very likely could lead to the global spread of sovereign money creation. Legislatures in countries having debt-based economies would find the attraction of taking back control of their currencies very compelling, particularly if it became clear that their financial sectors could become less prone to excess, their economies more stable and productive, their people more prosperous and their societies safer.

Financial Regulation

The key for all this to work would be to prevent the creation of new bank money through lending. Otherwise, commercial banks simply could circumvent the whole system, thereby undermining it and perpetuating the risk of volatility. The creation of private monies is not the issue. Banks, like any other private entity, should have an ability to become accredited seigneurs, and would do so in a future economy whose money were regulated by a global monetary authority. The risk to the financial system and to the economy at large comes from the creation of mountains of debt. Debt is destabilizing in and of itself. However, as Chap. 1 described, debt also fuels demand for consumption, which, absent any countervailing improvements in material efficiency, increases the material intensity of the economy. Debt exacerbates myopia on the part of economic actors, which hinders decision-making for the long term. It works against efforts to increase income equality. Debt, fundamentally, represents a crippling impediment to the long-term alignment of economic processes with natural ones.

Debt, in moderation, is acceptable. However, the economy of the past few hundred years has become so inebriated on debt that it cannot see beyond its nose and is in perpetual danger of falling over. If it wants a long and peaceful life, it needs to sober up.

A planetary economy would have a financial system allowing the creation of money through seigniorage but disallowing the creation of debt money. Specifically, only institutions would be able to register as seigneurs; individuals could not. Checks would be necessary to safeguard against fraud, such as to ensure that registered entities' operations were actual. Seigniorage brings responsibility and accountability for monetary stability, and for confidence in the money supply. A requirement for incorporation, whether on the part of private entities or public ones, would support these ends.

Although lending would be allowed, it would be possible only from operating funds. Banks as we know them today might cease to exist because the depository function would be so easily fulfilled through electronic means. Peer-to-peer depository functions may become the dominant pattern, to the point where anyone essentially could be their own 'bank'. Where today much economic activity is funded through bank-created debt, in this future economy it would be funded to a much greater extent through peer-to-peer investment or lending.

Would funds be harder to obtain? Not necessarily. Traditional debt would be superseded by a proliferation of other funding mechanisms, employing a profusion of currencies. The creation of new money through seigniorage would be debt-free. Of course, unchecked seigniorage could present a risk of instability to the financial system through inflation. The role of a global monetary authority would be to protect against inflation in any one, or all, of the new monies. Over time, as the world's population stabilized and the economy became materially more efficient, demand for the issuance of new money would decline: not to zero, but to a lower level than in the present growth economy. By the time this leveling-off were approached, the prevalence of debt within the economy probably would have become a modest percentage of all economic activity, as opposed to the multiples of it we see today. The concept of a government 'bail-out' of lenders—be they institutions or individuals—would be utterly alien, not only because the overall amount of debt would be low but also because lending inherently would not use leveraged assets, and because any risk of default would be spread among a large number of market participants.

One of the potentially very valuable uses of seigniorage, to be discussed in the next chapter, is to fund basic living standards. If debt money became relatively harder to obtain then individuals, families and households may need other means to cover living expenses. Peer-to-peer funding could not be relied upon for this. Its strength is as a convergence of interests, disintermediating the funding of projects, matching supply with demand. The creation of new money, on the other hand, has the potential to partially replace household debt, thereby strengthening the financial basis of the vast majority of households and narrowing the income gap.

With these considerations in mind, it is clear that legislation at the national level to separate the depository function from the lending function, as pure reserve banking would do, cannot be undertaken in isolation. It also cannot be phased in: a partial separation would be no better than no separation at all. It would have to be done in one fell swoop. Legislation would have to specify a future date on which the break would occur, far enough in the future for the financial sector to plan accordingly. A time-frame of ten years might be reasonable. Sunset provisions for existing loans would allow them to wind down naturally. New loans would be available through institutions operating in the capital markets, or from banks' operating funds. Other legislation establishing basic income schemes might be coordinated with this legislation so as to cushion the transition to a new world of reduced debt money. Finally, the concurrent enactment of legislation separating deposits from lending in multiple jurisdictions would reduce the risk of regulatory arbitrage on the part of finance multinationals.

In parallel, legislation at the national level to establish the parameters of monetary seigniorage would create clear guidelines for the creation of new monies that could fill many of the roles currently played by debt money. Registration, accreditation and regulation of seignieurs would form a clearly defined legal framework. It should be acknowledged, though, that public institutions might be subject to somewhat different requirements from private ones, simply because of their tax-exempt status and incorporation in the public interest. As with the separation of deposits from lending, there would be benefits in national governments coordinating with one another on legislation for monetary seigniorage, not only as a bulwark against capital flight but also for regulatory consistency.

A parallel legislative track would nationalize the financial rating process. Allowing private rating agencies to offer cozy credit ratings to private financial institutions is perverse and fundamentally not in the common interest. The past destruction wrought by private finance to society through its serial irresponsibility requires its institutions to be assessed independently in the future. Public entities, such as municipal and regional governments, would welcome this change, as their creditworthiness would be measured on a level playing field with private institutions. A national rating agency could be trusted not to show favoritism to anyone.

These kinds of legislations can be enacted in advance of the emergence of a global monetary authority, or similar institution; in fact, they would represent the first steps toward one. The proof of the concept would rest upon the success of the early adopters in stabilizing their financial systems, enabling long-term planning for material discipline, stimulating innovation, job creation and economic growth, and broadening the basis of prosperity among their people.

References

Dodd, N. (2014). *The Social Life of Money*. Princeton: Princeton University Press.

Hardt, A., & Negri, M. (2009). *Commonwealth*. London: Harvard University Press.

Hayek, F. A. (1978). *Denationalisation of Money: The Argument Refined*. London: Institute of Economic Affairs.

Huber, J., & Robertson, J. (2000). *Creating New Money: A Monetary Reform for the Information Age*. London: Report of the New Economics Foundation.

Jackson, A., & Dyson, B. (2013). *Sovereign Money*. London: Positive Money.

Joób, M. (2014). The Sovereign Money Initiative in Switzerland. *World Economics Association Newsletter, 4*(3), 6–7.

Kelly, M. (2012). *Owning Our Future: The Emerging Ownership Revolution*. San Francisco, CA: Berrett-Koehler.

Levitin, A. J. (2016). Safe Banking: Finance and Democracy. *University of Chicago Law Review, 83*(1), 357–455.

Mellor, M. (2016). *Debt or Democracy: Public Money for Sustainability and Social Justice*. London: Pluto Press.

Orrell, D., & Chlupatý, R. (2016). *The Evolution of Money*. New York: Columbia University Press.

Panayotakis, C. (2011). *Remaking Scarcity: From Capitalist Inefficiency to Economic Democracy*. London: Pluto Press.

Sigurjonsson, F. (2015). *Monetary Reform: A Better Monetary System for Iceland. Report Commissioned by the Prime Minister of Iceland*. Iceland: Reykjavik.

Striner, R. (2015). *How America Can Spend Its Way Back to Greatness: A Guide to Monetary Reform*. Santa Barbara, CA: Praeger.

Wray, L. R. (2011). *Keynes After 75 Years: Rethinking Money as a Public Monopoly*. Working Paper no. 658, Levy Economics Institute of Bard College, Annandale-on-Hudson, New York.

CHAPTER 12

Economic Controls 1: Principles and Requirements

The preceding chapters on norms, institutions, policies, accounting, money and finance lay the foundations for an examination of economic controls for a planetary economy. Controls enact policy through mechanisms designed to guide the economy toward one or more normative goals, progress toward them being measured through systems of accounts.

The meaning of 'control' here is not mechanistic but adaptive. Economic controls exert selective pressures upon economic agents, encouraging certain kinds of choices over others. Although the kinds of controls needed for a planetary economy are largely lacking in today's growth economy, the specific instruments outlined in this chapter should not be taken as an exhaustive set. They are instead examples of the kinds of controls that may help to initiate the process. Because the economy is an evolutionary system, its suite of social technologies, including economic controls, will evolve over time, in ways that cannot be anticipated today.

This chapter begins with general principles that economic controls must satisfy over the decades to come. It then summarizes the requirements for economic controls. From there, it dives into new currency creation, picking up from the previous chapter. The next chapter addresses fees, and the one after that taxation. The fourth and final chapter on controls examines subsidies, incentives and market instruments. The focus is on market economies, these predominating today and likely into the future.

F. Murison Smith, *A Planetary Economy*,
https://doi.org/10.1007/978-3-030-49296-0_12

General Principles

From the normative requirements of Chap. 3, it is possible to identify four general principles for economic controls for a planetary economy: (1) material discipline; (2) economic stability; (3) widespread prosperity and (4) encouragement of innovation and selection through markets, where applicable.

(1) Material Discipline

Material discipline is the principle of minimizing the economy's material intensity and maximizing its material recirculation. Economic controls conforming to this principle would encourage economic processes that: (a) minimize resource extraction from nature, or (b) maximize the material efficiency of economic production or consumption, or (c) minimize effluent flows to nature. The best controls, in principle, would accomplish all three.

The purpose of material discipline is to minimize the risk of unexpected effects of the economy and nature upon one other. Although the interactions of the economy and nature are unpredictable, all these exchanges are nonetheless material in nature. Irrespective of the specific kinds of interactions, reducing the material intensity of the economy would reduce the risk of deleterious feedbacks over time. The quantity of material outflows from the economy can be immediately reduced, leaving aside the question of the quality of outputs and their uncertain effects upon nature.

The effect of an instrument for material discipline would be to steer material flows toward recirculation within the economy. The instrument would operate as a kind of inoculation against a potentially deleterious material flow, analogous to the way a small dose of an antigen stimulates the body's production of antibodies. Once antibodies have been made, the body can produce them at any time in response to another infection by that same antigen. This is biochemical memory. Similarly, an economic instrument would redirect an initially small flow of effluent away from discharge into nature toward recirculation and processing within the economy, just enough to stimulate the necessary innovations. Once the economy has innovated the requisite technology, it has the capacity to apply it at increasing scales: a form of 'technological memory'. The size and scope of the instrument can be scaled up to process ever-larger flows of the effluent in this way. Not all new technologies scale easily, of course, but the analogy is not seriously weakened by this qualification.

Although some kinds of output will show stubborn resistance to the development of technologies for internal processing, this is not a critical issue as long as the overall volume of output declines over time. "Pollution", Stephen Marglin reminds us, "is not a commodity."[1] Material recirculated within the economy, as avoided pollution, can become a commodity. It has monetary or other economic value. That value can be enhanced relative to pollution through the application of economic instruments.

(2) Economic Stability

The principle of stability relates both to material flows and to financial flows. It does not mean that these flows would remain constant; otherwise, it would violate the principle of material discipline. It means that improvements in material efficiency over time would attempt to minimize volatility in material efficiency, as well as in such financial indicators as prices, inflation and income distribution. To seek a monotonic series of improvements is, of course, unrealistic: a degree of volatility is inevitable, and perhaps is to be encouraged, for stimulating innovation. However, excessive volatility hinders the kind of long-term view so essential to this program. On a calm sea, the horizon is distant, whereas on a rough sea, one can see only the next wave. By following a directive of material discipline, the economy will evolve toward relative material stability, as well as toward relative financial stability. Progress along this path would be accomplished all the more quickly if the path itself were kept as smooth as possible.

(3) Widespread Prosperity

The principle of widespread prosperity is similar to, but also somewhat distinct from, the conventional economic term of distributional equity. Distributional equity refers to a fair distribution of income, in money terms. What counts as 'fair' is, of course, a matter of judgement, and the subject of much political debate. 'Widespread prosperity' is a broader term. 'Prosperity' takes many forms: financial, material and temporal. Modernity, championed by conventional economics and the policymaking

[1] Marglin (2008, p. 283).

derived from it, asserts that financial success is a prerequisite for material and temporal prosperity. Temporal prosperity is having plenty of time to call your own. The problem, as the left has pointed out for decades, is that anyone who retires rich does so at the expense of many others who do not. Yet, as the right argues, if one's riches were taxed away, there would remain little incentive to make money, and the economic engine would slow.

Under a principle of widespread prosperity, financial, material and temporal prosperity would become to an extent decoupled from one another. Financial wealth would no longer be a prerequisite for material or temporal wealth. This means that even people on low financial incomes would have sufficient material and temporal resources to be economically productive, whether in the labor market, as family caregivers, or in any other capacity. The distribution of prosperity, however measured, could still be relatively wide, provided that a lower bound on prosperity—not only financially but also materially and temporally—were maintained. Provided this floor were kept in place, there would be no need for a formal upper bound on prosperity, however measured. The requirement to maintain the floor will, by itself, force the economy to establish a range of prosperity it can support.

(4) Innovation Through Markets

Innovation through markets is a fourth principle to which economic controls to encourage alignment with natural processes would conform. Markets are the manifestation of the adaptive, evolutionary nature of economic processes. Markets transact not only in money but also in other currencies. These may be quantitative, such as materials or energy, or qualitative, such as ideas or relationships. Diplomacy, for instance, operates as a market of goodwill among nation-states. Market operation is driven by a variety of forces, one of the most important being strong reciprocity, according to recent research. People make decisions based as much on fairness as on maximization.

Markets, being adaptive and evolutionary, have a built-in capacity to innovate, in the same way that genetic variety, in combination with differential survival and reproduction, produces novelty and change within natural populations. The selection pressures upon natural communities come mostly from the natural environment, whereas those upon markets come mostly from their economic, social and policy environments.[2] Economic controls to encourage alignment with natural processes would translate selection pressures from policymaking and social norms into changes in the way transactions are carried out, manifesting in aggregate changes to market behavior.

Innovation in response such to selection pressures would not only produce changes to market behavior but also could result in innovations to the controls themselves. An undesirable economic control would be one that stifles innovation, because it is through innovation that the economy would evolve toward alignment with natural processes. In the market of economic controls, one that stifles innovation would be selected out of existence in favor of those encouraging it.

Quid Pro Quo *with Nature*

All four of these principles represent facets of a *quid pro quo* with nature. Material efficiency minimizes the economy's material intensity upon nature; stability supports decision-making for the long term, resulting in greater congruence with natural timescales; widespread prosperity is essential for substantially full participation among economic actors in an economy aligned with natural processes; and innovation through markets mirrors the mechanisms of change in living systems. A *quid pro quo* with nature, then, can be seen as an overarching principle for economic controls encouraging alignment with nature. Economic controls that encourage natural regeneration, or at a minimum do not discourage it, while at the same time encouraging (or at least not discouraging) innovation within the economy, would be selected for within the technological space of all possible economic controls at any point in time. Further, innovations to economic controls themselves would be selected for if they satisfy the *quid pro quo* principle.

In essence, the modern market economy must treat nature well in order for the favor to be returned. It is an application of the pre-modern participatory worldview to the development of instruments for the bounding and guidance of the world's interconnected markets. For the economy to

[2] Mostly but not entirely. The essence of the coevolution of the economy and nature is that the human economy exerts artificial selection pressures upon natural communities, and nature exerts selection pressures upon the economy.

become well aligned with nature over the long term, the controls in operation must change from being exploitative to being generative.

Howard Odum offered some proposals along these lines. Beginning from the premise that natural services are not free, he asked how the economy might pay natural networks for the services they provide. "If man withdraws materials and energy [from a system] without paying for them in receivable currency", Odum wrote, "he has done the opposite of rewarding loops; he has broken loops and punished the loop he finds useful." The 'receivable currency' used to pay nature has to be material or energetic. Trees do not accept credit cards. The mechanisms at play within the economy must be able to convert between economic value and natural value. "If, however, man pays nature in a feedback reward to that loop, and the reward has as much energy-benefiting effect as he has withdrawn, and especially if he closes the loops, he will ensure the survival of the product of his interest."[3]

Managing the material or energetic interface between the economy and nature for the benefit of both is called stewardship. Closing resource loops necessitates "using man's waste flows to stimulate circuits of his interest"[4] or, if a waste flow cannot be transformed into a circuit of interest, eliminating it altogether and selecting for substitutes.

Odum's principle of coupling people and nature in reward loops using money flows is a powerful one for guiding the design of economic instruments in a crowded world. When people extract crude oil for motor fuels, for example, they do not reward nature for this extraction but instead punish it with the emission of pollutants from the fuels' combustion. Rewarding the loops would entail establishing flows of money in the opposite direction from the material flows, thereby placing a higher valuation upon them and encouraging the development of substitutes.

Common Capacity Ownership and Rent

The *quid pro quo* with nature, then, has two overlapping elements. One is the lowering of the economy's material intensity and the other is promoting and encouraging natural regeneration. Since natural networks are not free, rewarding them represents a payment of rent. The only difference between this natural rent and social rent is that natural rent is paid in a

[3] Odum (1971, pp. 289–290).
[4] Odum (1971, p. 290).

currency receivable by nature: materials or energy. Any diminution in the material intensity of the economy reduces the amount of rent payable.

In addressing institutional requirements for a planetary economy, Chap. 8 discussed how property ownership might evolve to recognize, encompass and maintain natural capacity. Well-developed property rights are essential to the delineation of ownership, and therefore of responsibility for natural capacity. A property owner's decisions about the use of biota within their property boundaries are in the interests of all people, even if in some nominal way. The same reasoning applies to the atmosphere above and the strata below. The natural capacity established by these natural resources therefore is a common asset.

Within the economy, other common assets exist, these being artificial rather than natural. Those with global extent include the internet and the financial markets. Sub-global ones include physical infrastructure, such as highways, railways, airspace and electrical grids, which may extend across national borders. Regional or local ones include water supply systems and electromagnetic spectra. They are all regulated by various levels of government, this being a proper function of government.

Ideally, such regulation should be conducted on behalf of all people, as the stakeholders in these common assets. In practice, this is not always the case. In countries ruled by autocrats, the infrastructure is regulated not on behalf of the people but on behalf of the régime in power. Infrastructure may still benefit the citizens of these nations but they have no claim of ownership over it, unless of course they were to topple the despot. The distinction is important because without a clearly defined role for government, there may be no way to establish a mechanism for the disbursal of commons rent. In countries where the government operates in the service of the people, or at least is constitutionally required to do so, such a mechanism can, in principle, be established. The common ownership of common resources or assets is predicated upon the establishment of democratic institutions requiring the government to recognize any rent extracted from their use as belonging to the people, and to disburse it accordingly.

Role of Government as Steward of Capacity

The idea of common resource ownership can be extended to allow a specific role for government to be defined. Economic infrastructure, in the form of communications networks, the electromagnetic spectrum, utilities and transportation networks, all are forms of artificial capacity. As such, they are able to accommodate a certain volume of business per unit time.

They rely on, and are complements to, Earth's natural capacity. Accepting that artificial capacity is commonly owned, the institution representing the common interest should be the one protecting and managing this commonly owned capacity. That institution is government. In other words, government would have jurisdiction over this capacity on behalf of the people, and would be responsible for its stewardship.

This is why Barnes and Murphey proposed their dividend schemes to be administered by government, as the Alaska Permanent Fund is today, or by a trust established through legislation. Because the State of Alaska constitutionally represents the people of Alaska, it has the ability to collect fees from corporations on the behalf of Alaska's people for the use of its commonly owned capacity. Having jurisdiction over this capacity gives government the necessary agency to levy the fees. As such, they constitute a form of rent.

From this line of reasoning, a general principle emerges for the role of government, not only on a crowded planet but also in just about any civilized economy operating within a democracy:

> A key role of government is to rent physical capacity upon which enterprise generates wealth.

Physical capacity could be artificial or natural, or both. The 'enterprise' in question may be private enterprise, public enterprise—such as municipal government—or individual enterprise. Wealth thus generated may be financial, physical or other non-physical wealth. When you pay a fee to enter a national park, you pay rent to the government for the experience of the park, which is a form of wealth.

Going one stage further, it is then possible to define what the wealth thus generated would do within a planetary economy:

> The characteristic of a planetary economy would be that the wealth generated by enterprise would maintain the physical capacity—in particular, the natural capacity—rented by government.

As the agreed-upon steward of natural capacity, government would rent access to it, even though it would not hold title to it. It would do so in its capacity as agent for nature. The incentive structures operating within the economy would enable government to continue offering natural capacity for rent essentially into perpetuity, because they would reward forms of enterprise that support natural capacity and penalize those that do not.

Although this outcome could be accomplished by a non-democratic government—if it served the interests of all rather than the people who run it—in practice, a non-democratic structure probably would be less stable in the long run than a participatory democracy. It probably would gravitate toward one.

Function of Economic Instruments

Economic controls are exerted through a variety of instruments, such as taxes, incentives and subsidies. The function of these controls is to provide signals of value, distilling the impractically complex into the actionable. Whether one is willing to pay a couple of dollars more at the supermarket for organic produce is a function not only of relative prices but also of other information about the relative merits of organic and non-organic produce, and of the prevailing norms within the community. What economic instruments can do is to tip the balance of price in favor of one type of good or service or another. Tipping the balance, such as by adjusting the taxes and incentives on organic and non-organic food production, is a policy decision.

Since the economy is a network of actors whose information about natural capacity is incomplete, and whose individual, short-term motivations may sometimes fall at odds with a long-term prime directive of material discipline, the instruments at work within the economy must be able not only to improve material efficiency and reward natural networks, but also to help align individual preferences with material discipline. In this way, preferences would not rely upon an individual's evaluation of natural capacity, which is virtually impossible, but only upon the signals being transmitted through economic instruments into actionable information.

Price is the most obvious of these but is not the only one. Preferences are expressed also through such other kinds of information as certifications, actual or perceived quality, reusability, recyclability, corporate reputation, and availability of spare parts, service or support.

Functional Criteria

It is not a single instrument but suites of instruments working in concert that can address the various attributes of a good or service, such as price, certification or reusability. A well-functioning suite of instruments should deliver a selection of products or services that meets consumers' needs and supports a *quid pro quo* with nature. Neither a consumer's selection nor the outcome thereof need be 'optimal' in a Pareto sense; they only need to be *adequate* to adapt to the prevailing selection pressures within the economy. If consumer selection or market outcomes could be improved, then a suite of instruments would face economic selection pressures to evolve in that direction. Such evolution either could take the form of a change to the value of an instrument or a wholesale 'speciation' into a new instrument—that is, an innovation of a new social technology.

Seen through the lens of adaptive evolution, the development of economic controls for material discipline as collections of economic instruments is a process of iteration under selection pressures imposed by policy. The adaptive criterion of adequacy in the face of selection pressures allows instruments to be simple and robust. The system would be set up with a beginning suite of rules and be allowed to iterate on those rules. As the system changes, the rules would be adjusted or old ones discarded and new ones introduced.

With new social technologies emerge new selective measures. By analogy to nature, the evolution of birds from dinosaurs resulted in the emergence of a whole new family of selective measures related to feather structure, wing profile, thoracic musculature, and so on. Within the economy, the emergence of the worldwide web in the 1990s resulted in the emergence of such new selective measures as website visits, mouse-overs and click-throughs.

In general, then, any adaptive, evolutionary system can be described as consisting of populations of actors whose decision-making is inductive—selection followed by evaluation—under a collection of rules, which also may change over time.[5] The instruments having the greatest effect in the service of some overarching directive would satisfy certain functional criteria, such as simplicity, robustness and adaptability. They also should be implementable within their intended economic community, and they should be integratable with other instruments acting upon complementary attributes, thereby forming a coordinated suite exerting the desired selection pressure.

[5] A foundational text in this area is Holland et al. (1986).

Summary of Requirements

The controls summarized below necessitate a great deal of additional research, particularly in their interactions. By introducing them, this chapter and the next three seek to prompt the development of potentially better ones, from which a coordinated suite of instruments would be put into action. Table 12.1 summarizes the instruments and their applicability to end purposes.

If these particular instruments were integrated into a suite of measures in support of the four principles, then they might have the characteristics listed below. The instruments apply primarily to OECD countries and other high-income nations with well-developed institutions, these being the ones that can most immediately turn the global economy toward long-term stability and prosperity:

Table 12.1 Matrix of instruments described in this chapter and the three to follow, showing their applicability to government expenditure, direct natural capacity improvement (such as ecological restoration), and the four principles of controls for a planetary economy

Instrument	General government expenditure	Direct natural capacity improvement	(1) Material discipline	(2) Economic stability	(3) Widespread prosperity	(4) Innovation through markets
New currency creation						
Common capacity fees on artificial assets	(N/A)					
Natural capacity fee						
Material intensity tax						
Reductions in income & corporate taxes	(N/A)					
Property tax reform for natural capacity	(N/A)					
Subsidies / incentives for natural capacity / material efficiency	(N/A)					
Tradable quotas	(N/A)					
Natural capacity share-ownership	(N/A)					

Source: Author's creation

- Sovereign money creation, giving government control of the money supply for government programs and reducing the burden on society of financial market instability.
- Common capacity fees charged to corporations for the use of artificial capacity, the proceeds from which would be predicated to a basic living program.
- A natural capacity fee charged to corporations, and on financial trades for the use of natural capacity, the proceeds from which would be predicated to government-funded programs to improve natural capacity.
- A material intensity tax levied on all goods and services passing through the economy, calculated from reported, auditable, material intensity.
- Reductions in income and corporate taxes, along with an elective redistribution scheme applied to households on higher incomes.
- Reform to property taxes, scaling them according to assessed maintenance or improvement of natural capacity within property lines.
- Shifting of subsidies and incentives from extractive to generative industries and practices.
- Expansion of tradable quota markets to global scale, followed by eventual phase-out.
- Establishment of markets for natural capacity share-ownership, paying dividends in tradable natural capacity units.

Beyond these general requirements, any given instrument must be implementable within the economic environment prevailing at any time. The measures above can be put into action within the economy of the early twenty-first century. Chapter 16 will outline how they might become established and evolve over the long term to steer the economy toward alignment with nature.

Each instrument examined in this chapter and the next two addresses one or more of the four principles to varying degrees, as Table 12.1 summarizes. To save having to refer back to Table 12.1, each section provides a miniature reference table summarizing that instrument's emphasis relative to the four principles.

References

Holland, J. H., et al. (1986). *Induction: Processes of Inference, Learning and Discovery*. Cambridge, MA: MIT Press.

Marglin, S. (2008). *The Dismal Science: How Thinking Like an Economist Undermines Community*. Harvard University Press.

Odum, H. T. (1971). *Environment, Power & Society*. New York: Wiley-Interscience.

CHAPTER 13

Economic Controls 2: Currency and Fees

New Currency Creation

New currency creation holds promise as a vehicle for debt-free government investment and control of the money supply. As Chap. 11 discussed, the investment could take a variety of forms. Although new currency could be invested in government programs, it equally could be invested directly in citizens as basic income. The concept is astoundingly simple: create new money and give it in equal amounts to all people. Most versions of the basic income idea define 'all people' as every adult and child living within a monetary area who satisfy certain residency or citizenship requirements (Table 13.1).

This definition usually has implied that the money-issuing authority would be a national government and that the money in question would be the national currency as legal tender. Orrell and Chuplatý suggest instead that the government could issue a complementary currency. Additionally, a handful of privately funded basic income schemes have been piloted, such as in Oakland, California. Neither of these ideas is particularly compelling. A private scheme, when scaled up, would not achieve any public benefit if it were exclusionary. It would have to fund everybody. Yet it would offer no obvious return to a private funder, in contrast to a public one, which could be returned through taxes. If a private scheme issued a private currency, then that currency would have to be able to buy all the things the national currency could. The national currency is already accepted everywhere within national borders, and it is the currency the

F. Murison Smith, *A Planetary Economy*,
https://doi.org/10.1007/978-3-030-49296-0_13

Table 13.1 Approximate conformance of new currency creation to the four principles of economic controls for a planetary economy

Material discipline	
Economic stability	
Widespread prosperity	
Innovation through markets	

Source: Author's creation

Note: Darker shading denotes relatively greater conformance

government uses to collect taxes. Issuing a private or complementary currency for basic income does not appear to offer any advantages over the national currency, whereas issuing it in the national currency offers clear advantages over a complementary currency. A national basic income scheme funded through the creation of national currency could lay the foundation for a future global one, if or when a global money-issuing authority were established. Within that framework, other private or complementary basic income schemes might be feasible.

Sovereign money creation would require the government to take back control of the public money on behalf of the public, precluding commercial banks from creating new public money as debt. Government expenditure would have a stimulative effect, putting people to work and placing money into their pockets as wages. Economic activity would grow, both on the supply side and on the demand side. Basic income, on the other hand, would stimulate only the demand side of the economy. Either way, the resulting economic growth would increase the economy's material intensity in the absence of any countervailing improvements in material efficiency.

Sovereign money-funded basic income could be inflationary. For goods whose supply is relatively price-inelastic, an increase in demand for consumption would drive prices up. Could basic income become self-defeating if it raised prices proportionately? In a world of higher prices, would low-income people on basic income be just as poorly off as before?

The question cannot be considered in isolation. An economy on a pathway toward material efficiency would be different in several ways from the current economy. A much greater amount of production would come from the recirculation of material rather than from its extraction from nature. Other controls, such as taxes and incentives, would operate within the economy to stimulate this change. It would result in a great deal of technological innovation which, it is fair to say, would release a significant amount of new latency in production. Therefore, production can be expected to be fairly elastic in response to increases in demand for consumption. Additionally, as discussed below, taxes on households would shift away from income onto material intensity. People living at or just above the basic income level would pay no income tax at all. However, prices on certain goods would be higher, depending on the material intensity of their production. One might expect food also to be pricier, although agricultural incentives could offset this effect.

Sovereign money creation can improve people's net incomes in ways other than simply providing basic income. It also can pay for services used and needed by everyone, particularly healthcare and education, discussed later. If these services were taken as a public right, then it would become in the public interest to fund them through the creation of new public money. The removal of the costs of these services from household budgets amounts to an indirect form of additional income which, of course, would deliver the greatest benefit to those least able to pay for them.

Counterbalancing the inflationary effects of sovereign money creation is the removal of public money from circulation. This the government would accomplish through taxation, as Mellor and others have described. Having created new money, and placed it into the hands of the people, the government would recoup most of it after it has circulated within the economy. If basic income at a level of, say, $15,000 per year for every adult and child in the US were funded solely through sovereign money creation, then the money-issuing authority would have to issue roughly $4.5 trillion in new money each year. Most of this new money subsequently would have to be mopped up through taxation. It does not really matter what form of taxation the government used: the collection itself would have a deflationary effect. Any excess money not returned by tax would be a benefit to the economy as a whole.

Another related economic control to consider within this context pertains to income distributions. Basic income, at a stroke, would eliminate absolute poverty by bringing everyone up to a level at which needs are met and a minimum of personal dignity is assured. However, if incomes remain widely distributed and, in particular, the median income is many multiples of the basic income, then an upward pressure on consumption would be created. A prime directive of material discipline implies an increase in prosperity, as distinct from an increase in consumption. In the current economy, the two have become conflated: prosperity partly is a function of consumption. However, as earlier chapters have argued, material consumption would be only one facet of prosperity in an economy operating for the purpose of material discipline. To the extent that prosperity becomes decoupled from consumption, the economy overall would have a greater chance of attaining the material efficiency required for long-term alignment with nature. For people, this translates into a *sense* of prosperity: of not feeling an urgent need to strive upward in material consumption to have a sense of belonging within society. It is itself a form of personal dignity. Leaving aside all the other non-material facets of prosperity, such as more time, better education and free healthcare, the sense of already belonging to society, even at or just above basic income, can be enhanced by establishing economic controls distributing incomes appropriately. It is perhaps less important for the upper tail of the income distribution to be brought down than for the median and mean incomes to be maintained within a few multiples of the basic one. This is not a matter of taxing middle incomes but of maintaining income tax rates on higher ones, possibly also instituting some form of elective redistribution scheme.

The combination of these various policies and controls would have to be modeled in detail to gain a clearer picture of where the gains and losses would lie. It is a rich seam for research. Sovereign money creation may be a feasible funding source for basic living, provided it is established as part of a suite of measures. However, for reasons discussed below, a basic living scheme might more feasibly be funded through corporate fees on commonly owned assets.

Common Capacity Fees

Table 13.2 indicates the approximate conformance of common capacity fees to the four principles of economic controls for a planetary economy. Common capacity fees are divided into artificial capacity fees and natural capacity fees.

Table 13.2 Approximate conformance of common capacity fees to the four principles of economic controls for a planetary economy

Material discipline	
Economic stability	
Widespread prosperity	
Innovation through markets	

Source: Author's creation

Note: Darker shading denotes relatively greater conformance

Artificial Capacity Fees

Peter Barnes' proposal to charge corporations small fees to use commonly owned assets, introduced in Chap. 8, was made for the purpose of paying a dividend to individuals as a form of basic income. Barnes provides some rough estimates of the scale of the annual income that these fees could generate in the US for selected assets, summarized in Table 13.3.

All these assets constitute types of capacity. Because they are commonly owned, the fees charged on them may be called common capacity fees.

The transaction fees for use of financial infrastructure are defined by Barnes as payable by an asset owner on a trade made less than one year since the previous trade. A possible refinement to this fee structure could account for differences in temporality between different kinds of financial assets. Derivatives, for instance, tend to trade over much shorter intervals than stocks and bonds. A progressive fee structure, although not as simple, would be more finely tuned to the properties of the assets in question, as Table 13.4 suggests.

Although the fees themselves would be small, this progressive structure would discourage short-term speculation, particularly in derivatives, and would encourage longer-term investment. The fees would have to be modeled and trialed before they could be implemented, of course, so as to arrive at workable values. According to The Money Project, the world's 60 major

Table 13.3 Summary of annual income estimates from fees on co-owned wealth, for selected assets in the US

Asset	*Mechanism*	*Rate*		*Estimated total*
Air	Tradable permits			$198 billion
Financial infrastructure	Transaction fees	Stocks	0.5%	$62 billion
		Bonds	0.15%	$170 billion
		Derivatives	0.05%	$120 billion
Intellectual property protection	20% value-added fee			$320 billion
Electromagnetic spectrum	20% value-added fee			$84 billion
Total				$954 billion

Source: Author's creation, using data from Barnes (2014, pp. 143–148)

Table 13.4 Illustrative structure of a progressive financial infrastructure fee levied on trades of different intervals since the previous trade in the same asset by the same owner

Time interval	*Financial infrastructure fee (%)*		
	Stocks	*Bonds*	*Derivatives*
≥ 10 years	0	0	0
10 years or less	0.1	0.03	0
1 year or less	0.5	0.15	0.01
1 month or less	1	0.5	0.05
1 week or less	1	1	0.1
1 day or less	1	1	0.5
1 hour or less	1	1	1
1 minute or less	1	1	5

Source: Author's creation

Note: Fee is expressed as a percentage of the value of the trade

stock exchanges traded a total of $69 trillion in 2015.[1] If Barnes' total of $230 billion from fees on stock and bond trades in the US represented roughly half of the potential worldwide revenue—say, $500 billion—then that amount would equate to a weighted average fee of about 0.7 percent, consistent with the range of the values in Table 13.4. The global volume of derivative trading on registered exchanges in 2016 totaled more than 25 billion trades.[2] If the global annual revenue from financial infrastructure fees on derivatives were, say, $300 billion—two-and-a-half times the US estimate—then the per-trade fee would be about $12.

Provided the fees were small enough, the risk of financial firms relocating to countries where the fees were not charged would be minimal, because the cost of relocation would outweigh the cost of the fees. The spread of adoption of the fees to more and more jurisdictions would set the stage for supranational administration. A planetary trust, in conjunction with a supranational operating partner, such as the International Monetary Fund, could accomplish this. In the more distant future, a global monetary authority could take on this role, perhaps as a successor to the IMF.

Barnes' revenue analysis does not include many other kinds of capacity, such as the internet, publicly owned utility infrastructures, roads, railways, other transportation assets, or natural capacity. Estimating such revenues,

[1] http://money.visualcapitalist.com/all-of-the-worlds-stock-exchanges-by-size/.
[2] Acworth (2017).

Table 13.5 Estimate of annual income from common capacity fees on assets not reviewed by Barnes (2014)

Infrastructure asset	*Fee levied on*	*Rate*	*Unit*	*Revenue ($ bil.)*
Water delivery	Volume delivered	$0.01	Gallon	58
Wastewater systems	Volume processed	$0.01	Gallon	58
Electric grid	kWh delivered	$0.01	kWh	234
Gas pipelines	Cubic feet delivered	$0.001	CF	227
Telephone lines & cell networks	Call minutes	$0.01	Minute	39
Internet (fixed & mobile)	Data use	$0.001	MB	2099
Roads	Miles traveled	$0.08	Mile	24
Railways	Freight car miles	$1.00	Mile	36
Waterways and ports	Tons of cargo	$10.00	Short ton	23
Air traffic control	Per outbound flight	$500.00	Departure	4
Total				2805

Source: Author's creation

he claims, is complex.[3] Yet a brief survey of the volumes of traffic on these infrastructures suggests that small charges levied on their use could yield close to an additional $3 trillion in the US alone, as Table 13.5 shows. The Appendix to this chapter describes how the totals are derived.

An advantage of charging a common capacity fee is that it helps to secure the integrity of the underlying asset. The asset becomes something valued because its use carries a cost, even a small one. In turn, the revenue derived from the fee would serve one or more common purposes, whether the support of natural capacity or the establishment of a broad basis of prosperity. People would acquire a vested interest in preserving the integrity of the underlying assets because the fees now provide a vital income stream. It is a form of what one could call common asset security.

Public ownership of these infrastructure assets would not be a prerequisite for the implementation of common capacity fees. It is commonplace for federal or state regulators to impose small fees on customers' utility bills where private utilities hold a natural monopoly on delivery infrastructures. These fees cover costs of regulation and of state or federal programs in the public interest. Their existence recognizes that the operation of a natural monopoly is both a privilege and a responsibility. Similarly, the revenue from common capacity fees would cover the government's

[3] Barnes (2014, pp. 147–148).

administrative costs which, given the scale of the potential revenue, would amount to a tiny fraction.

The possibility exists, of course, that businesses might pass on the cost of these fees on to their customers. The selection of fee rates would have to strike a balance between raising revenue and enabling companies to absorb the fees within their overall costs of doing business. If necessary, legislation would require companies to absorb the fees.

Natural Capacity Fee

One type of common capacity not listed in Table 13.5 is natural capacity. Mechanisms for collecting fees on artificial common assets are tractable, in the sense that the value of a financial trade, a kilowatt-hour of electricity or a megabyte of data can easily be measured. Natural capacity, however, does not have a single unit of measure. Nonetheless, a natural capacity fee still can be collected, such as from corporate income and from financial trades, the latter mirroring the artificial capacity fee. It would be predicated specifically to improving natural capacity, such as through pollution abatement and ecological restoration. Any funds predicated to improving natural capacity would benefit it so, as long as the fees were small enough to be absorbed into operational costs, they would promote common asset security.

As with artificial capacity fees, a natural capacity fee would be administered by a planetary trust. However, the two flows of revenue would be kept distinct, to maintain predication. The trust could readily be established at the national level by legislation and then, once enough countries had established one, expanded globally.

Natural Capacity Fees on Corporate Income

Because larger businesses, on average, should be expected to have greater effects upon natural capacity than smaller ones, a natural capacity fee on corporate income should scale with income. It also could be slightly progressive. The rule of diminishing marginal returns implies that larger businesses will have greater leeway to pay larger fees. Table 13.7 in the Appendix to this chapter illustrates a schedule of fees and potential revenue using data on US businesses. The fees range from 0.5 percent of net income for small businesses up to about 1 percent for large ones. Total revenue, based on 2013 data, would be about $18 billion per year.

Natural Capacity Fees on Financial Trades

If natural capacity fees were levied on financial trades, then any given financial trade would be subject to two fees: an artificial capacity fee predicated to a basic living program and a natural capacity fee predicated to the maintenance and improvement of natural capacity. Although financial trading does not make direct use of natural capacity, the justification for levying such a fee on financial trades is that underlying assets—corporations, whose stock is traded—make use of natural capacity. Within today's economy, growth in the valuation of financial assets is built upon the exploitation of natural capacity. Charging small fees on financial asset trades would provide a mechanism for a small percentage of that value growth to be returned to nature.

Fees can be designed to deliver a comparable scale of revenue to that obtained from artificial capacity fees, in the low hundreds of billions of dollars a year for the US on present volumes. This is an order of magnitude larger than the potential natural capacity fee revenue from corporations. The total value of stock trades on the New York Stock Exchange and the Nasdaq, which host most of the stock and bond trades in US markets, was about $26 trillion in 2015.[4] For the sake of illustration, suppose that the average fee rate on a stock or bond trade were 0.2 percent of its value. Natural capacity fee revenue from stock and bond trades then would be about $52 billion. On the derivatives side, about eight billion trades took place in US markets in 2016.[5] A per-trade fee of $10, comparable with the above estimate for artificial capacity fees, would yield revenue of about $80 billion, bringing the total to $132 billion.

Uses of Common Capacity Fees

What distinguishes a fee from a tax, generally speaking, is that a fee is predicated, whereas a tax is not. Although revenue from all the various fees could in principle be pooled to fund a combination of common benefits, it would be more appropriate to predicate the natural capacity fee to the improvement and maintenance of natural capacity, and the various artificial capacity fees to social prosperity.

Natural Capacity

A growing number of studies during the 2010s highlight the benefits of ecological restoration and carbon capture upon natural capacity. For instance, Griscom et al. calculate that reforestation and other forest-related actions "offer over two thirds of cost-effective natural climate mitigation" to

[4] http://money.visualcapitalist.com/all-of-the-worlds-stock-exchanges-by-size.

[5] Acworth (2017) reports 8.59 billion for North America, of which the US would be the lion's share.

meet the goals of the 2015 Paris Climate Agreement, the remaining one-third coming from the protection of carbon-storing peatlands and improved management of soils and grasslands. Many of the measures can be delivered at or below US$10 per tonne of carbon dioxide removed. Other measures between US$10 and US$100 per tonne of CO_2 would be cost-effective by 2030, the authors note. They estimate that as much as 11 trillion tons of CO_2 per year could be mitigated through a variety of bioenergy and carbon-capture methods[6] which, if implemented at US$10 per tonne, would cost upward of US$110 trillion per year. This amount is more than the total global GDP. Studies of this kind illustrate the scale of what is possible in material terms and the scale of the economic challenge ahead.

Natural capacity fees also could be predicated to subsidies and incentives for material efficiency. While not directly benefiting natural capacity, they indirectly benefit it by reducing the economy's material intensity upon nature. The funding of subsidies and incentives thus could be taken out of the taxation process and instead be passed directly from the treasury to the government agencies responsible for implementing them. In the US, these include the Environmental Protection Agency and the Department of Energy. Natural capacity fee revenue flowing to these agencies would be used both for in-house programs and for eligible outsourced programs, whether in the form of project funding or as subsidies. In this way, private landowners, for instance, could apply for financial or technical assistance to implement ecological restoration, carbon capture or any other relevant improvement. The tie-in to reductions in property taxes should be evident.

Even though the funds themselves would not pass through the budget process, their disbursal still would have to be subject to legislative oversight, to ensure funds were spent appropriately. This oversight might present an opportunity for citizen participation, which would be a democratic enhancement.

Basic Living Program: Income, Health and Dependent Care, Education

If fees on artificial common assets could yield close to $4 trillion per year in the US, then the per-capita value of that revenue would be close to $12,500 per year. Basic income is one of three principal benefits this dividend could fund, the others being health and dependent care and education.

The dividend of $12,500 per person is close to the median US household income of $50,000 to $60,000 during the 2010s, assuming an average household size of four. Yet paying the dividend only in basic income

[6] Griscom et al. (2017).

might not yield the greatest overall benefit. Households would have to purchase their own health care, dependent care and possibly also education through markets. A simple basic income scheme, by itself, might fail in its intended goal because these services, being fundamental and necessary for widespread prosperity, are relatively price-inelastic. Households with special needs in particular might face costs far exceeding their ability to pay, even with a basic income scheme. The policy also could risk collusion among providers to drive up costs.

A more integrated approach would establish a basic living program, incorporating basic income, free health and dependent care, and free public education. The health and education services would be free, or substantially free, at the point of consumption. Whether the government provided them directly or contracted them is a question for implementation. The difference, however, between the government paying for them on households' behalf and households paying for them directly is a difference in bargaining power. Service providers would find it much harder to drive prices up against a government, as a single, monolithic customer, than against millions of small customers.

Although the United Nations Universal Declaration of Human Rights includes health care and education as 'basic human rights',[7] it does not include a minimum financial income. The present thesis proposes that it should. Moral justifications aside, the reason is because a large economy on a crowded planet can only become materially stable if substantially all members are able to participate in its material efficiency. By subsidizing the basic necessities upon which all households depend—staying healthy, caring for loved ones and educating the family—the purpose of widespread prosperity would be minimally satisfied. Only once these necessities were provided for could households address a prime directive of material discipline. A basic living program also acts as a form of insurance against surprises from nature, until such time as the economy becomes aligned with it. The vulnerability of so many within society during the coronavirus pandemic of 2020 has thrown this advantage into sharp relief.

In such countries as the US, making the change from a system of private health insurance to one of government-subsidized health and dependent care would be substantial yet achievable. In countries with little or no institutional health care, the challenge would be far greater. However, elective redistribution, discussed earlier, could begin to bridge that gap.

[7] United Nations (1948) Articles 25 and 26.

The establishment of non-tax funding streams to offer free health and dependent care and education has three other potential benefits. One is a simplification of the tax code in countries operating under a health insurance model, such as the US, where a complex system of tax benefits offsets health and dependent care costs.

Another benefit, broader and less quantifiable but potentially more significant, is that removing the burdens of health and dependent care and education from families frees them not only to engage in a drive for material efficiency but also to innovate and contribute to the economy in other, unforeseen ways. A base level of prosperity, in finance, health and education, should deliver numerous societal benefits, and therefore economic ones, as studies by Wilkinson and Pickett and others have suggested.[8] Crime rates would fall, health care costs borne by providers would fall, educational standards would increase and, with them, economic competitiveness. Individuals who might not have had an opportunity to bring an innovation, a new business or a new service to their community, or to society as a whole, now would have a platform from which to do so. A virtuous cycle would become established, in which prosperity begat prosperity, economy-wide.

If societies that are more equal are indeed more economically competitive, then a third, macroeconomic, benefit should accrue, which is comparative advantage. Early adoption of a basic living program could place a national economy at a short-term comparative advantage over others. This hypothesis that deserves rigorous, quantitative examination.

Chapter 16 will examine the likely budgetary requirements for a basic living program. If common capacity fees by themselves are insufficient to fund those requirements then other mechanisms, such as sovereign money creation, can be employed. Chapter 16 also will examine the potential for natural capacity fees to fund programs benefiting natural capacity.

Appendix: Potential Revenue from Common Capacity Fees

Potential Revenue from Artificial Capacity Fees

In addition to the sources of revenue from artificial capacity fees proposed by Barnes (2014), this chapter lists a number of others. Table 13.6 provides the revenue estimates, along with data sources and notes on

[8] See Chap. 8.

Table 13.6 Summary of potential revenue from artificial capacity fees on common infrastructure assets

Infrastructure asset	*Fee levied on*	*Rate*	*Unit*	*Revenue ($ bil.)*	*Total US (commercial)*	*Year*	*Sources*	*Notes*
Water delivery	Volume delivered	$0.01	Gallon	58	5.8×10^{12}	2015	1	1
Wastewater systems	Volume processed	$0.01	Gallon	58	5.8×10^{12}	2015	1	2
Electric grid	Energy delivered	$0.01	kWh	234	2.3×10^{13}	2017	2	3
Natural gas pipelines	Volume delivered	$0.001	Cubic feet	227	2.3×10^{14}	2017	3	4
Telephone lines & cell networks	Call minutes	$0.01	Minute	39	3.9×10^{12}	2012		5
Internet (fixed & mobile)	Data use	$0.001	MB	2099	2.1×10^{15}	2017	4	6
Roads	Miles traveled	$0.08	Mile	24	3.0×10^{11}	2015	5	7,8
Railways	Freight car miles	$1.00	Mile	36	3.6×10^{10}	2014	5	8,9
Waterways and ports	Tons of cargo	$10.00	Short ton	23	2.3×10^{9}	2014	6	10
Air traffic control	Per outbound flight	$500.00	Departure	4	9.0×10^{6}	2014	7	11
Total				2805				

(*continued*)

Table 13.6 (continued)

Infrastructure asset	*Fee levied on*	*Rate*	*Unit*	*Revenue ($ bil.)*	*Total US (commercial)*	*Year*	*Sources*	*Notes*

Source: (1) Author's creation based on USGS: (a) https://waterdata.usgs.gov/nwis; (b) https://water.usgs.gov/watuse/wudo.html; (2) US Energy Information Administration: https://www.eia.gov/electricity/annual/html/epa_02_05.html; (3) US Energy Information Administration: https://www.eia.gov/dnav/ng/ng_cons_sum_dcu_nus_a.htm; (4) Cisco Systems (2018), Table 8; (5) US DoT (2017), Table 1–35; (6) US DoT (2017), Table 1–56; (7) US DoT (2017), Table 1–37

Notes: (1) Public, non-residential supply only. Groundwater and surface water withdrawals would be addressed through a material intensity tax. Total public supply in 2015 was 39 billion gallons per day (bg/d; source 1a) of which 23 bg/d was domestic (source 1b), leaving 16 bg/d non-domestic; (2) actual data on wastewater volumes are spotty, so the volume of wastewater processed is assumed to equal the volume delivered; (3) non-residential supply calculated as total delivered energy minus residential delivered energy. Includes commercial, industrial and transportation; (4) non-residential supply calculated as total supply minus residential. Includes commercial, industrial, vehicle fuel and fuel for electric generation; (5) 'business' calls means calls made on phones whose accounts are billed to businesses. Data on total call minutes across all platforms are not comprehensive or systematic. A web article on Quora estimates the number of cellular calls in the US each day at around 6 billion (2012). Statista estimates an average cellular call duration of 1.8 minutes in 2012. These data include all sectors (commercial, residential etc.) but they exclude landline and VOIP calls. As such, the estimate in the table might be within the right order of magnitude for total business calls, but it is very uncertain; (6) business fixed internet and mobile traffic calculated by business proportion of total fixed internet and mobile traffic. Managed IP (intranets, VPNs etc.) omitted; (7) total miles in two-axle, six-tire or more trucks, combination trucks and buses. Light-duty, two-axle, four-tire trucks may be a mix of private and commercial. These are omitted; (8) blended rate as weighted average of assumed rates and miles traveled. Assumed rates by vehicle class are: truck, single-unit, two-axle, six-tire or more: $0.05/mile; truck, combination: $0.10/mile; bus: $0.10/mile; (9) class 1 freight, car miles. This is the number of miles traveled by all freight cars, as distinct from all freight trains. Thus, a lumber company, for example, using a freight car to ship a load of lumber would pay $1.00/mile for that shipment; (10) collected at point of loading or unloading; (11) unlike other fees, this one would be charged not to the end-use customer but to the carrier, meaning that it could be passed on to the customer, whether their trip is for business or pleasure. Many trips combine both. It should equate to a few dollars per seat or about $1/100 lb. of cargo: therefore, it might vary according to equipment. About $500 per departure is a rough approximation of an average assumptions and methods. The fees would be charged to end-use business customers rather than to carriers, with the exception of the air traffic control fee, for reasons explained in the notes.

Illustrative Schedule of Natural Capacity Fees Using Asset Brackets for US Businesses

Table 13.7 Illustrative schedule of natural capacity fees using asset brackets for US businesses

Asset brackets ($ mil.)		*Average net income*	*Annual natural capacity fee*	*Percentage of net income*	*Number of reporting entities*	*Revenue ($ bil.)*
From	*To*					
0	0	$42,297	$200	0.5%	1,095,574	0.22
0.001	0.499	$28,299	$200	0.7%	3,825,213	0.77
0.5	0.999	$70,792	$400	0.6%	374,436	0.15
1	4.999	$142,020	$1000	0.7%	415,997	0.42
5	9.999	$382,176	$2000	0.5%	70,737	0.14
10	24.999	$787,825	$4000	0.5%	48,639	0.19
25	49.999	$1,359,729	$10,000	0.7%	19,031	0.19
50	99.999	$2,073,411	$20,000	1.0%	11,674	0.23
100	249.999	$3,609,822	$40,000	1.1%	10,344	0.41
250	499.999	$7,557,527	$80,000	1.1%	5580	0.45
500	2500.00	$28,496,034	$300,000	1.1%	7312	2.19
2500		$393,561,849	$4000,000	1.0%	3266	13.06
Total	18.43					

Source: Author's creation, using 2013 data from the US Internal Revenue Service

Notes: Annual natural capacity fees are imposed at between 0.5 percent of average net income for small businesses up to about 1 percent for large ones. Revenue is derived from the fees multiplied by the number of filing entities

References

Acworth, W. (2017) *2016 Annual Volume Survey.* Futures Industry Association, London, Singapore, Washington DC. www.marketvoicemag.org.

Barnes, P. (2014). *With Liberty and Dividends for All.* San Francisco, CA: Berrett-Koehler.

Griscom, B. W., et al. (2017). Natural Climate Solutions. *Proceedings of the National Academy of Sciences, 114*(44), 11645–11650.

United Nations. (1948). *Universal Declaration of Human Rights.* New York: United Nations.

CHAPTER 14

Economic Controls 3: Taxation

In addition to use fees, corporations could be charged larger amounts for material throughput, the revenues thus collected going to the treasury as tax. Some households would be charged similar taxes as well. Any tax increase on one factor should be balanced with a tax decrease on others, so that the tax burden on most taxpaying entities would not increase. Increasing the total tax burden invites political resistance.

The golden rule of taxation policy is to tax social 'bads' while not taxing social 'goods'. Probably the most significant example of taxation on a 'good' is income tax, whether individual or corporate. Human productive labor is a good. Therefore, a tax on productive labor, in principle, is a bad and should be minimized. It is not only a social bad, in the sense of being economically inefficient, but also more broadly an environmental bad because it reduces spending power on materially efficient goods and services. In the economy of the early twenty-first century, these still tend to be more expensive than materially inefficient ones. Herein lies another perversion of taxation policy: consumption that pollutes is not taxed enough. As Robert Frank and many others have pointed out, taxes on polluting consumption "would encourage not only the development of cleaner technologies but also a shift to cleaner forms of consumption".[1]

Taxing material inefficiency while rewarding material efficiency would provide a very strong motive force within the economy toward material efficiency. It also would generate revenue for government programs, albeit

[1] Frank (1999, p. 102).

F. Murison Smith, *A Planetary Economy*,
https://doi.org/10.1007/978-3-030-49296-0_14

a declining flow over time. Such a decline is not a reason to avoid taxing material inefficiency, especially in the presence of other sources of revenue, such as sovereign money creation. As the economy evolves toward greater material efficiency, governments may come to rely progressively less on income taxation as a source of revenue.

This chapter outlines a taxation structure for a planetary economy, conforming to the principle of taxing 'bads' rather than 'goods'. It suggests some broad implications for revising tax codes. It also addresses distributional implications, identifying risks to the principle of full social participation in material discipline.

From an equity point of view, argued Julian LeGrand, it would be preferable to establish a tax basis independent of preferences. LeGrand was referring to income tax, although the requirement could apply to any form of tax. Individuals or other taxable entities having identical choice sets would be taxed the same, a situation he called 'horizontal' equity. Taxable entities with different-sized choice sets would be taxed according to the degree of difference, a situation he called 'vertical' equity.[2] For example, since people on higher incomes have more options to improve their material efficiency than people on lower incomes, some form of progressive taxation or incentivation would be justified so as to give the disadvantaged a comparable range of opportunities as the more advantaged. A broader social benefit would be high participation rates in the economy, which conforms to the principle of strong reciprocity.

Progressive Consumption Tax

A progressive consumption tax is a tax on economic consumption measured in currency (Table 14.1). It is designed to address distributional inequities by disincentivating high consumption. It also is designed to encourage savings. The tax has a long history of endorsements by various economic and political illuminati: Adam Smith, John Stuart Mill, Alfred Pigou, Alfred Marshall, Milton Freidman, Kenneth Arrow and Ronald Reagan being a few examples. "Liberals and conservatives may disagree on many details concerning the consumption tax, including the extent to which it should be progressive", continued Frank, "but there is little

[2] Provided that the smaller choice set is a subset of the larger one (LeGrand 1991, pp. 71, 155–156).

Table 14.1 Approximate conformance of a progressive consumption tax to the four principles of economic controls for a planetary economy

Material discipline	
Economic stability	
Widespread prosperity	
Innovation through markets	

Source: Author's creation

Note: Darker shading denotes relatively greater conformance

disagreement on the principle that it is better to tax consumption than income."[3]

Frank's concern in particular was what he called the 'consumption arms race' that can take hold in market economies, where consumers compete for access to the best services, such as good school districts or hospitals, and for status. The solution he advocated was a steeply progressive consumption tax. Since the 1970s, income taxes, particularly in higher tax brackets, have been drastically cut, and many basic services funded by income taxes have been cut accordingly. Frank suggested a consumption tax rate of 20 percent on the lowest taxable incomes, rising to 70 percent on taxable incomes representing roughly the wealthiest 1 percent of households.[4]

Frank built upon a 1942 proposal by Irving and Herbert Fisher, which defined taxable income as gross receipts minus what the Fishers termed 'investments'—taxes paid during the current year—and 'proper exemptions' for a taxpayer and their dependents. The remaining taxable income represents what would actually be spent. The term 'investments' was used in its traditional sense, meaning 'waitings'. An asset, such as cash, is set aside for a certain period before it can be liquidated: it is 'vested'. These investments would be rather like putting pre-tax funds into a retirement account, except that one would not have to wait until retirement to liquidate them.[5] Instead, the vesting period would be shorter, perhaps a few years.

A progressive consumption tax naturally encourages saving because the investment described above reduces the amount a household would pay in tax. Higher household savings rates likely would result in lower

[3] Frank (1999, p. 223).

[4] Frank (1999, p. 215).

[5] Frank (1999, pp. 213–214), and Fisher and Fisher (1942) cited therein.

indebtedness and lower rates of default or bankruptcy. Funds invested in one year would become liquid some years later. At an aggregate level, the economy likely would become more stable over time. The scheme, Frank argued, would bring individual and collective interests better into line.[6] Because income equality is linked with relatively higher economic growth, greater social harmony and wider prevalence of democratic values,[7] a progressive consumption tax that improved income equality would be expected to improve these other measures as well.[8]

This kind of progressive consumption tax would be preferable to regular income tax in that it would not tax a social good. Whether high consumption, in money terms, is a social bad depends on the degree to which it correlates with high material intensity. In today's growth economy, it does, although the linkage may be weakening. In today's economy, then, a tax on consumption in money terms should be expected to reduce overall material intensity. However, it would not in and of itself create an incentive to increase material efficiency.

To the extent that high income inequality is a social bad, a progressive consumption tax would be beneficial in narrowing the income gap from the top end. However, in the absence of a guaranteed income floor, the poor would remain poor, even if they were not subject to a progressive consumption tax. Participation in an economy-wide drive toward material discipline would suffer. A progressive consumption tax in isolation probably would not satisfy the requirement for widespread prosperity.

Other ideas for consumption taxes, which focused on specific classes of 'luxury goods', date back to times when a monarch sought to retain an appearance of greater wealth than even the richest merchants. The problem with these variants of the consumption tax is that they are arbitrary ('Who is to decide what constitutes a luxury?') and do not make much of a dent in society's material intensity. A buyer in the market for a Rolls-Royce is not going to be very much concerned whether the car costs $400,000 or $800,000, and the market for these kinds of products is very small anyway.

[6] Frank (1999, p. 213).

[7] See in Frank (1999, pp. 243–245) and in Wilkinson and Pickett (2009); also Friedman (2006).

[8] Frank (1999, p. 247).

Taxes on Material Flows

Unlike a progressive consumption tax levied on financial income, a tax on material flows will create an incentive for greater material efficiency. In the aggregate, it can be expected to drive the economy toward greater material discipline. However, a tax on material flows cannot be applied in isolation. Businesses and households face capital costs in switching to new, more efficient practices and technologies. Overcoming these capital costs would require bridging funds of some kind, such as in the form of an incentive. A tax on material flows also would have to be considered within the context of a broader tax-shifting policy so as to keep the taxpayer relatively indifferent to, or attracted to, the change.

In principle, a tax could be applied to any material flow declared a 'bad'. Examples of taxes that have been established on material flows in different parts of the world as of the early twenty-first century are listed in Table 14.2. Taxes on material inputs include resources extracted from nature. Taxes on outputs are those imposed on effluent flows, such as atmospheric emissions or liquid and solid waste streams.

During the early years of the twenty-first century, the so-called carbon tax has gained increasing acceptance as an instrument to drive down carbon emissions. The tax specifically targets carbon emissions, as distinct from carbon inputs. On the input side, carbon typically is bound up with many other substances, principally in the form of biomass, so it becomes effectively impossible to tax separately. Instead, as Table 14.2 shows, taxes are levied on such inputs as lumber from old-growth forests. The combustion of carbon-based fuels yields a readily determinable mass of carbon emitted in bound form as one of a small number of chemical species: carbon dioxide, carbon monoxide or soot. The same line of thinking applies to other varieties of atmospheric pollutant: oxides of sulfur, oxides of nitrogen and fine particulates. These reliable quantities provide a basis for taxation, so as to encourage reductions in the flow in question, and the innovation of alternatives.

Some outputs are less well defined. Industrial processes produce all kinds of waste sludges, which are differentially toxic to aquatic systems and to people. For these kinds of effluents, an undifferentiated per-tonne tax may be the only feasible form of regulation. Solid waste streams to landfills also are undifferentiated. Their disposal through typically is charged as fees, levied per unit of weight or for discrete disposal events.

Table 14.2 Selected existing or attempted taxes on material flows as of the early twenty-first century, organized into input and output categories

Category	*Quantity taxed*	*Jurisdiction(s)*	*Effective date*	*Rate(s)*	*Notes*	*Sources*
Input	Biodiversity	Many OECD and other countries	N/A	Hunting or fishing as percentages of sales; also habitat alteration	As distinct from hunting license, which are fees	1
Input	Lumber	California	2013	1 percent sales tax on lumber and engineered wood		2
Input	Minerals	Many, mostly regional and local	N/A	May be levied at a few US dollars per ton extracted or as percentage (typically 10–20 percent) on mineral sales	Exploration costs typically excluded from taxable amounts Some countries charge land-use fees or royalties	3
Output	Carbon	Australia	2012–2014	A$23—A$25.40 per tonne	Repealed in 2014	4
Output	Carbon	British Columbia	2008	C$10 (US$7.60) per tonne	Increased by C$5 per year to C$30 by 2012	4
Output	Carbon	Canada (8 provinces)	2018	C$10 (US$7.60) per tonne	Increases by C$10 per year until C$50 by 2022	4
Output	Carbon	Chile	2018	$5 per tonne	Applies to 55 percent of emissions	4

(*continued*)

Table 14.2 (continued)

Category	*Quantity taxed*	*Jurisdiction(s)*	*Effective date*	*Rate(s)*	*Notes*	*Sources*
Output	Carbon	Finland	1990	€66.20 per tonne (mid-2010s)	First-ever national carbon tax	4
Output	Carbon	Ireland	2010 2012	€15 per tonne €20 per tonne	Tax on fuels. Replaced an increase in income taxes	4
Output	Carbon	Sweden	1991	$150 per tonne (mid-2010s)	Industries required to pay only 50 percent of tax. Separate taxes levied on electricity and motor fuels	4
Output	Carbon	UK	2013	£18 per tonne	Carbon price floor: any time EU carbon price drops below UK price, the difference is paid to the British Treasury	4
Output	Municipal solid waste	Many, mostly regional and local	Late nineteenth century onward	Various, typically US$1–100 per tonne, depending on the material	Also changed as a percentage of garbage rates, e.g., 10–20 percent in the US.	5

Source: (1) Author's creation, from the OECD Policy Instruments for the Environment Database: https://pinedatabase.oecd.org; (2) State of California: www.cdtfa.ca.gov/industry/lumber-products.htm#Overview; (3) Otto et al. (2006), Guj (2012), PricewaterhouseCoopers (2012); (4) Carbon Tax Center: www.carbontax.org.; (5) Wikipedia; various US states' departments of revenue

Above all, Table 14.2 illustrates how the efforts to tax material flows have been piecemeal and rather scattershot. A simpler, all-encompassing approach might be to levy a blanket material intensity tax.

Material Intensity Tax—Introduction

Like a common-resource fee on a business's material throughput, a material intensity tax would be levied on every tonne of material passing through the hands of an economic actor (Table 14.3). However, the two instruments differ in key respects. A material throughput fee, predicated to basic income, would never pass through the exchequer. A material intensity tax, on the other hand, would pass through the exchequer, and could be used for whatever the government decided in its budget process.

Like the material intensity fee, a material intensity tax would distinguish among material inputs that are extractive in nature, such as from depletable mineral sources; those that are regenerative in nature, meaning living systems; and those that are recirculated within the economy. Extractive inputs would be subject to a high rate of tax, as they could not be replaced. Regenerative inputs would be subject to a medium rate of tax because of their regenerative nature, setting aside for a moment questions related to ecological structure, diversity and stability, discussed later. Recirculated inputs probably would not be taxed at all and may be incentivated.

The certification of material flows through businesses would be carried out by a material efficiency rating agency. Businesses would self-report, as they do for taxes, and the rating agency would have the ability to audit, investigate or penalize, just as government tax agencies do for tax reporting. A material intensity report would be used by the tax agency as the basis for a business's material intensity tax within a given reporting period.

A tax on primary extraction from nature would be assessed at the point of extraction. From there, raw materials passed on to primary economic

Table 14.3 Approximate conformance of a material intensity tax to the four principles of economic controls for a planetary economy

Material discipline
Economic stability
Widespread prosperity
Innovation through markets

Source: Author's creation

Note: Darker shading denotes relatively greater conformance

production would be treated as equivalent to material recirculated within the economy. In this way, a given tonne of material would not be double-counted for taxation. The cost of higher taxes paid by primary extraction would be passed, in whole or in part, on to primary production as material costs. These costs then would be passed, in whole or in part, through the supply chain all the way to the point of final consumption.

It is important to understand that it would not be necessary to track the proportions of extractive, regenerative and recirculated resources comprising any given good or service as it passes through the economy, provided that the cost of the tax at the point of extraction from nature were passed on to the downstream activities within the supply chain. However, it is useful to track these quantities for the purpose of accounting for material performance. Being able to demonstrate high material efficiency ought to confer value or competitive advantage within the markets of an increasingly materially efficient economy.

Material outflows from final consumption would be taxed similarly to material inflows to the economy. Outflows expelled into nature as effluent would be taxed at a high rate. Outflows recaptured and recirculated within the economy would not be taxed.

Figure 14.1 illustrates the concept.

James Baker and others have proposed a hybrid plan for the US. A carbon 'tax' would be levied at the point where fossil fuels first enter the economy. Revenues from this tax would be paid out directly as dividends to all citizens. Since the authors do not state that the revenues in question would pass through the Treasury, is it fair to assume that the instrument in question is not a tax as such but closer to a fee levied on the use of a common resource.[9] However, a small but growing number of governments during the 2010s have instituted carbon taxes levied at the point of consumption as distinct from production. The European Union is considering implementing a carbon tax alongside its emissions trading program, as are a number of other jurisdictions.[10]

The overall effect of a material intensity tax would be to drive the recirculation of as much material as possible within the economy, minimizing resource extraction and effluent flows. A couple of examples serve to illustrate the concept.

[9] Baker et al. (2017, p. 1) and Americans for Carbon Dividends (www.afcd.org)
[10] World Bank (2016).

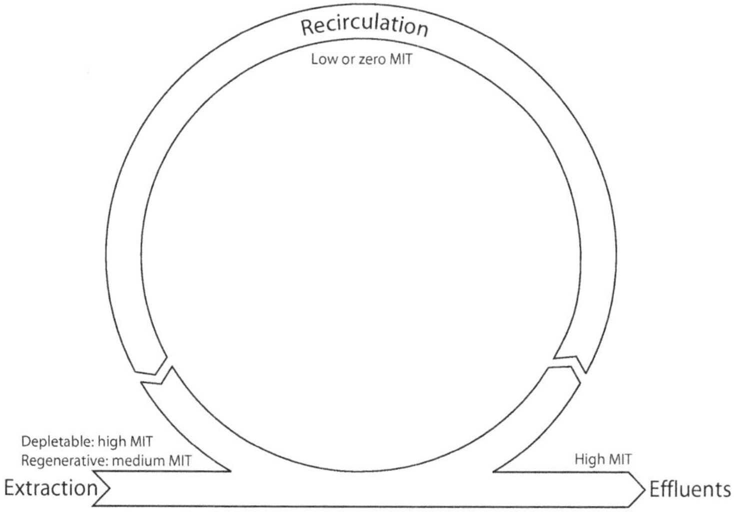

Fig. 14.1 Schematic of the application of a material intensity tax, illustrating the distinction between extractive, regenerative and recirculated flows, as well as effluent, with respect to tax rates. Source: Author's creation

The supply chain for motor fuels consists of several stages. During the twentieth and early twenty-first centuries, the scale of this supply chain has become immense. Crude oil is extracted from deposits in Earth's crust and stored at or close to the point of extraction. From there, it is transported, by pipeline, rail, road or ship, to a refinery where it is stored prior to refining. A variety of intermediate products are produced through refining—for example, feeder stocks for other industrial processes, such as plastics manufacturing—as well as finished motor fuels. Motor fuels are transported by rail or truck to the point of delivery: the filling station. Once inside a vehicle, the motor fuel is consumed, releasing gaseous effluents and fine particulates through the tailpipe.

A material intensity tax applied to this supply chain would levy a high tax on the extraction of crude oil. The company extracting the oil would be charged according to the volume of oil it extracted and stored. This tax would significantly increase the cost of extraction. Consequently, the price of crude oil sold to the distributor would become significantly higher, and this cost would be passed along the supply chain to the refiner and ultimately to the filling station. These entities could all be subsidiaries of a single conglomerate providing end-to-end services, or they could be distinct companies; it does not matter as far as the tax is concerned. A second high rate of tax would be applied to each gallon of fuel pumped into a customer's vehicle.

Thus, two taxes would be applied: one at each end of the supply chain. The tax at the input end to the supply chain would create an incentive to reduce the rate of extraction. The tax at the output end would put downward pressure on the rate of consumption. Any of the multitude of other products made from crude oil as a feeder stock would be similarly taxed for the mass of effluent their consumption released into nature, whether in gaseous, liquid or solid form. The effect would be to reduce the demand for crude oil and its derivatives in fuels, plastics, chemicals, paints and many other products—subject, of course, to the price elasticity of demand for these products.

Since the price elasticity of demand depends upon the availability of substitutes at the point of demand, the tax also should be expected to stimulate the emergence of substitutes, particularly for those products whose demand is comparatively price-inelastic. In the case of motor fuels, alternate, cleaner means of delivering energy to vehicles would appear relatively more attractive. Demand for them would increase.

A second example pertains to the use of regenerative, or ecological, resources at the point of supply into the economy. These would be subject to a lower tax than depletable resources. Yet here the situation appears more complex, because there is a difference, for instance, between extracting wood from old-growth forests, which would take hundreds of years to regenerate, and harvesting crops on land that may have been converted long ago to agriculture. Should the tax differentiate between monocultural, industrial agriculture on the one hand and biodynamic forms of agriculture, such as agroforestry, which aim to preserve a degree of ecological diversity on the other?

A monocultural crop and a virgin ecosystem represent two ends of a spectrum of ecological diversity. Since it is not possible to draw clear lines demarcating industrial agriculture from biodynamic agriculture, nor, in turn, a 'gardened' forest from a relatively untouched ecosystem, it would be more straightforward simply to levy a flat tax on all extracted regenerative resources. At the same time, as the discussion of subsidies will articulate, the application of a subsidy to operations enhancing natural capacity would offset the effect of the tax for those operations incorporating ecological practices. In this way, landowners would have an incentive to preserve old-growth forests and to manage other forests for lumber in a way that mimics natural treefall. Even without such a subsidy, an agricultural operation relying heavily on industrial chemicals would find those inputs much more expensive. It would either have to pass these expenses on to its customers or eliminate them by changing its operations.

If the tax were flat, why apply a tax at all? Why not simply apply subsidies for ecological practices? The reason is because there are technological substitutes to land-based agriculture, ranching, lumber harvesting, fisheries and other uses of natural capacity. Many have yet to be invented, although some already exist. Various forms of urban agriculture, for instance, employ techniques to grow food crops at high density in small spaces, whether on reclaimed urban land or hydroponically in controlled environments—so-called vertical farming. Biotechnologies to grow animal protein in vitro are in commercialization. Wood substitutes made from recycled plastics have been a growing industry for many years. These and other substitutes for regenerative inputs from nature would not be subject to a material intensity tax, at least until better substitutes came along.

Tax-Shifting on Income

The amount of revenue generated by a material intensity tax would depend on how much of the tax burden could be shifted away from income onto material intensity (Table 14.4). The question, however, is not quite as simple as an aggregate estimate would suggest. A key prerequisite is being able also to shift the tax burden up the income scale. Low and middle earners would pay virtually no income tax whereas the wealthy—those with annual incomes in the millions—would pay more. The rule of diminishing marginal utility implies that increased consumption costs arising

Table 14.4 Approximate conformance of tax-shifting away from income to the four principles of economic controls for a planetary economy

Material discipline	
Economic stability	
Widespread prosperity	
Innovation through markets	

Source: Author's creation

Note: Darker shading denotes relatively greater conformance

from a material intensity tax could be extremely burdensome on the 90 percent whose earnings are less than three times the median income, whereas the 1 percent whose annual earnings are in the seven figures would barely notice. A similar line of argument applies to corporate incomes.

To illustrate how much of the tax burden could be shifted, income tax data from the US Internal Revenue Service are used as a basis for comparing current income tax structures to a possible future one, for both individual and corporate filers. The example presented here examines how much of the tax burden could be shifted off income onto material intensity and how much of it could reasonably be shifted up the income scale.[11] IRS data organize individual filers by income brackets, whereas corporate filers are sorted into asset brackets. This analysis uses the same brackets.

Income tax rates for individual filers apply at different income levels depending whether they are married filing jointly, married filing separately, head of household or single. Ninety percent of all returns come from filers who are married filing jointly or single, so the schedules for these two are a good approximation of all individual returns.

In 2013, the IRS received about 111 million returns, of which nearly 54 million were from married couples filing jointly and close to 69 million from single filers. It collected slightly less than $1.2 trillion from these two classes, $880 billion from the former and $279 billion from the latter. As Fig. 14.2 shows, most of that revenue came from filers in higher income brackets. The figure indicates that if the tax code were modified to eliminate income tax for people in lower-income brackets then it should not be expected to affect the overall tax revenue very much.

In this example analysis, seven tax rates that existed in 2013 are replaced with a simpler four-tiered system summarized in Table 14.5.

The thresholds are selected for convenience to conform to thresholds in the IRS income data. Incomes below the median are not subject to tax. Those between the median and roughly three times the median pay a marginal rate of 10 percent on incomes within those thresholds. Above three times the median up to 20 times the median, incomes are taxed at 40 percent. Incomes above 20 times the median—that is, the very wealthy—are subject to a high rate of 60 percent. Only about 0.2 percent of filers reported incomes above 20 times the median. Incomes between 3 and 20 times the median numbered about 7 percent of the total. The vast

[11] Source data file available upon request.

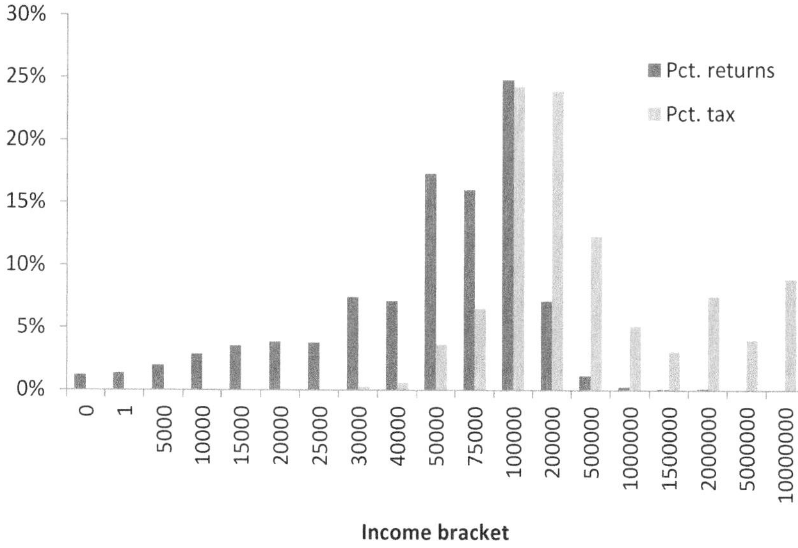

Fig. 14.2 Percentages of returns and income tax received from married couples filing jointly, organized by income bracket. Source: Author's creation, using income brackets from US Internal Revenue Service Data Table 1.2: all returns: adjusted gross income, exemptions, deductions, and tax items, by size of adjusted gross income and by marital status. Data are from 2013. Note: The pattern for single filers is similar

majority of filers, then, would pay effective, or weighted average, tax rates well below 10 percent.

The rates are chosen to encourage broad prosperity: that is, to minimize the tax burden on middle incomes and to maintain a moderate rather than excessive degree of income inequality. Placing a high rate of marginal income tax on incomes above 20 times the median helps to correct a social 'bad' which is excessive income inequality. The purpose of this high marginal rate is not only to return revenue to the government but also, equally, to constrain the income distribution. No social benefit is derived from allowing more billionaires, any more than requiring everyone to be equal.

The analysis shows that for filers reporting income up to three times the median, the effective tax rate would drop from between 5 and 10 percent to below 4 percent. For those reporting income between three and 20 times the median—affluent filers—effective tax rates would be a few

Table 14.5 Current (2013) and proposed US income tax rates for tax-shifting analysis

Current (2013) structure					
Income brackets				Rate	Notes
Married filing jointly		Single			
Min	Max	Min	Max		
$-	$17,850	$-	$8925	10%	
$17,851	$72,500	$8926	$36,250	15%	
$72,501	$146,400	$36,251	$87,850	25%	
$146,401	$223,050	$87,851	$183,250	28%	
$223,051	$398,350	$183,251	$398,350	33%	
$398,351	$450,000	$398,351	$400,000	35%	
$450,001		$400,001		39.6%	
Proposed structure					
Income brackets				Rate	Notes
Married filing jointly		Single			
Min	Max	Min	Max		
$-	$74,999	$-	$29,999	0%	
$75,000	$199,999	$30,000	$99,999	10%	1
$200,000	$1,499,999	$100,000	$499,999	40%	2
$1,500,000		$500,000		60%	3

Source for rates: The Tax Foundation, https://taxfoundation.org/selected-federal-rates-2012-2013/

Notes: (1) Lower thresholds close to median incomes in the two classes. Married filing jointly (MFJ): $76,509; single: $31,178 (DeNavas-Walt and Proctor 2014, Table 1, p. 6.); (2) lower threshold: MFJ: around 2.7 x median; single: around 3 x median; (3) lower threshold: MFJ: around 20 x median; single: around 17 x median

percent less near the lower end of this income bracket and a few percent higher near the upper end of it. For the wealthiest 0.2 percent, effective tax rates would increase from between 25 and 30 percent to between 40 and 60 percent. Table 14.6 shows how the income tax burden would shift among the tax brackets under this scheme.

Figure 14.3 shows how the effective tax rates would shift among IRS income brackets, for married couples filing jointly. A similar pattern is found for single filers.

Table 14.6 Effects of changes on the income tax burden by bracket, in aggregate and per capita, for the two tax classes in the analysis

Bracket	*Married filing jointly*		*Single*	
	Aggregate ($m)	*Per capita*	*Aggregate ($m)*	*Per capita*
Zero to median	(39,677)	$(1461)	(18,952)	$(437)
Median to 3 x median	(179,170)	$(8147)	(79,603)	$(3575)
3 x median to 20 x median	(5935)	$(1290)	3205	$1147
Above 20 x median	173,979	$1,110,690	46,189	$390,163

Source: Author's creation

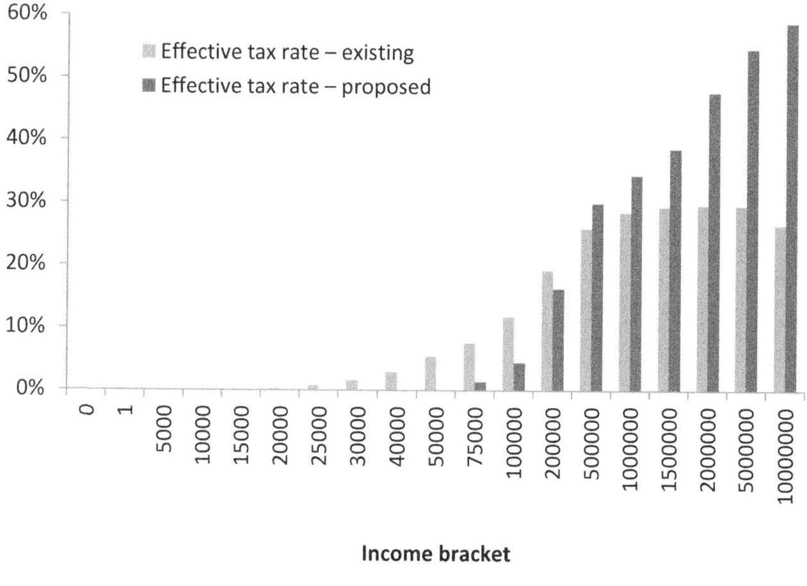

Fig. 14.3 Existing (2013) effective tax rates for married couples filing jointly, by income bracket, and new effective tax rates under a proposed tax structure described in the text. Source: Author's creation, using data from IRS Data Table 1.2

The total amount of income tax revenue generated under the new scheme is slightly less than under the current system. Compared to $1.2 trillion collected by the IRS in 2013, income tax receipts under the new scheme would be around $111 billion, or 9 percent, lower. Yet significant beneficial effects upon the net incomes of the '90 percent' would be realized.

Fine-tuning the thresholds and the rates could yield an additional $100 to $200 billion without needing to add tax brackets. The idea would be to maintain or slightly increase the effective tax rate on the affluent, who would be able to afford the higher consumption costs arising from a material intensity tax. As Chap. 16 details, a suite of economic instruments would be required to ensure that lower and middle earners are able to afford the higher consumption costs of material intensity. Although the tax savings in this example would help, they alone would not be sufficient.

On the corporate side, income tax data are organized by company size, measured in the dollar value of a company's assets. As Fig. 14.4 shows, more than 75 percent of all corporate income tax revenue comes from a relatively small number of behemoth companies, with assets over $2.5 billion.

In 2013, the IRS applied no less than eight corporate income tax brackets, ranging from 15 percent to 39 percent. For all corporations with income above $75,000, tax rates were in the range of 34 percent to 39

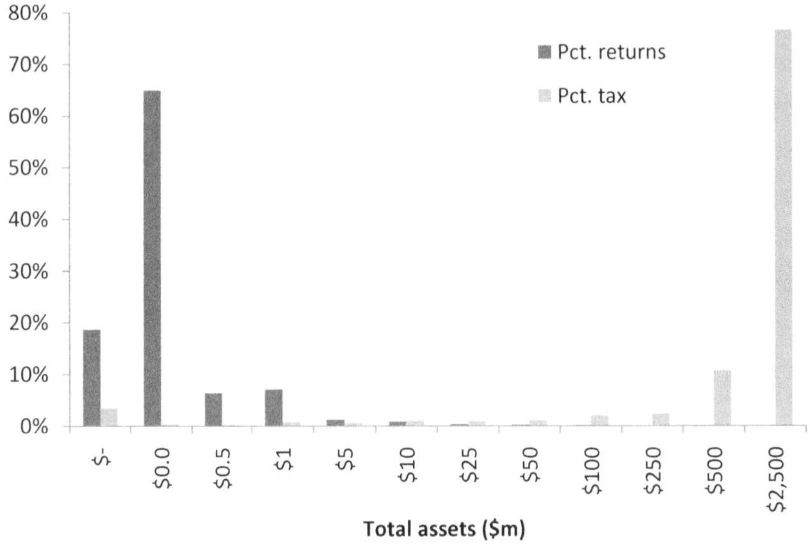

Fig. 14.4 Percentages of returns and income tax received from corporations, organized by asset bracket. Source: Author's creation, using data from the IRS Income Tax Returns Complete Report, Fig. A, p. 2. Data from 2013

percent. Effective tax rates therefore also were in roughly the same range as the statutory rates.

Across all corporations, the average income subject to tax was about $214,000. For this reason, the present analysis sets the deductible, below which no income tax would be payable, to $250,000. Given the skewness in the distribution of assets and in incomes among corporations, the threshold for the 30 percent tax rate is set at ten times this amount, or $2.5 million. The results are in Fig. 14.5.

Corporations would benefit across the board, but small ones in particular would save relatively the most. The total tax shifted away from income under this structure would be about $113 billion, or about 26 percent of the 2013 total of about $442 billion.[12]

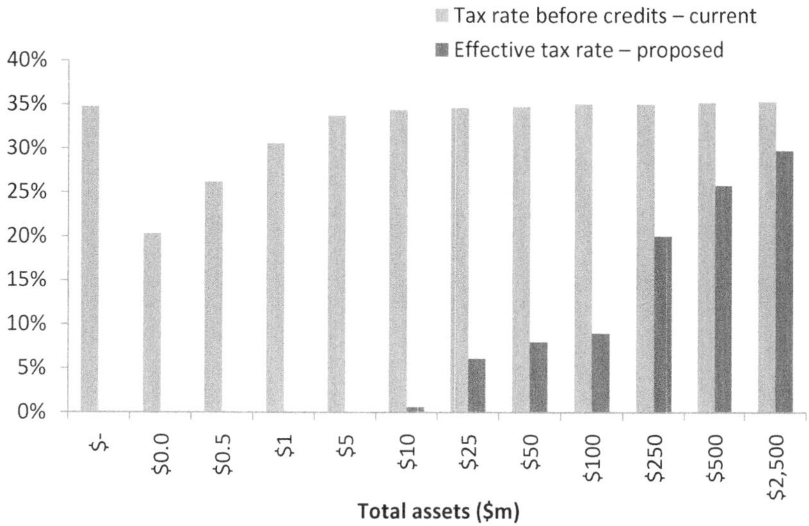

Fig. 14.5 Existing (2013) effective tax rates for corporations, by asset bracket, and new effective tax rates under a proposed tax structure described in the text. Source: Author's creation, using data from the IRS Income Tax Returns Complete Report, Fig. A, p. 2. and IRS Form 1120

[12] This is the total income tax before credits rather than after credits. This income tax is used to calculate effective rates because any new rates established also would be before credits.

Because a material intensity tax would affect consumption choices in the market, individuals and corporations would be able to reduce their total tax burden over time. Thus, the total revenue to the IRS would fall. As argued earlier, though, this decline would be offset by a relatively lighter budgetary burden if other non-tax sources of revenue were established, such as common capacity fees and sovereign money creation.

Material Intensity Tax Rates and Revenue

With $111 billion lost from reductions in individual income tax, and $113 billion lost from reductions in corporate income tax, the IRS would institute material intensity taxes to make up these shortfalls.

In the US economy, as in most of the world, almost all material extraction is performed by corporate entities. Even farming, which extracts the products of regenerative resources, is carried out in the US mostly by incorporated farms.

As sources cited in Chap. 1 describe, the US was responsible in 2013 for 13.3 gigatonnes (Gt) of material extraction from nature and about 7 gigatonnes of effluents in all forms.[13] Out of the total extraction, about 10.8 gigatonnes came from depletable sources (fossil fuels, mineral ores) and about 2.5 gigatonnes came from biomass, which is regenerative.[14]

These categorizations line up with the IRS distinctions between primary extractive industries and secondary and tertiary industries in its data on corporate tax returns. The IRS conveniently divides primary extractive industries according to the North American Industrial Classification categories of 'Mining' and 'Agriculture, forestry, fishing and hunting'.

For simplicity, let us divide the $113 billion tax burden proportionately by material flow so that 66 percent (13.3 Gt out of 20.3 Gt) goes onto extraction and 34 percent goes onto effluents. Then roughly $74 billion would be raised from the companies on the extractive side and $39 billion from all US corporations, assuming all businesses have positive net effluent flows. Dividing the extractive activities into depletable and renewable categories by proportion of material flows, the 'mining' industries

[13] Material extraction figures from Wirtschafts Universität Wien (2013), available at www.materialflows.net. Effluents figure estimated from total world effluent flows in EU EDGAR database (http://edgar.jrc.ec.europa.eu) and US CDIAC data estimating that the US is responsible for about 14 percent of the world total (http://cdiac.ornl.gov/trends/emis/tre_glob_2013.html)

[14] Wirtschafts Universität Wien (2013): www.materialflows.net

extracting depletable resources would be responsible for about $60 billion (equivalent to 10.8 Gt out of 13.3 Gt) and the industries responsible for harvesting regenerative resources would pay the remaining $14 billion.

According to the IRS, approximately 25,000 companies were engaged in the extraction of depletable resources (oil, gas, minerals, etc.) in the US in 2013. On average, each of these companies would be responsible for about $2.4 million in material intensity taxes, although for the larger ones, of course, the tax burden would be considerably higher. Approximately 110,000 companies were engaged in the harvesting of regenerative resources in 2013, which includes all farms.[15] These companies each would be responsible for about $127,500 in material intensity taxes, on average. To a farmer, this might seem a significant burden. However, a great deal of it would be passed on to distributors and wholesalers in the form of higher prices. Some of it would be offset by subsidies and incentives.

On the effluents side, not only corporations but also individuals would be responsible for material intensity taxes on the roughly 7 billion tonnes of effluents. It is assumed that all corporations would produce effluents. Because all individuals are, first and foremost, economic consumers rather than direct consumers from nature—the odd hunting or fishing trip notwithstanding—their material intensity tax burden would come only from effluents. They already would experience higher prices for goods and services from economic production.

Among corporations, the $39 billion in material intensity taxes would be borne not only by for-profit companies but also by non-profits. Just because a non-profit is exempt from paying income tax does not make it exempt from paying a material intensity tax. According to the IRS, close to 1.5 million tax-exempt organizations filed returns in 2013, in addition to the roughly 5.9 million for-profit corporations.[16] Dividing the $39 billion corporate material intensity tax burden into these 7.4 million business entities produces an average material intensity tax on effluents of about $5700 per year.

Individuals or households would be responsible for $111 billion in material intensity taxes. The IRS recorded approximately 146 million individual income tax returns in 2013.[17] Therefore, the average individual

[15] IRS aggregate data on corporate tax returns, table 1 (2013).
[16] Internal Revenue Service (2014, p. 4).
[17] Internal Revenue Service (2014, p. 4).

material intensity tax burden would be about $760 per year if it had been implemented by 2013.

What is not known with any certainty is how much of the 7 gigatonnes of effluents would be produced by corporations compared to households. Therefore, it is not possible to reliably infer a rate for the material intensity tax paid by corporations on effluents, compared to that paid by households. However, by combining the data for corporations and households, it is possible to derive an aggregate rate. That rate is computed as $39 billion plus $111 billion in material intensity taxes, divided by 7 gigatonnes, which is about $21 per tonne. This amount is about half the $40 per tonne proposed by Baker et al. for a carbon tax.

On the extraction side, the rate works out to $74 billion divided by 13.3 gigatonnes, which is $5.56 per tonne. As Chap. 16 describes, these rates would be starting-points for periodic increases.

It might be tempting to try to distinguish among different kinds of material in terms of the tax rate. This could open a can of worms. There is no reliable information on the relative effects of different kinds of material extraction upon natural capacity, nor on the relative effects of different kinds of effluent upon it. Granted, we can say that certain chemical species have different effects on specific natural processes. Methane, for example, is reliably estimated to be roughly 40 times as potent a greenhouse gas as carbon dioxide. But how does one set a material intensity tax rate on the combustion products of gasoline, for example, compared to the dumping of waste construction materials? Is mineral extraction worse for natural capacity than ecological degradation? No-one really knows. In the face of this radical uncertainty, a simple, adjustable tax structure would be preferable. Simply levy a tax by tonne of material intensity. Other economic instruments are available to steer specific activities toward beneficial practices and away from detrimental ones. The material intensity tax would, irrespectively, provide an impetus for economic activity to become materially more efficient. Any increase in material efficiency will be desirable.

While it might be tempting to levy the same rate of tax on extraction and effluents, combining the entire tax burden divided by total material intensity, there is something to be said for having a higher tax rate on effluents. It will stimulate demand for more materially efficient supplies of products and services. On the supply side, it is possible to impose material limits upon extraction in addition to financial ones, such as setting aside certain areas or imposing quotas. Emissions, on the other hand, are harder to control.

Returning to the gasoline example, the imposition of a $31-per-tonne material intensity tax upon the extraction of crude oil would increase the price of a barrel of oil from about $54 in 2017 prices to about $58, there being about seven barrels in a tonne. At current prices, then, a material intensity tax on extraction would increase oil prices by about 7 percent. Let us suppose that the industry passed this entire increase on to the end customer. In addition, the end customer would pay a material intensity tax upon tailpipe emissions of $109 per tonne, levied at the pump. Each gallon of gasoline, when burned, emits close to 18 pounds of carbon dioxide, as well as other by-products of air combustion, such as nitrates and ozone. A rough approximation would be about 20 pounds of effluents for every 5.7 pounds of gasoline in a gallon. A metric tonne, therefore, represents 110 gallons'-worth of effluents which, if taxed at $109 per tonne, would add about $1 to a gallon of gas.

This may not seem an unduly heavy burden, perhaps not one significant enough to drive sizable increases in material efficiency in one of society's largest polluting industries. However, as material intensity began to drop, the Treasury would have an ability to raise the material intensity tax rate accordingly, thereby stimulating further increases in material efficiency. After a number of years, it should not be surprising to find gasoline at the pump costing $10 a gallon, with much less of it being consumed than before. Automakers and their customers by then would have switched to alternate sources of energy. The consumer, paying much less income tax than before, would have relatively more disposable income to pay $10 a gallon at the pump, if they so wished.

Effects

In summary, the expected effects of a material intensity tax may be as follows:

- New goods will become more expensive.
- Consumers will reuse and resell more used goods.
- New products may become designed for greater durability, rather than for planned obsolescence, in part to justify their higher price tags.
- Relatively more new products, particularly large-ticket items, may be leased to consumers. Then, upon retrieval by the manufacturer, they may be repurposed, or re-equipped and re-leased, or recycled.

- Product design could become more modular to accommodate technological improvements to subsystems.
- Assuming the tax burden is shifted from income onto material intensity, people and businesses will have relatively more disposable income.

One important difference between a material intensity tax and income tax is that in the case of a material intensity tax the activity being taxed is geographically specific. Whereas corporations, particularly multinationals, have an ability to offshore revenue to avoid paying taxes within a given jurisdiction, there is no obvious way to offshore material intensity. Assuming primary extraction is taxed at the point of extraction then that point is geographically known. Similarly, a material intensity tax on effluents paid at the point of consumption also can be geographically known, even if a purchase is made online.

This localization has implications for international trade. As Baker et al. identify in relation to a carbon tax, border adjustments may be necessary to "protect American competitiveness and punish free-riding by other nations, encouraging them to adopt carbon pricing of their own".[18] Ultimately, if or when most of the world economy came to operate under a material intensity tax or similar mechanism, such cross-border adjustments could become unnecessary, provided the tax burdens were roughly equivalent across all jurisdictions.

Taxes on Ownership

Shifting the tax burden from income onto material intensity raises a question about how other common kinds of tax might be treated in a materially efficient economy. Two in particular, which are widespread within many national economies, are property taxes and capital gains taxes. They are both forms of taxes on ownership, as distinct from income. Ownership of both real property and capital assets, such as financial stocks, may result in unearned income through appreciation in financial value.

Conventionally, real property and capital gains are taxed. However, there is no overarching economic principle justifying their taxation other than local and national governments' needs to raise revenue. The question that follows, then, is whether the ownership of real property or the receipt of capital gains is generative (good) or extractive (bad). Here, it is the

[18] Baker et al. (2017, p. 1).

ownership structure that should be examined rather than the type of income.

Property Taxes

In the present economic system, title to real property incurs a cost to the title holder in the form of property taxes. It also incurs a benefit, which is an ability to extract rent. An ancillary benefit may be an appreciation in value over time.[19] Property ownership is not at all evenly distributed. Much of it is concentrated in the hands of the wealthy.

As Chap. 8 discussed, property taxes historically have been calculated based on the assessed value of a property in financial terms. It has never occurred to jurisdictions levying such taxes to assess them any other way. However, an opportunity exists to redesign property taxes so that they conform to the generative principle of maintaining or improving natural capacity.

A property tax structure for a planetary economy would shift emphasis from the assessed financial value of land and structures—these being the conventional components of property tax assessments—onto the degree to which a property is managed to support natural capacity. Even though natural capacity cannot be directly measured, a variety of proxies can be used. Natural capacity assessments that could result in property tax increases include such actions as clearing naturally occurring habitats for the building of structures or for agriculture. Conversely, the restoration of formerly developed land to a natural state would reduce property taxes. Restoration would not necessarily have to return a habitat to its original state to qualify. Examples would include the regeneration of urban wastelands or the restoration of wetlands formerly drained for agriculture. Similarly, the construction of housing on vacant urban lots could benefit from lower property taxes relative to new developments on the edge of a town. The effect would be to encourage regeneration of urban centers on land already developed, rather than promoting outward sprawl. Generally speaking, urban densification will benefit natural capacity. Property tax benefits also would be available for agricultural land on which ecological function were encouraged.

[19] On average property values should be expected to increase while the population continues to rise, although it will depend upon the specific market.

Table 14.7 Approximate conformance of a property tax structure supporting natural capacity to the four principles of economic controls for a planetary economy

Material discipline	
Economic stability	
Widespread prosperity	
Innovation through markets	

Source: Author's creation

Note: Darker shading denotes relatively greater conformance

Table 14.7 summarizes approximately how a property tax structure to support natural capacity conforms to the four principles of economic controls.

A side effect of this change to property taxes would be to increase demand for the services of ecologists on behalf of taxing jurisdictions. The office of a county assessor-recorder, as it is known in the US, would include not only traditional property evaluators but also professionals with experience in evaluating the condition of natural habitats.

In this case, the tax burden is shifted from a financial quantity onto one or more proxies for natural capacity, the intent being to encourage generative rather than extractive uses. While the shift would not necessarily encourage broader property ownership, it would at least hold property owners, large and small, accountable for the natural capacity within their property lines. This measure would dovetail with the proposed changes to property titles themselves outlined in Chap. 8, in which title would be granted contingent upon a commitment to maintain or improve natural capacity.

Adjusting property taxes according to stewardship of natural capacity should not be confused with the so-called Georgian economics of taxing the economic value derived from land—named after the nineteenth-century economist Henry George. George maintained, correctly, that the taxation of productive labor was inefficient, perverse and unfair, arguing that government should raise revenue instead from a land value tax, which was an appropriate tax on economic rent.[20] Indirectly, a land value tax would have brought distributional benefits, and could have been extended to other forms of private capital. George's arguments might have made sense in the 1800s but they did not anticipate the global scale of the cur-

[20] George (1879).

rent economy's material intensity. A material intensity tax is a more appropriate instrument for a global, materially significant economy, since it exerts its greatest effects upon materially inefficient economic actors, whether they hold land or not. Adjusting property taxes according to the degree of stewardship of natural capital implicitly defines the 'value' of land as its natural capacity.

Capital Gains Taxes

Theoretically, it is perverse within the milieu of a market economy to tax capital gains. A market economy, by definition, treats capital gains as good, since they are an outcome of the competitive processes operating within a market. However, in the present economy, those competitive processes are exploitative upon both natural capacity and—to varying degrees—human capital. In today's growing, increasingly unequal economy, the issue is not with capital gains in and of themselves but with who benefits from them. Those holding capital to begin with stand to gain at the expense of those who do not.

Today, capital gains tax receipts flow into the treasury or the exchequer. Arguments for an increase in capital gains taxes are, for all intents and purposes, arguments for governments to raise additional revenue. They miss the point that the capital gains enjoyed by large market players are not distributed. Capital gains receipts from the top 1 percent could be disbursed as cash to the bottom 1 percent, and from the next-highest 1 percent to the next-lowest 1 percent, and so on, similar to the mechanism proposed for elective redistribution. However, some of the other measures proposed here, particularly relating to money and finance, might render such reform unnecessary because they would eliminate much of the outlandish capital gains from the financial markets in the first place.

In a planetary economy, there might be no need for a capital gains tax at all. The rate of gain in financial capital within an economy approaching long-term material stability would be relatively low. Gains in financial capital would come through innovations in the use of a relatively constant material recirculation rather than through material increase. Provided the incentives within such an economy were set up to reward increasing material efficiency, capital gains would be encouraged rather than discouraged. Taxes on them could be abolished, provided other mechanisms already outlined maintained an income floor and constrained the distribution of incomes.

Taxation Policy

To summarize, the taxation policy for a planetary economy would incrementally shift the tax burden from goods, such as income and capital gains, onto bads, such as material intensity and other practices detrimental to natural capacity. Table 14.8 summarizes this policy.

This kind of taxation policy can be adopted today. By introducing it incrementally, the risk of economic surprises arising from unintended consequences would be minimized, and any unintended effects mitigated. Its success would be predicated upon the adoption of other economic controls for economic stability and widespread prosperity. The taxation policy for a planetary economy would necessarily work in concert with other economic policies for widespread prosperity.

Under this taxation policy, households and businesses would find goods and services flowing through the economy more expensive. However, by paying significantly lower income taxes and capital gains taxes, they would have more disposable income with which to select more materially efficient goods and services within the market, thereby rewarding innovations in material efficiency. Households and businesses owning real property would be rewarded for supporting natural capacity by paying relatively lower property taxes.

The power for change would be placed to increasing degrees into the hands of economic actors, who would be able to make decisions within markets structured for material discipline. Substantially all economic actors would have this ability: virtually none would be excluded.

Tax Code

The tax code for a planetary economy could be significantly simplified and shortened, compared to present codes, provided it reflected the four principles outlined above. The US Internal Revenue Code, for example, is nearly 10,000 pages long in its entirety. The parts dealing with income and employment taxes comprise about 2500 pages.[21] Although much of this text documents legislative changes over time, the substance of the tax code could be reduced in size and complexity through the enactment of a material intensity tax, reductions in income and capital gains taxes, and the establishment of a basic income scheme. Further complexity could be

[21] US GPO (2009), Subtitles A to C.

Table 14.8 Summary of taxation policy for a future planetary economy, showing changes from current (early twenty-first century) policies prevalent within most market economies

Type of tax	*Relative tax burden*	
	Current	*Future*
Income tax	High	Low
Material intensity tax	Non-existent	High
Property tax	High	Low if supportive of natural capacity; high if not
Capital gains tax	Medium	Abolished

Source: Author's creation

eliminated if, as a companion to basic income, government were to provide health care and education largely free of charge, services which could be funded through sovereign money creation or common-resource dividends.

Table 14.9 summarizes how selected Sections from the US Internal Revenue Code could be modified or replaced under a tax code for a planetary economy. Similar modifications to other countries' tax codes would be possible.

Sections of the tax code dealing with the economy's interface to nature—such as Sections 30, 45, 169, 175, 179A and 194—could be phased out over time as a material intensity tax took effect, and as regeneration of natural capacity were rewarded through reductions in property taxes. Certain itemized deductions would become unnecessary as materially intense activities became more expensive and taxpayers acquired relatively more disposable income.

A basic living scheme would reduce or eliminate the need for various low-income tax credits, such as relate to housing, education and health care, as well as unemployment insurance. Chapter 23 of the IRC could be eliminated in its entirety as long as basic income provided at least the same level of support as unemployment insurance.

There would be little need for government intervention in the capital gains and losses of households or business, provided such gains and losses were realized legally. Government oversight thus would extend to the legality of the transactions rather than adjustments to the tax basis, which in the US falls under the Securities and Exchanges Commission rather than the Internal Revenue Service.

Table 14.9 Selected Sections of the US Internal Revenue Code (26 USC) related to income and employment taxes showing how such taxes might be treated under a modified tax code for a materially stable US economy

US Internal Revenue Code Section	Tax code for planetary economy
Subtitle A—Income Taxes	
Chapter 1—Normal Taxes and Surtaxes	
Subchapter A—Determination of Tax Liability	
Part I—Tax on Individuals Part II—Tax on Corporations	Significantly reduced rates over time as tax burden shifted onto material intensity.
Part IV—Credits Against Tax	
Subpart A—Nonrefundable Personal Credits	
§21. Expenses for household and dependent care services necessary for gainful employment §22. Credit for the elderly and the permanently and totally disabled §23. Adoption expenses §24. Child tax credit	Might not be needed if covered by basic income or if dependent care covered by government in place of a portion of basic income (see under Subsidies in Chap. 15).
§25A. Hope and Lifetime Learning Credits	Might not be needed if covered by basic income or if education provided free by government (see under Subsidies in Chap. 15).
Subpart B—Other Credits	
§30. Credit for qualified electric vehicles	Replaced with material intensity tax and repealed.
Subpart C—Refundable Credits	
§35. Health insurance costs of eligible individuals	Might not be needed if covered by basic income or if health care provided free by government (see under Subsidies in Chap. 15).
Subpart D—Business-Related Credits	
§42. Low-income housing credit	Might not be needed if covered by basic income.
§45. Electricity produced from certain renewable resources	Replaced with material intensity tax and repealed.
§45F. Employer-provided child care credit	See Sections 21 to 24 above.
Subchapter B—Computation of Taxable Income	
Part VI—Itemized Deductions for Individuals and Corporations	

(continued)

Table 14.9 (continued)

§169. Amortization of pollution control facilities	Represents an incentive for pollution control. May be retained in near term, repealed in long term under low-income-tax régime.
§174. Research and experimental expenditures	Could be expanded in near-to–medium term to stimulate innovation.
§175. Soil and water conservation expenditures	An incentive to improve natural capacity. May not be needed in long term as other incentives take hold.
§179A. Deduction for clean-fuel vehicles and certain refueling property	Replaced with material intensity tax and repealed.
§194. Amortization of reforestation expenditures	An incentive to improve natural capacity. May not be needed in long term as other incentives take hold.
Subchapter P—Capital Gains And Losses	All Sections (1201 to 1298) eventually could be repealed under shift of tax burden onto material intensity, provided such repeal did not produce adverse distributional effects.
Chapter 2—Tax on Self-Employment Income	Significantly reduced rates over time as tax burden shifted onto material intensity.
Subtitle C—Employment Taxes	
Chapter 21—Federal Insurance Contributions Act	Basic income would supplement social security for those individuals whose work may not be compensated by social security. Homemakers and people who do occasional work for cash are examples.
Chapter 23—Federal Unemployment Tax Act	Could be replaced in its entirety by basic income.

Source: Author's creation, from US GPO (2009), Subtitles A to C

Tax credits pertaining to health and dependent care and to education could be eliminated if such services were largely free. Not only would the tax code become simplified but so would the lives of millions of households and businesses.

References

Baker, J. A., et al. (2017). *The Conservative Case for Carbon Dividends.* Washington and London: Report of the Climate Leadership Council.

DeNavas-Walt, C., & Proctor, B. D. (2014). *Income and Poverty in the United States: 2013. Current Population Reports.* Washington, DC: United States Census Bureau.

Fisher, I., & Fisher, H. W. (1942). *Constructive Income Taxation*. New York: Harper Collins.

Frank, R. H. (1999). *Luxury Fever: Why Money Fails to Satisfy in an Era of Excess*. New York: The Free Press, Simon & Schuster.

Friedman, B. M. (2006). The Moral Consequences of Economic Growth. *Society, 43*(2), 15–22.

George, H. (1879). *Progress and Poverty: An Inquiry into the Cause of Industrial Depressions and of Increase of Want with Increase of Wealth: The Remedy*. Garden City, NY: Doubleday, Page & Co..

Guj, P. (2012). *Mineral Royalties and Other Mining-Specific Taxes*. Perth, Western Australia: Report of the International Mining for Development Centre.

Internal Revenue Service. (2014). Internal Revenue Service Data Book, 2013. Publication 55B, IRS, Washington, DC.

LeGrand, J. (1991). *Equity and Choice: An Essay in Economics and Applied Philosophy*. London: HarperCollins Academic.

Otto, J., et al. (2006). *Mining Royalties: A Global Study of Their Impact on Investors, Government and Civil Society*. Washington, DC: Report of the World Bank.

PricewaterhouseCoopers. (2012). Corporate Income Taxes, Mining Royalties and Other Mining Taxes: A Summary of Rates and Rules in Selected Countries. PricewaterhouseCoopers global mining industry update, June 2012.

US GPO. (2009). *United States Code, 2006 Edition, Supplement 3, Title 26: Internal Revenue Code*. Washington, DC: US Government Publishing Office.

Wilkinson, R. G., & Pickett, K. (2009). *The Spirit Level: Why More Equal Societies Almost Always Do Better*. London: Allen Lane.

World Bank. (2016). State and Trends of Carbon Pricing, 2016. Washington DC.

CHAPTER 15

Economic Controls 4: Subsidies, Incentives and Market Instruments

Subsidies and Incentives

In addition to taxes, which can render undesired behaviors less attractive than desired ones, governments have at their disposal instruments to render desired behaviors relatively more attractive. These are subsidies and incentives (Table 15.1).

The difference between a subsidy and an incentive is that a subsidy helps a household or business to pay for something, constituting a form of financial aid, whereas an incentive offsets the price of a good or service in a market as a way to influence consumer behavior. A subsidy is somewhat broader than an incentive, in that it can support the cost of certain needs for which markets may be ill-suited.

Indeed, a great deal of political debate in modern market economies revolves around the extent to which such services as health care and education can or should be provided through markets. The fashion since the rise of market liberalism in the 1970s has been to impose market doctrine upon these areas, as evidenced by the health care market in the US and such concepts as charter schools in education. However, as Julian LeGrand argued, market liberalism tends to conflate the two. Preference is a choice typically made within a market. As such, it may be tunable through the application of incentives. Need, by definition, is not a choice. People have needs for health care and education, independent of any choices that may be on offer. Those needs may be subsidized if people are unable to afford

F. Murison Smith, *A Planetary Economy*,
https://doi.org/10.1007/978-3-030-49296-0_15

Table 15.1 Approximate conformance of subsidies and incentives for material efficiency or natural capacity to the four principles of economic controls for a planetary economy

Material discipline	
Economic stability	
Widespread prosperity	
Innovation through markets	

Source: Author's creation

Note: Darker shading denotes relatively greater conformance

any of the choices. Or, put another way, their needs could be satisfied through preference if more than one of the available options were affordable.

Subsidies

Subsidies for a planetary economy would, where applicable, support needs for material efficiency or for the improvement of natural capacity in situations where markets may not exist or be poorly suited to the allocation of available resources. In the interest of widening prosperity, other services also may be subsidized, such as through sovereign money creation or capacity use fees. These include health and dependent care and education, as discussed earlier.

Material Efficiency Subsidies

In today's growth economy, subsidies can help to accelerate the transition toward material efficiency. In the long term, there may be no need for material efficiency subsidies, since a material intensity tax would do the work of pricing material efficiency favorably relative to material intensity.

A recent study by David King and others estimated the global total of subsidies for renewable energy production in 2015 at about one-fifth of the total of subsidies for fossil-fuel-based energy: $101 billion compared to $500 billion.[1] These subsidies are paid mostly from taxes. This means that any taxpayer concerned about atmospheric emissions would continue to subsidize fossil-fuel emissions through their income taxes, even if they were to divest themselves of all fossil-fuel-related stocks. Taxpayers have a choice about where their taxes are spent only through the ballot-box

[1] King et al. (2015, p. 14).

which is, at best, a weak medium through which to have a voice in government spending.

The balance of subsidies between fossil fuel energy and renewables, as King et al. note, seems perverse in the face of mounting evidence of fossil-fuel-driven climate change. In any case, it violates principle 1 above (material discipline) and possibly principle 2 (economic stability). The question King et al. do not ask is what might happen if the subsidies were flipped, so that $500 billion would go to renewables and $101 billion to fossil-fuel-based energy.[2] In order to answer that question, the scale of these respective industries, in terms of their costs, needs to be known.

Conveniently, King et al. provide the data to answer this question, at least approximately. Table 15.2 shows how much energy is produced worldwide from various sources, by percentage and by the average levelized cost of energy (LCOE) from those sources. LCOE is the cost of energy produced from a generating source over its typical operating lifetime.

The data in Table 15.2 allow an estimate of the weighted average annual cost of all energy produced worldwide from renewable sources (solar,

Table 15.2 Composition of electrical energy produced worldwide in 2012, showing also total production in terawatt-hours, and approximate, 2013 average levelized cost of electricity (LCOE) in dollars per megawatt-hour

Composition of electrical energy produced	*Pct. (2012)*	*LCOE, \$/MWh (2013)*
Solar photovoltaic	0	150
Onshore wind	3	80
Biomass	3	130
Hydro	16	70
Nuclear	13	90
Gas	18	90
Coal	46	90
Other	1	200
Total	100	
Production/TWh	20,725	

Source: Author's creation, from data in King et al. (2015, figures A1 and A3)

[2] A recent study by Bridle et al. (2019) asks a similar question and comes to similar conclusions as here.

Table 15.3 Subsidies to renewables and fossil fuels, worldwide, 2013

Source	*Subsidy ($billion)*	*Total annual LCOE ($billion)*	*Notes*	*Total annual LCOE w/o subsidy*	*Total annual LCOE if subsidies flipped*
Renewables	101	363	Excludes nuclear	464	(80)
Fossil fuels	544	1194	Excludes other	1738	1637

Source: Author's creation, from data in King et al. (2015, table 2), based on data from the International Energy Agency, cited therein

Note: The total levelized cost of energy from these sources comes from Table 15.2, renewables being the sum of solar photovoltaics, wind, biomass and hydropower, by proportion of production, and fossil fuels being the sum of gas and coal, by proportion of production. The table shows the total LCOE without the subsidy, then with the subsidy flipped, i.e., if the fossil-fuel subsidy were applied to renewables, and vice versa

wind, biomass and hydropower) compared to that from the fossil sources listed by King et al. (gas and coal). These costs are shown in Table 15.3.

The difference in size of the two industries is significant. The total annual levelized cost of producing fossil-fueled energy is about $1.2 trillion per year, compared to about $0.36 trillion for renewables. If the subsidies were flipped then the total annual levelized cost of producing energy from renewables would turn negative: in other words, renewable energy producers would be paid by governments simply to exist, which is perverse. The total annual levelized cost of producing energy from fossil sources would increase to about $1.6 trillion per year.

A reasonable middle ground would eliminate subsidies altogether from fossil-fuel production and increase subsidies to renewable production to drive the average annual LCOE close to zero. As an illustrative example, if subsidies for renewable production totaled $450 billion and subsidies for fossil-fuel production were zero then about $95 billion in subsidies would be saved worldwide. Because the renewable basis would grow relative to the fossil-fuel one, that growth probably would soak up the annual savings over time. At some point, the renewable energy industry would cease to be fully subsidized. As economies of scale took hold and the renewable basis began to displace the fossil-fuel one, its subsidies would be scaled back.

Energy costs in fossil-fuel-dependent regions would increase, while those in regions able to access renewable sources would decline significantly. Such an approach may increase geographical inequities in energy costs over the short term. However, over the medium term, it also would drive the development of renewable sources of production or procurement in regions formerly dominated by fossil-fueled generation. The effect would eventually even out any geographical inequities in the distribution of energy costs, particularly within countries having integrated energy grids.

Over time, the subsidy-free LCOE of renewable energy sources would fall below that of fossil sources. Some time after the energy industry ceased investing in new fossil plant, there would follow a tipping point at which the industry would accelerate the decommissioning of existing plant. At or around that point, the subsidies would have done their job and market forces pure and simple could take over, driving the industry toward a new state dominated by renewables. If the subsidies were flipped early in the twenty-first century, then the new state could be achieved in the second half of it. The timing of the application and progressive removal of the subsidies, and their effects upon the energy industry, are topics for exploration using scenario models.

Natural Capacity Subsidies

Natural capacity fees levied on corporations could subsidize the maintenance and improvement of natural capacity in a variety of ways. One would be payment to leave ecosystems intact, as Norway does to Guyana. A non-exploitation subsidy of this kind could, in principle, be applied at many institutional scales, from the nation all the way down to the household or business. However, at the level of a private landowner, a discount on property taxes might be a more appropriate instrument. It could be applied into perpetuity, whereas a subsidy should be designed to eventually terminate.

A second, perhaps more appropriate vehicle for natural capacity subsidies is the regeneration of previously degraded habitats. This is an activity with a finite schedule: once a habitat is restored in some shape or form, the subsidy would go away. This category of activity also includes pollution mitigation. The clean-up of polluted areas, whether the removal of contaminants from soils or plastic from the oceans, reduces the amount of harmful material making its way into food chains. Subsidies for natural capacity regeneration could work in concert with discounts to property taxes by providing financial support for regeneration, at which point discounted property taxes would take over to encourage long-term maintenance.

A third vehicle for natural capacity subsidies is the encouragement of regenerative natural resource management. Perhaps the greatest opportunity in this area lies in replacing exploitative agricultural practices with those that regenerate natural systems, such as organic, biodynamic or polycultural practices. A parallel line of development is to partially or wholly replace field agriculture altogether with vertical farming and the in vitro production of animal protein.[3] Subsidies to support this 'factory' agriculture, as it might be called, would shift demand for food away from field-based agriculture, reopening land to ecological restoration.

Data on agricultural subsidies serve to illustrate the potential scale of the shift from exploitative to regenerative practices. Generally speaking, the single largest regenerative practice in today's early-twenty-first-century economy is organic farming, although it is only partially regenerative. Organic farming does away with synthetic herbicides, pesticides and hormones, and it encourages the use of natural pest controls. However, it does not specifically require the use of polycultures or mixed-ecological cultures. Nonetheless, the data that exist pertain mostly to organic farming which, for purposes of illustration, can be treated as a proxy for other, more fully regenerative practices.

World Bank data from 2014 report that the value-added percentage of GDP from agriculture averages about 3.9 percent, globally. Given a total global GDP of about US$78 trillion, this means roughly US$3 trillion of that comes from agriculture.[4] According to the Worldwatch Institute, about US$486 billion was paid in agricultural subsidies worldwide in 2012,[5] which equals about 16 percent of agricultural GDP.

The organic agriculture market was estimated at around US$81.6 billion in 2015,[6] which is less than 3 percent of all agriculture. Subsidies for organic agriculture, as of the early twenty-first century, are few and far between. Given the vastly greater scale of non-organic farming, the relative amount of subsidies to organic farming therefore is negligible.

If organic agriculture were fully subsidized as of 2015 then roughly US$404 billion would be saved in agricultural subsidies annually. Since agricultural subsidies are raised principally from income taxes, mostly

[3] Referred to by some of its proponents as 'cellular agriculture'.

[4] http://data.worldbank.org/topic/agriculture-and-rural-development.

[5] http://vitalsigns.worldwatch.org/vs-trend/agricultural-subsidies-remain-staple-industrial-world.

[6] Willer and Lernoud (2017).

within OECD countries plus a handful of other Asian countries, they are funded from the income taxes of about two billion people. Assuming these two billion people belong to roughly 500 million households, a US$404 billion savings would represent an average of about US$800 in annual income tax savings per household, if they were passed along to the taxpayer.

These numbers are very approximate, of course, although they illustrate the scale of the perversity of subsidizing exploitative agriculture. Shifting subsidies onto organic agriculture would tilt the playing field in favor of regenerative practices. The organic agriculture market would grow, along with factory agriculture and urban small-scale agriculture. Farmers in poorer countries would find themselves more competitive with multinational agribusinesses. With US$404 billion per year to soak up, there would be a great deal of headroom to fully subsidize regenerative practices for some time. It would be appropriate for some of those subsidies to be funneled to farmers in poorer countries, through a multilateral agreement among OECD countries and other high-subsidy Asian countries, provided the subsidies could be ensured to reach the intended recipients.

As in the energy sector, the shift in subsidies would encourage exploitative producers to modify their practices, not only to avail themselves of the subsidies themselves but also to ensure their long-term competitiveness within a market moving inexorably toward regenerative practices. Prices at the store might increase for a number of years until regenerative agriculture came to dominate. However, as discussed earlier, the concomitant shifting of tax burdens and the establishment of an income floor would offset the effects of these increases for the consumer.

An argument could be made that a transition to new field practices would result in lower yields, a point of concern within the context of a growing population. Several reasons exist why this might not be the case. While mechanized monoculture likely will produce a higher total yield than an agroforested plot, much mechanized farming grows crops only for cattle feed. Reducing the consumption of beef, or growing it in vitro, allows the land in question to be turned over to food crops. A material intensity tax would encourage this shift. Beef raised on the land could become as expensive as caviar and, if cellular agriculture took off, it would become out-competed on price and go extinct. Agricultural yields might continue to increase anyway with the spread of factory agriculture, which already produces much higher yields per unit area than conventional farms. The factory environment can be finely controlled, with racks of crops stacked vertically, producing crops year-round.

A transition from exploitative to regenerative practices could occur more quickly in agriculture than in the energy industry because it is not constrained to the same extent by large, fixed, or stranded physical assets. If the subsidies were flipped today, the transition could be made in a few decades. The natural landscape would become significantly altered over a relatively short period, just as happened during the transition to mechanized agriculture in the first place.

Incentives

Like subsidies, incentives are intended to operate over a finite lifetime. They are established to move a market in a particular direction, such that the price of a more-desired good or service becomes competitive with the price of a less-desired good or service. Whereas subsidies may be applied before a cost is incurred—such as in agricultural subsidies—incentives typically are applied at or after the point of sale. Generally, a subsidy is applied to production within the economy, whereas an incentive is applied to consumption. Popular examples in early-twenty-first-century markets are incentives for plug-in hybrid or electric vehicles, which typically are applied at the point of sale, and rebates for solar photovoltaics, which, as the name implies, are disbursed after the product has been purchased.

The mirror instrument of a material intensity tax is a material efficiency incentive. The vendor of a good or service would receive an incentive, usually from the government, to lower its price based upon the material efficiency of its production and supply, the measurement and accounting of which would follow the same outline as for a material intensity tax. The intent would be to stimulate consumer demand for goods and services having greater material efficiency. Ideally, an increase in demand should more than compensate for the reduction in unit price as it is passed up the supply chain.

Administratively, it would be cumbersome to apply the incentive to consumers themselves, and in any case that approach might defeat the object of lowering the price at the point of consumption. It would be more practical to provide the incentive to the vendor, who passes it on to the consumer through lower prices. However, if the value of the incentive went to the consumer then the unit price of the good or service from the wholesaler to the vendor would fall accordingly. A risk might arise that the wholesaler, and producers further up the supply chain, might be squeezed to a point where the price leaves no margin for costs and overhead, at which point it would become uneconomic to produce.

A solution would be to apply a material efficiency incentive in concert with the material intensity tax and the subsidies on materially efficient production. The cost of production thus would be lowered for materially efficient goods and services, allowing the supply chain to cover its costs and overhead. A material intensity tax would enhance the effect of attracting consumers to relatively more materially efficient goods and services. Revenue to the government from the tax could be used to fund the subsidies and incentives. In this way, economic policy would steer markets toward greater material efficiency.

Over time, as the volume of the materially efficient supply chain grew, and the materially intense one shrank, the subsidies and incentives would be phased out. The material intensity tax would remain in place to maintain the price differential within the market. The purpose of the subsidies and incentives would be to catalyze and accelerate the process to a point of inflexion where material efficiency became economically advantageous. Although revenue to the government from a material intensity tax would fall over time, concurrent expenditures on the subsidies and incentives also would decline.

There is theoretically no limit to the range of goods and services subject to these instruments. In fact, the approach probably should be applied to all goods and services within the economy, to forestall substitution. The challenge, of course, is to establish the rates of these various incentives so as to guide markets as smoothly as possible toward material efficiency. This is why policymaking itself needs to be iterative and adaptive.

Market Instruments

A final category of economic controls for a planetary economy is those operating directly on the functioning of markets. The distinction between these controls and taxes or incentives is not hard and fast, as taxes and incentives themselves influence decisions made in markets. However, while taxes and incentives generally influence purchase decisions at the point of consumption, certain other instruments may create entirely new markets. New markets having the potential to guide the economy toward greater material efficiency would be encouraged under a prime directive of material discipline. This section examines a few possibilities.

The creation and operation of a new market inherently involves the establishment of one or more instruments, which are priced within that market, along with the rules of market operation. A new market thus requires both instrumental and regulatory components in order to come into existence.

Tradable Quotas

The most conventional and well-known form of market instrument to promote material efficiency is the tradable quota. In general, a quota of material intensity is established by a regulatory agency and is updated on a regular basis, such as annually. Market participants whose operations use the material in question would be subject to an upper limit on its consumption. Those whose operations are expected to exceed the limit would be able to purchase quotas. Those whose operations fall under the limit would be able to sell them (Table 15.4).

This 'cap and trade' system has been applied in a number of industries. Emissions markets, for example, have become established in several US states, Canadian provinces, the European Union, New Zealand, China, South Korea and Japan as of the late 2010s. The markets differ from one to another in the details of their implementation, although they all regulate carbon dioxide emissions. US EPA emissions trading markets additionally regulate sulfur dioxide and oxides of nitrogen, while a regional market in Illinois regulates volatile organic compounds in the Chicago area. A well-established market also exists in the US for tradable renewable energy certificates in the electric utility industry. These certificates give energy producers flexibility in meeting renewable energy quotas for electric generation. China has two programs, one for carbon dioxide and the other for pollution permits on other chemical species. In 2017, an energy

Table 15.4 Approximate conformance of tradable quotas to the four principles of economic controls for a planetary economy

Material discipline	(dark shading)
Economic stability	
Widespread prosperity	
Innovation through markets	(light shading)

Source: Author's creation

Note: Darker shading denotes relatively greater conformance

consumption trading market was added for industrial customers in several provinces. Tradable quotas also have been explored for fisheries, with mixed success.

The success of tradable quotas in commercial and industrial sectors stimulated experimentation, beginning in the 2000s, in a handful of consumer markets. In 2007, for instance, trials of tradable carbon quotas began in Manchester and Cardiff, UK. Consumers using less than their quota could sell unused allocations back to a carbon bank for purchase by high-carbon users. The quotas paid for such commodities as fuel, power and air travel.[7] However, a national system was not implemented. Other jurisdictions, such as British Columbia, have explored the concept,[8] although this sector remains very much in its infancy.

Peter Barnes suggested a variant that combines elements of both systems, in which 'pollutees' (i.e., consumers) are represented by a trust, which sells pollution rights to polluters. The trust gradually reduces the number of permits available until a safe level is reached and polluters have developed alternate technologies. Revenue from the sale of pollution permits benefits the commons. It could, for instance, be divided among a cash dividend to all citizens, improvements to education, and ecological restoration.[9] As yet, no such scheme has been implemented.

The Kyoto Protocol and the European Union Emissions Trading System (EU ETS) recognize the importance of expanding emissions trading beyond national frontiers to allow nations to trade quotas off against one another. Because nations have different marginal abatement costs—the cost of abating an additional increment of pollution—the total abatement in theory could be optimized by enabling international trade in quotas. For this reason, the EU ETS incorporates both quotas and offsets,[10] the latter behaving rather like the tradable renewable energy certificates mentioned above. However, existing systems do not always account adequately for technology costs in abatement.[11]

Nonetheless, cap-and-trade systems on atmospheric pollution have been shown to drive down overall emissions within the countries where they have been implemented. This has been accomplished through improved efficiencies, technological innovations and adoption of

[7] Wilkinson and Pickett (2009, p. 222).

[8] Guzman and Clapp (2017).

[9] Barnes (2006, pp. 90–91).

[10] Braun et al. (2015).

[11] Golombek and Hoel (2006).

substitutes. A question remains whether cap-and-trade has been, or could be, the most effective instrument to drive rapid, large-scale migration from polluting industries to non-polluting ones. Since the introduction of cap-and-trade, emissions have fallen within the participating jurisdictions, although worldwide they have continued to rise, an issue known in the field as leakage. Given a choice of comparable jurisdictions in which to do business, companies can be expected to select those with the least costs. A material intensity tax, or a carbon tax, is very simple and clear. It also raises revenue for government. However, it does not impose a physical limit as such. A cap-and-trade market is only as effective as it is comprehensive: if certain companies or industries are exempt then its effectiveness may be weakened. It has the advantage, however, of imposing absolute and declining physical caps on emissions, something a tax does not do.

The most effective solution likely is not one or the other but both. Combining a material intensity tax with cap-and-trade markets for material intensity should combine the best of both instruments, if implemented synergistically. The quantity of material removed from nature or sent back to it would decline at the same time as its market price increased. Selection pressure like this, exerted on both material and monetary quantities, would drive industry into rapid innovation, provided the pressure is not beyond industry's ability to adapt. However, implementation may have to be effectively global for the greatest reductions in material intensity to be seen.

Financial Instruments

Markets for tradable quotas, being markets for physical commodities, shape the financial valuation of material intensity. Any investment in a new project or venture made under a régime of capped and traded quotas, a material intensity tax, and subsidies and incentives for material efficiency, would appear more attractive financially the more materially efficient the project was projected to be.

All well and good for direct financial investments; however, as noted earlier, most of the volume of trade within the financial markets is speculative. Although the secondary market would be indirectly influenced by these instruments, they alone would not curb any excess of speculation. As yet, no instrument has been developed to harness financial speculation to improve natural capacity. The earlier section on natural capacity fees suggested one: fees on financial trades. This section discusses another: natural capacity share-ownership.

Natural Capacity Share-Ownership

What if the very natural systems on which all economies, rich and poor alike, depend were valued directly within the financial markets? What if people living within any one of the twenty-five internationally recognized ecotypes[12] owned shares in those systems, shares that had value on the financial markets? What if everybody on Earth owned shares in the atmosphere or the oceans, those systems also having value on the world's financial markets, like multinational corporations? What if one could purchase futures or options in the value of these shares, and make, or lose, money by speculating on movements in value? Could these kinds of markets become established, and would they help to bind flows of money within the economy and flows of material through it into mutually reinforcing loops?

Although everybody already is a stakeholder in the world's natural systems, owning an equal stake in their functionality and long-term stability, there does not yet exist a structure for people to have agency through which to exercise these interests. Legal attempts to establish the foundations for this agency rested on suits brought against governments for failure to act over such issues as climate change. In the corporate world, shareholders of some wasteful companies have begun to exercise their agency, voting to require management to disclose risks to share value of any exploitative operations, along with plans to mitigate those risks. Both kinds of measures are weak, being more likely to deliver too little change too late than the rapid, systemic change to the economy that is needed (Table 15.5).

The difference between a stakeholder and a shareholder is that a shareholder owns a specific quantity of shares in an asset, and has access to an institution through which to represent their interest in the asset, whereas a stakeholder holds only an informal interest. In the case of a publicly traded corporation, the asset owned by a shareholder is the corporation itself and the institution is the corporate structure, consisting of shareholders, directors and management, and their interrelations. In the case of

[12] These include such types of ecosystem as lowland rainforest, cloud forest, savannah and boreal tundra.

a natural system, the asset is the system itself; however, there does not exist a corresponding institution as yet to formalize it.

Being a shareholder presumes a market of some kind through which to buy or sell shares in an underlying asset. For natural capacity, there would need to exist a market or markets for natural systems indicating the value of those systems to their shareholders. These markets would have to be accessible to everybody in order for everybody to have agency and for their shares to have value. If the shares were tradable then a risk of consolidation of share-ownership would arise. People who sold their shares would lose agency within the market and, with it, their stake in natural capacity.

Notwithstanding these questions, an opportunity exists for an institutional structure to establish a market for natural capacity. If ever a perfect role for a planetary trust existed, this would be it. A government-established trust for natural capacity share-ownership would fulfill the role of government for a planetary economy articulated in Chap. 12, which is to rent capacity upon which enterprises generate wealth, the rent serving to maintain this capacity.

As part of a well-established practice of assessing the health of natural systems, scientists regularly sample and monitor the atmosphere, soils and oceans. Ecologists also regularly assess the state of the world's ecosystems. As noted earlier, this work has served as a foundation for estimates of the economic value of nature in money terms.[13] However, a money estimate of nature's value would be unnecessary for a functioning market in natural capacity. Instead, all that would be needed is a rating, analogous to the existing financial ratings of corporations and of the material efficiency rating proposed earlier. This would be a natural capacity rating, applied to the 25 ecotypes, as well as soils, atmosphere and water systems. It would

Table 15.5 Approximate conformance of natural capacity share-ownership to the four principles of economic controls for a planetary economy

Material discipline	(dark shading)
Economic stability	(no shading)
Widespread prosperity	(light shading)
Innovation through markets	(light shading)

Source: Author's creation

Note: Darker shading denotes relatively greater conformance

[13] For example, Costanza et al. (1997).

be reviewed and amended by scientific consensus. It could be applied at a variety of scales, from the local to the regional to the global. Obviously, a great deal of interaction takes place between an ecotype and its substratum, the air and water around it: the boundaries are fuzzy rather than crisp. The natural capacity ratings of all these systems would influence one another. These interactions, however, would not invalidate the ratings: they would reinforce the necessity to maintain and improve the interconnected whole. Just as in the corporate world, the ratings would influence the share prices. If generative management of a regional biome caused its natural capacity rating to rise, then the value of shares in that biome also would be expected to rise.

In keeping with Berg's bioregionalism, share-ownership in ecosystems and soils might be regional, whereas share-ownership in the atmosphere and oceans more appropriately would be global. The bioregional nature of ecological ownership means that people from multiple neighboring jurisdictions would share common interest in the health of a given ecosystem, and also that people within the same province or country might hold shares in different ecosystems, depending where they live. Share-ownership of African tropical rainforest, for instance, would be held by people living in central Africa. People from the same country might hold shares in different ecotypes depending where they lived. Angolans in the north of the country might hold shares in tropical rainforest, whereas those living in central or southern Angola would hold shares in tropical scrubland or savannah.

These bioregional considerations inform the institutional structure and administration. The trust would have to be supranational, not only because biomes cut across national borders but also for consistency among natural capacity ratings and neutrality in the administration of shares. Good census data would be essential, to protect against fraudulent acquisition of shares, and to ensure new shares are issued to people born, and existing shares retired upon death. Regional or national jurisdictions would pass census data up to the planetary trust, which would review the data, reconcile any anomalies, and pass down new shares and records of retired shares for distribution. As with elective income redistribution, a significant challenge would be to ensure the benefits made it into the hands of their intended recipients. The challenge would be greatest in regions with weak internal institutions.

Each person would be allotted one share in their resident biome, including its soils, along with one global share in each of the atmosphere and oceans. The shares would be non-tradable, instead returning a dividend when the health rating of any of the systems in question improved from one year to the next. As with common capacity fees, this system would promote common asset security, the common asset in this case being natural capacity.

Valuing the dividend in money terms would introduce the now-familiar, intractable problem of placing a money value on the intrinsic value of natural systems. A workaround exists, however. Assuming the program would operate on a global standard, a parallel currency could be introduced, which we could call a natural capacity unit (NCU). An increase in the rating of a given biome or other natural capacity would translate into a dividend valued as a specific number of NCUs. These NCUs would not be redeemable as conventional money; however, individuals would be able to sell them as offsets to corporations whose operations diminish natural capacity. A market would be established along the lines of an emissions trading market or a renewable energy credit market in which corporations needing to improve their own material efficiency rating, or to meet certain environmental standards for projects, could purchase NCUs to contribute toward these goals. Agencies in charge of material efficiency ratings or environmental standards would set exchange rates for NCUs in relation to the specific conditions of their localities or projects. It would be left to the market to discover the exchange value of NCUs for money. In this way, individuals could convert NCUs into liquid currency, in return for helping corporations to meet compliance requirements. The idea bears some similarity to Barnes' idea of administering pollution quotas through a commons trust.

Would the NCUs be worth anything? Certainly, given the natural system degradation of the twentieth and early twenty-first centuries, one might expect an NCU to decline in value into the future. Yet the change in value would depend upon the initial rating. If the highest rating were assigned to a natural system in its pre-human state, then the rating given to any system at the onset of the program obviously would be much lower. If other economic instruments then created incentives to improve natural capacity, the rating would increase from its starting value, triggering the payment of dividends.

This structure probably would stimulate the emergence of a secondary market in NCUs, trading on expectations about the movement in value of a given type of NCU over time. Many of the same kinds of derivative instruments found in the financial markets, such as futures, options and swaps, could equally emerge for NCUs. As long as it were not funded through leveraged assets, this kind of speculation would be unlikely to cause market instability.

Natural capacity share-ownership could be implemented today in regions having the requisite institutional and regulatory apparatus, along with widespread internet access to enable trading in NCUs. However, its success, even in developed economies, probably would be predicated upon the establishment of other policies and controls in support of the four principles outlined above, so as to increase the likelihood of share-ownership paying a dividend sooner than later. Natural capacity share-ownership could accelerate the transition toward economic alignment with nature, but it probably could not initiate it.

References

Barnes, P. (2006). *Capitalism 3.0: A Guide to Reclaiming the Commons.* San Francisco, CA: Berrett-Koehler.

Braun, N., Fitzgerald, T., & Pearcey, J. (2015). Tradable Emissions Permits with Offsets. In B. Hintermann & M. Gronwald (Eds.), *Emissions Trading as a Policy Instrument: Evaluation and Prospects.* MIT Press.

Bridle, R., et al. (2019). *Fossil Fuel to Clean Energy Subsidy Swaps: How to Pay for an Energy Revolution.* Winnipeg, MB, Canada: International Institute for Sustainable Development.

Costanza, R., et al. (1997). The Value of the World's Ecosystem Services and Natural Capital. *Nature, 387*, 253–260.

Golombek, R., & Hoel, M. (2006). *Endogenous Technology and Tradable Emission Quotas.* Working Paper of Fondazione Eni Enrico Mattei, Milan, Italy.

Guzman, L. I., & Clapp, A. (2017). Applying Personal Carbon Trading: A Proposed 'Carbon, Health and Savings System' for British Columbia, Canada. *Climate Policy, 17*(5), 616–633.

King, D., et al. (2015). *A Global Apollo Programme to Combat Climate Change.* London School of Economics and Political Science.

Wilkinson, R. G., & Pickett, K. (2009). *The Spirit Level: Why More Equal Societies Almost Always Do Better.* London: Allen Lane.

Willer, H., & Lernoud, J. (Eds.). (2017). *The World of Organic Agriculture: Statistics and Emerging Trends, 2017.* Bonn, Germany: Research Institute of Organic Agriculture (FiBL) and IFOAM Organics International.

PART III

How Will We Get There?

Part III articulates a pathway from today's growth economy toward a future planetary one. Chapter 16 analyzes how the various policies and instruments described in Part II may be combined to get the ball rolling over the next 20 years. The outlook is promising: households, businesses and government, all stand to benefit. The analysis forms the basis of a manifesto in Chap. 17 for an economic doctrine of 'market planetarianism', which could guide development farther into the future.

Chapter 18 imagines a planetarian society of the future as an outcome of evolutionary processes set in motion today, offering some concluding thoughts about our chances of success. It proposes that the economic worldview known today as 'capitalism' will be replaced by a new one, 'capacitism', resulting from a normative, social and political evolution from exploitation to regeneration.

CHAPTER 16

Pathway Toward a Planetary Economy

It's time to upgrade the world.
—Farallon Murison Smith, *age 10*

Introduction

The five years it took to write this book spanned a comment from a five-year-old boy about how nature keeps us safe to another by his twin sister about humanity's part of the bargain. This chapter describes how we might keep our part of the bargain to upgrade the world. It outlines a path along which the economy can be steered toward long-term material stability and widespread prosperity, by combining the norms, institutions, policies and instruments proposed in the previous chapters. These measures together form a platform called 'market planetarianism', a manifesto for which is offered in the next chapter.

The Economy and Economics of the Future

Some time around seven generations, or 200 years, from now, the human economy should be materially efficient, most of its material needs being met through internal recirculation. Rates of extraction from nature and flows of effluents to it should be relatively small, within nature's capacity to support them, such as that is known. The economy of the future will be materially disciplined.

F. Murison Smith, *A Planetary Economy*,
https://doi.org/10.1007/978-3-030-49296-0_16

For this to happen, certain internal adjustments will be necessary, these arising from the establishment of new social technologies and the abandonment of old ones. Economic instruments for alignment with nature should conform to the four principles articulated in Chap. 12: material efficiency, economic stability, widespread prosperity and innovation through markets.

There needs to be a way for substantially everybody within society to participate in material discipline. The impetus to accumulate financial wealth, and the inequalities it generates, must be replaced with a broadening of economic prosperity, specifically, the establishment of a hard floor below which no-one may be allowed to fall. Part 2 proposed for this floor to consist of three necessities: income, health care and education. The basis of prosperity would mean basic income for all; free, or substantially free, health care at the point of consumption; and free public education at all levels. These measures would be funded not though income taxes but through fees levied on corporations for the use of commonly owned artificial capacity: electric grids, the internet, the financial system and so on. The system would be administered by a planetary trust. Income taxes below the median income would be substantially eliminated. No upper bound would be placed upon prosperity, although income tax rates for high earners would be no less than they are today and, for the very wealthy, probably higher. High earners might opt to electively redistribute some of this tax burden into designated programs to help people on low incomes, thereby exercising a degree of influence over the way their earnings were spent.

The overall effect would be to substantially narrow the income distribution, mostly from the lower end, and to ease the burden of worry among low and middle earners about such essentials as health care and education. People would have more time and more inclination for activities adding value to society. These measures, along with reductions in corporate income taxes, would enable the implementation of instruments to encourage material discipline, including, for example, a material intensity tax and subsidies and incentives for material efficiency. A separate natural capacity fee levied on corporations at a rate of about 1 percent of annual income would be predicated to government investment in improving and maintaining natural capacity.

The revenue base for government investment would be solidified by its taking back control of the national currency through sovereign money creation. This in turn would significantly reduce speculation within the financial markets and eliminate fractional reserve banking, thereby stabilizing the economy against excessive market volatility.

This future economy also might operate a form of natural capacity share-ownership, in which a supranational authority would issue shares in ecosystems, the atmosphere and oceans, whose value would depend upon regular health ratings produced by scientific consensus—similar to today's IPCC climate assessments. Dividends in natural capacity units (NCUs), a parallel currency, would be paid out to all individuals, who would be able to sell them to corporations on an exchange. Corporations would use them to offset natural capacity commitments, much the same way as emissions trading markets work today. In this way, every person alive would hold a lifelong stake in the health of the world's natural systems.

Governance of the global economy would fall upon a global governance council, a hierarchical institution providing representation at all social levels, from individuals to nations. The council would not pass laws but instead would establish policies and goals for member institutions to legislate and implement. It also would oversee the planetary trust. A parallel institution, a global monetary authority, would provide hierarchical governance over the world's monies, setting and enforcing currency standards, establishing macroeconomic policies, overseeing elective redistribution, and ensuring currency stability.

These or similar institutions can only come into existence under a system of norms that rejects individual self-maximization and nationalism, embracing instead collective persistence and planetarianism: a worldview of global interconnection between society and nature, and within society itself. The tightening coevolution between the economy and nature is anticipated to stimulate a normative shift from exploitation toward generation, which ultimately would bring a planetarian worldview into the mainstream.

An economics of this society would reflect these norms. The economy would be viewed as a wholly contained material subsystem of nature, subject to the laws of thermodynamics. Economics would be, first and foremost, a teleological undertaking. Today's emerging theory of markets as evolutionary, agent-based systems, will, in a future planetary economy, have become the mainstream. The analytical tools of economics would reflect this change, being iterative and simulation-based, rather than mechanical and optimality-based. Certain core concepts, such as supply and demand, along with aspects of marginality theory, would persist as useful heuristic tools, although their representation in introductory texts would be very different from today's. Economists would play a central role in pragmatic policymaking for long-term economic stability, a role in which they would collaborate closely with natural scientists. The days of economics as a theoretical abstraction, or of being perceived as one, would be long gone.

This economy of the future is not a utopia but a positive model. "The energies of a system cannot organize themselves around a negative", writes Marjorie Kelly. "We need a positive vision ... that is so strong, it exerts its own magnetic power."[1] As such, it is a description of a desirable future that can be debated, refined and converted into pragmatic steps taken today. It is a guiding point, not merely for a way of living but for a way of being. "Some type of holistic, or participating, consciousness and a corresponding sociopolitical formation have to emerge if we are to survive as a species", wrote Morris Berman. "The implication is that a way of life is slowly coming into being which will be vastly different from the epoch that has so deeply colored, in fact created, the details of our lives."[2]

Harnessing Markets for Collective Ends

By placing primacy upon the system and upon a livable baseline for all, the normative environment within a planetary economy would represent a return to the kinds of community-based norms characteristic of early societies. Yet at the same time the economy would be globally more interconnected than it is even today. This state is characterized by a collective awareness and a collective ethos, yet it will be market-based.

Civilization brought anonymity—or, to be more precise, the opportunity for relative anonymity—rather than being known to everybody within a small community. Not knowing everyone else within a city dissipates a sense of the collective, and perhaps heightens individuality. It might not come as a great surprise to learn that the individualism of the modern era arose among Enlightenment thinkers living in the largest cities in the world at the time, such as London and Paris. Since then, the world's population has swelled, people have migrated to cities in ever greater numbers, and the once rather isolated economies of different regions have become connected together. Little wonder that the collectivism of the village progressively gave way to the individualism of the urban consumer.

[1] Kelly (2012, p. 160).
[2] Berman (1981, p. 23).

Yet there are signs of a new perception emerging, one that recognizes the global implications of individual consumer decisions. This recognition is a recognition of a new collective: not the small collective of the village or of the nomadic tribe, but the collective of everybody on Earth. This is altogether a different kind of collective from that of the small group, because communication occurs among individuals not known to one another. One cannot gather billions of people around a campfire. The global collective exists at a much higher level of organization than historical ones. It likely will display certain emergent properties characteristic of very large groups: diversity, structure, conflict, a herd mentality. Some of these properties will be enabled by advances in global communication. Yet, following Maturana and Varela, the act of distinction calls this collective into existence. If people within the civilized world begin to recognize how their and others' self-motivated actions affect everybody on Earth, then this acknowledgment distinguishes the global collective.

This distinction represents a first step along a path from an economy whose agents were assumed to be motivated by self-maximization toward one organized by a system-stabilizing norm. The way along the path will be iterative, with advances in the implementation and adoption of economic instruments stimulating normative changes, these in turn prompting new institutional and political questions. The distinction of a global collective, however, is a tacit acknowledgment of the unity of society upon planet Earth. It is a planetarian act. The deployment of the market as the principal agent of economic change for the collective departs from the old notion of central planning, recognizing the global collective as a diverse collection of societies rather than a single homogeneous one.

Market planetarianism is both an economic philosophy and a policy platform. It can be a doctrine against which alternate ideas for moving forward may be evaluated. It would succeed by being broad, manifesting in different ways at different times and places. However expressed, its central criterion is the implementation of market solutions to serve the needs of the planet: both the alignment of the economy with nature and the establishment of substantially full social participation in this alignment. The overarching requirements of minimum living standards and relatively high distributional equity are not in conflict with the adaptive, selective dynamics of the market, provided the market is not assumed to be ideologically fundamental but instead an engine of teleological change.

Market planetarianism should not be confused with the economic ideology referred to as 'market socialism'. Market planetarianism creates a clear distinction with negative historical connotations of 'collectivism'. Market socialism, in contrast, argues that so-called free markets are not completely 'freed' until relieved of capitalist exploitation, private property and alienated labor. Under market socialism, all property or capital would be public, all entrants to a market would do so on an equal basis, and prices would be centrally controlled in some way. Market socialism essentially is one great, big barter system among more or less equal stakeholders. Market socialism attempts to formalize the collective ownership of wealth. Market planetarianism is ideologically agnostic and globally focused. It accommodates public and private ownership. It allows for uneven distributions of well-being, within certain bounds. Market planetarianism simply recognizes the fundamental, natural constraints upon human activity, and deploys the power of the market to find solutions to meet them. It asserts a critical role for government, to provide normative bounding and guidance.

We no longer live in a world dominated by a war over capitalist and communist ideologies. The present and future situation is much more fluid. To the extent there exists a conflict among economic ideologies, it is between the old, capitalist economics of the nineteenth and twentieth centuries, and an emerging holistic, regenerative economics of the future.

Combining Instruments for Economic Redirection

Although it is impossible to speculate about the long-term pathway toward a materially stable, prosperous future, it is possible to project how the economy might embark upon the journey by embracing a market-planetarian approach. This section outlines the possible effects of combining the measures presented in previous chapters over 20–50 years, starting from the early twenty-first century. It is a social-technology road-map. The actual building, testing, release and refinement of the social technologies in question will be for the talents of the coming decades to address.

None of the instruments described in the previous chapters can, by themselves, satisfy all four principles of material discipline, economic stability, widespread prosperity and innovation through markets. They have to act synergistically. This chapter examines how the implementation of a coordinated suite of instruments might affect individual and corporate incomes, as well as the national budget. The geopolitical focus is on the

US, using US data.[3] It illustrates how similar changes could unfold in other countries and regions. The results represent only one of a number of possible scenarios. Additionally, uncertain, longer-term changes, such as political diffusion among nation-states, are omitted for simplicity.

The scenario imagines the establishment of a basic living program, consisting of basic income, free health care and free education, funded by small fees on the use of commonly owned artificial capacity. Basic income would replace most of the US's social security and income security programs. Medicare would become free or be replaced with a new, free system. A natural capacity fee would be levied upon corporate income and financial trades, the revenue from which would be used to fund programs to improve natural capacity within US borders and support subsidies and incentives for material efficiency. Taxes would be shifted from income onto material intensity, along the lines described earlier. Government operations would be funded through sovereign money creation, with new money later being reabsorbed through taxes. This mechanism would be used also to pay down the national debt and to provide capital for commercial lending. Fractional-reserve banking would be eliminated.

For the purpose of the analysis, a rough projection is needed for a rate of decline in material intensity of the US economy due to these measures. Although such a rate is virtually impossible to predict, a reasonable assumption would lie in the range of 1 to maybe a few percent per year. An illustrative rate of 2 percent year-on-year is assumed.

Basic Living Program Through Common Capacity Fees

Legislation enacted in year 0 would establish a basic living program to take effect in year 1. The legislation would establish a planetary trust to administer the program, maintain rates, collect payments, carry out audits and apply penalties for program violations, in much the same way as taxation institutions already do. However, the trust would operate independently from the government administration. Funds collected for a basic living program and revenue collected by the Internal Revenue Service would never commingle.

In the scenario, every resident adult and child in the US would be paid a starting annual amount of $10,000 up to a family maximum of four people.

[3] The data are baselined to 2013, which is year 0 in the analysis, this being the latest year for available tax data at the time of writing.

How residency was demonstrated is an open question, not addressed here. An advantage of national legislation, though, is that it would adopt the same definition of 'residency' in all states. National legislation need not supervene any existing state programs: Alaska residents, for instance, could enjoy a double benefit. Per-capita basic income is projected forward according to a growth function detailed in the Appendix to this chapter.

The funding requirement for the health and dependent care part of the program is determined according to the OECD average per-capita cost of health care. An adder of 30 percent covers dependent care. It increases the US amount slightly above the OECD average. The starting value is $2153 per year.

The per-capita education requirement is a rough estimate of annual per-student costs, adjusted for an expected proportion of an average person's life spent in education. Its starting value is $2500 per year. Both health and education per-capita requirements are projected forward using a similar growth function to basic income.

Figure 16.1 shows the first 20 years of the per-capita basic living dividend.

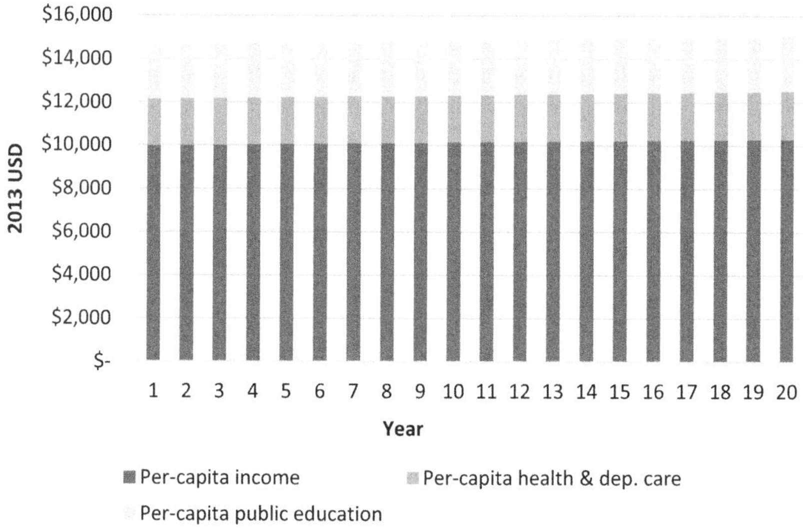

Fig. 16.1 Projection of dividends for a basic living program consisting of basic income, free health and dependent care, and free education over 20 years. (Source: Author's creation, based on data and assumptions in the text and the Appendix)

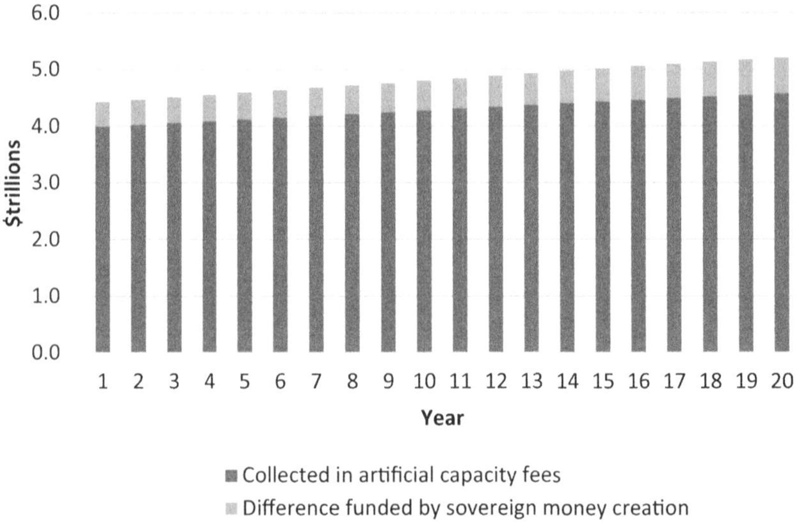

Fig. 16.2 Projection of artificial capacity fee revenue and sovereign money creation to meet the basic living program funding requirements in Fig. 16.1. (Source: Author's creation, based on data and assumptions in the text and the Appendix)

The total funding requirement for the basic living program is the per-capital total in each year multiplied by the eligible population. In year 1, this is about $4.4 trillion, rising to about $5.3 trillion by year 50. Following the estimates in Chap. 13, the potential revenue from artificial capacity fees in year 1 is set to $4 trillion. The difference between the funding requirement and the fee revenue is made up through sovereign money creation. Figure 16.2 shows the first 20 years.

Taxes and Redistribution

Income and Payroll Taxes

Following the tax-shifting analysis of Chap. 14, this analysis assumes a restructuring of income tax, beginning in year 1. Then, every four years, rates are adjusted as shown in Table 16.1. Income tax lost from these rate reductions is shifted onto material intensity.

The tax-shifting analysis in Chap. 14 did not consider certain tax implications of instituting a basic living program. This analysis incorporates them. Not only does basic income alter the distribution of incomes but

Table 16.1 Individual and corporate income tax rates and thresholds assumed for tax-shifting scenario

Years	*Individual*				*Corporate*		
	Below median (%)	*At median (%)*	*3 × median (%)*	*20 × median (%)*	*Below median (%)*	*At median (%)*	*10 × median (%)*
1–4	0	10	40	70	0	10	30
5–8	0	8	37	70	0	8	30
9–12	0	6	34	70	0	6	30
13–16	0	4	31	70	0	4	28
17–20	0	2	28	70	0	2	26

Source: Author's creation

Note: All rates below the median income are zero

also it alters households' pre-tax income. If a basic living program was funded through artificial capacity fees, then there would be no need for tax-funded social and income security and healthcare. Since these benefits are funded in today's economy through payroll taxes, those taxes would be eliminated under a new basic living program. Employees' pre-tax income would become correspondingly greater: in other words, they would enjoy a double benefit. The effects on the government budget also would be significant, discussed later.

To compensate for the loss of payroll taxes, the tax schedule in Table 16.1 increases the top marginal rate of income tax from the 60 percent used in Chap. 14 to 70 percent. This increase would more or less neutralize any gains the wealthiest half-percent of earners might realize from removing payroll taxes. As the Appendix explains, payroll taxes on earned incomes in the millions can be substantial.

Table 16.1 also shows how marginal income tax rates between the median and three times the median decline earlier and more steeply than rates between three and 20 times the median. The intent here is to create headroom in the disposable income of middle earners to accommodate increases in the material cost of goods and services arising from a material intensity tax. The same reasoning applies to the reductions in corporate income taxes.

The amounts collected in income tax are then fed into an accounting of total tax revenue. The effects on incomes by bracket also are captured, as described below.

Material Intensity Tax

The starting material intensity tax is calculated according to the method outlined Chap. 14. A tax on material extraction is levied only on corporations, whereas a tax on effluents is levied on both corporations and households. Using the material intensity estimates of 13.3 gigatonnes of extraction and 7 gigatonnes of effluents for the US, along with the estimate of the tax burden that could be shifted from individual and corporate incomes, Chap. 14 estimated a material intensity tax of $5.56 per tonne on extraction and of $21 per tonne on effluents. These become the starting values for the scenario.

The taxes rise steeply, using a logistic function described in the Appendix, so that, by year 20, they are both over $100 per tonne. The logistic property of the function ensures they would eventually level off. In tandem, the scenario assumes a 2 percent year-on-year decline in material intensity. The rate of tax multiplied by the material intensity gives the total tax collected.

Other Taxes

The Congressional Budget Office (CBO) categorizes other taxes collected by the US government into estate and gift taxes, excise taxes, customs duties and miscellaneous receipts. The scenario projects these taxes forward according to growth rates established using CBO data, described in the Appendix.

Total Tax Revenues

Total tax revenue is the sum of income tax, material intensity tax and other taxes. Figure 16.3 shows how the tax burden would shift from income taxes onto a material intensity tax under the assumptions of the model.

The dip in total tax revenue in year 1 is due to the shifting of income and health care support from payroll taxes onto the basic living program.

Effects on Corporate Incomes

Under the 2013 tax structure, small businesses, with assets up to $1 million, paid between 20 percent and 30 percent effective tax rates before credits. Medium and large businesses all paid close to 35 percent in effective tax rates. Under the proposed structure, small businesses by year 20 would pay no income tax; medium businesses, with up to $250 million in assets, would pay a few percent in effective tax rates; and large businesses would pay substantially less income tax than under the existing structure, as Fig. 16.4 illustrates.

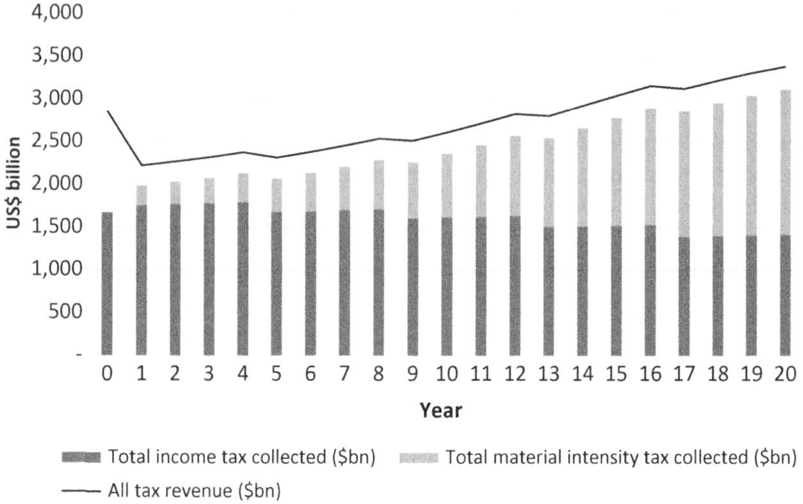

Fig. 16.3 Projection of income tax revenues, material intensity tax revenues and total tax revenues, using 2013 tax data as a baseline. (Source: Author's creation, using 2013 tax data as a baseline. Data from US Internal Revenue Service and Congressional Budget Office. Note: Assumptions and method explained in text and the Appendix)

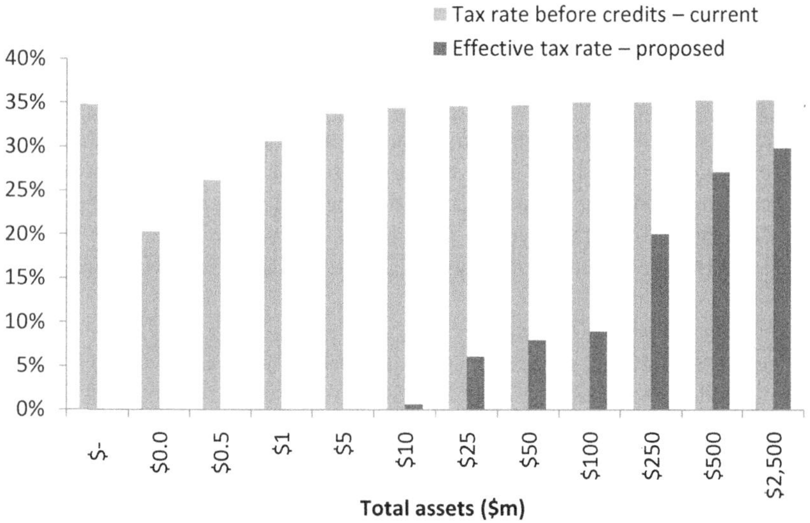

Fig. 16.4 Effect of tax-shifting on corporate effective income tax rates by year 20. (Source: Author's creation, baselined from 2013 IRS corporate income tax data for US corporations)

This may appear to be a great deal for businesses. However, by year 20 they also would be paying substantially more in material intensity taxes, the amounts varying substantially according to the type and size of business. Businesses' discretionary income will have increased, and so will their costs of doing business. The key benefit to this tax-shifting is that the power to change business practices would be placed into the hands of business managers themselves. If the material intensity of their operations is stiffly taxed, then it would create a selection pressure to reduce the material intensity of their operations.

Elective Redistribution

In addition to tax-shifting, the scenario incorporates elective redistribution among individual filers. Those whose taxable income exceeds three times the median would have the option to redistribute a portion of pre-tax income above that threshold. The scenario assumes that this portion, on average, would be 10 percent. For example, a married couple filing jointly with a combined income of $300,000 would be subject to a 40 percent marginal income tax rate on the income above $200,000, this being the threshold in the model approximating to three times the median. Under elective redistribution, they would disburse $10,000 of that income into redistribution and $30,000 into income tax.

Under each quadrennial adjustment to income tax rates, the total electively redistributed income above three times the median is calculated. This total is divided by the number of filers below three times the median to derive a per-capita dividend. This dividend is added to those filers' post-tax income along with the basic income. The number of filers eligible for the dividend will change over time. The model assumes, for simplicity, that it increases in line with population increase. A more fine-grained program would disburse dividends from the very top percentiles to the very bottom ones.

Effects on Individual Incomes

Figs. 16.5 and 16.6 illustrate the combined effects of the income and payroll tax changes, basic income and elective redistribution on mean net incomes for married couples filing jointly and single filers, respectively, in years 1, 5, 10, 15 and 20. Because the mean income is below three times the median, it is eligible for the redistribution dividend.

Reductions in income tax rates and the elimination of payroll tax have significant effects upon net incomes. For single filers, the effect is

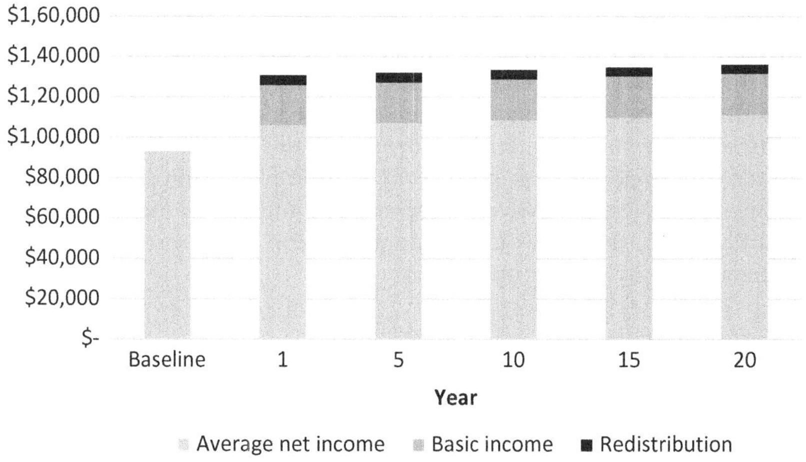

Fig. 16.5 Average (mean) net incomes in the baseline year and years 1, 5, 10, 15 and 20 in the scenario, showing the effects of the income tax reductions, basic income and elective redistribution: Married couples filing jointly. (Source: Author's creation, using 2013 IRS data. Note: Assumptions and method explained in text and the Appendix)

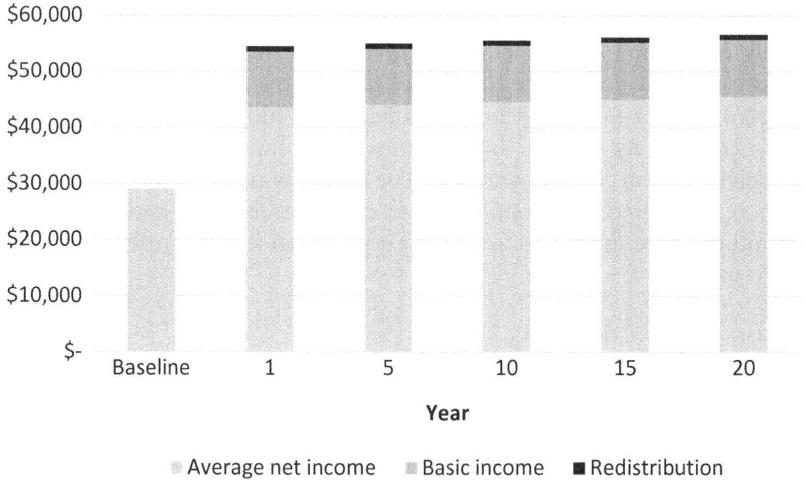

Fig. 16.6 Average (mean) net incomes in the baseline year and years 1, 5, 10, 15 and 20 in the scenario, showing the effects of the income tax reductions, basic income and elective redistribution: Single filers. (Source: Author's creation, using 2013 IRS data. Note: Assumptions and method explained in text and the Appendix)

dramatic. The addition of basic income has an equally pronounced effect, whereas the effect of elective redistribution is relatively small. Overall, the effects on average net income would be very beneficial. Individuals and households would have between 50 percent and 100 percent extra disposable income to be able to afford the higher material costs of living associated with the establishment of a material intensity tax.

How those higher costs of living might eat into consumers' disposable incomes is a question for future research. The present analysis serves merely to illustrate the bounds of what might be possible. The important benefit that applied to businesses also applies to consumers: if they have more discretionary income then they would have the ability to make choices in the marketplace to save money and, in the process, reduce their material intensity. It is not sufficient for households to have the *desire* to shift their preferences: they also must have the *ability* to do so. If a material intensity tax drove up the cost of living, then it is imperative that lower-income consumers would not be priced out of the market but that they would have sufficient discretionary income to put their preferences into action.

To illustrate how the measures described above would narrow the income gap, Figs. 16.7 and 16.8 plot the average net income by IRS income bracket for married couples filing jointly and single filers, respectively, in the baseline year and in years 1, 5, 10, 15 and 20.

For both categories of filer, the baseline distribution (black bars) is highly skewed. In 2013, the lowest average net income among married couples

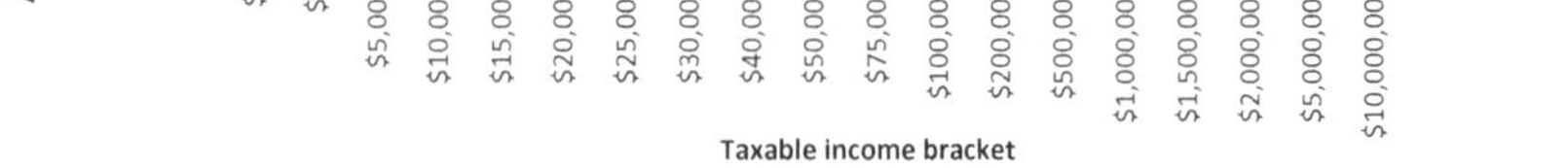

Fig. 16.7 Distribution of average net income by IRS taxable income bracket in the baseline year and in years 1, 5, 10, 15 and 20: Married couples filing jointly. (Source: Author's creation, using 2013 IRS data. Note: Assumptions and method explained in text and the Appendix)

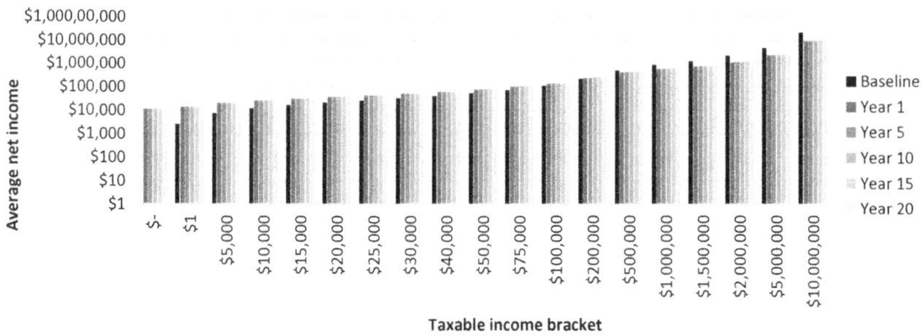

Fig. 16.8 Distribution of average net income by IRS taxable income bracket in the baseline year and in years 1, 5, 10, 15 and 20: Single filers. (Source: Author's creation, using 2013 IRS data. Note: Assumptions and method explained in text and the Appendix)

filing jointly was $2562. This income fell within the $1–$5000 taxable income bracket. For single filers, it was $2577. The highest average net income was $21.2 million among married couples filing jointly in the taxable income bracket above $10 million. For single filers, it was $21.6 million.

In years 1–4, under the new structure, average net incomes in the lower brackets jumped into the $30,000–$50,000 range among married couples filing jointly, and into the $15,000–$40,000 range among single filers. At the other end of the scale, average net incomes among the highest earners dropped, due to higher effective income tax rates and elective redistribution.

Subsequent income tax adjustments serve to flatten the distribution further, as the progressively lighter bars indicate. Married couples with taxable incomes around $1 million sit roughly at the inflexion point in the distributional changes, being left pretty much indifferent to them. Single filers in the $250,000–$500,000 range are largely unaffected by the changes.

The analysis suggests that a basic income scheme and—to a lesser extent—elective income redistribution are powerful enablers of a shift of the tax burden from income onto material intensity. The reason is because reductions in income tax for low earners save relatively little money for them, as they pay relatively little income tax to begin with. Facing higher living costs arising from a material intensity tax, they rely upon the other sources of income incorporated into the scenario, which are not taxable. That those sources of income do not come exclusively from the taxation of the rich should render them all the more politically palatable.

Natural Capacity Fees

As described in Chap. 13, the government would raise revenue for natural capacity investments and for material efficiency incentives and rebates through small fees charged on financial trades. Chapter 13 estimated the potential starting amount of revenue that could be raised this way using data on trading volumes in stocks and derivatives. An illustrative average rate of 0.2 percent on the value of a stock trade, along with a fee of $10 per derivative trade, could raise about $132 billion. Natural capacity fees charged to corporations at around 0.5 percent to 1 percent of net income could raise about an additional $18 billion.

The scenario begins with the dollar-value trade volumes used in Chap. 13 and projects them forward using known rates of increase in market volumes. The long-term average year-on-year growth rate in trading volumes on the New York Stock exchange is about 6 percent, and on the Nasdaq about 8 percent.[4] The growth rate in derivative trading volume used for the projection is a global average of about 1.7 percent per year.[5] For North America, this rate might be somewhat conservative. In the scenario, it is reduced incrementally to 1.5 percent by year 20, on the assumption that the natural capacity fee on derivative trades might have a slight dampening effect on volume. The elimination of fractional reserve banking could have a much greater dampening effect, although there is no simple way to capture that in this projection.

The natural capacity fee on stock trades is increased incrementally in the scenario from 0.2 percent in year 1 to 0.3 percent in year 20. The fee on derivative trades is increased from $10 in year 1 to $15 in year 20. The revenue from corporate natural capacity fees is increased from $18 billion in year 1 at a rate of 0.5 percent per year, roughly in line with long-term average US economic growth.

The revenue from these fees over the 20 years is depicted in Fig. 16.9.

The total amount collected from natural capacity fees is incorporated into the government budget projection, predicated to natural capacity improvements and material efficiency measures.

Sovereign Money Creation

Legislation is enacted to separate commercial deposits from lending and to establish the public sector as the sole issuer of the public currency. This currency is issued annually by a national monetary authority in response to

[4] Source: DJIA_Nasdaq Excel workbook (Yahoo Finance).

[5] This rate rather than the North American rate is used because changes in trading volumes vary considerably from year to year and from region to region (see in Acworth 2017).

Fig. 16.9 Projected revenue from natural capacity fees on financial trades and corporate net income. (Source: Author's creation, based on assumptions and data described in the text)

budget requirements from the government, subject to macroeconomic constraints. New funds are spent into circulation, subsequently to be recouped as government revenue in the form of taxes and fees as detailed above. Net public revenue from one year becomes available the following year for quantitative adjustment to meet the needs of commercial lending and investment. The government debt is paid down during the early years, which infuses the financial sector with additional funds, offsetting the elimination of fractional reserve banking. The government's long-term aim is to run a modest year-on-year deficit.

The analysis begins with a government budget, incorporating both existing programs and new ones, as set out below. The budget creates a requirement for sovereign money creation for government operations. This amount is the current budget adjusted by the previous year's net revenue. If the previous year's net revenue is a surplus, then this would reduce the current year's SMC requirement for the government. If the previous year ended with a deficit, then it would add to the current year's SMC requirement. The model sets a target long-run government operating deficit of $100 billion. The national monetary authority also commits to paying down the government debt. According to CBO data, the US paid $221 billion in interest in 2013 on a national debt of just under

$12 trillion. Little wonder the commercial banks love government borrowing. The model assumes $1 trillion per year going into debt service until the debt is paid off. The budget, the debt service and the balance of the basic living program not covered by artificial capacity fees comprise a total SMC requirement for the year in question. Debt payments are assumed in the model to flow into the financial sector. In years where government revenue exceeds expenditures, any excess SMC also flows into the financial sector as investment. A more sophisticated analysis would adjust this flow for macroeconomic criteria, such as to adjust the money supply; however, in this illustrative scenario, it is simpler to assume all such monies flow into the financial sector. It allows us to see how much money could be made available for private lending.

Government Budget Projection and Capital Flows

Under the scenario, the government budget in year 1 is radically different from the baseline budget in year 0. A basic living program, funded by artificial capacity fees, replaces tax-funded social and income security and healthcare. In practice, the replacement of these programs, which are 'mandatory' as distinct from 'discretionary' in the CBO categorization, probably would occur in phases. However, since the scenario is illustrative, and a phased replacement would introduce unnecessary complexity to it, the switch is accomplished in a single year.

A description of the various budget categories and their relations to one another is provided in the Appendix to this chapter. Suffice to say that, among the outlays, Social Security and Medicare are eliminated. Among the revenues, offsetting receipts, which are the portion of payroll taxes collected from employers, along with the balance of payroll taxes collected from employees, also are eliminated. Discretionary programs, other mandatory programs and interest paid on debt, are retained.

The budget projection method is described in detail in the Appendix. In summary, the total budget minus the total revenue in any year gives the net revenue as a surplus or a deficit. If a deficit, it is added to the national debt. In years where revenues exceed the budget, the excess is fed into the following year's investment in the financial sector. Total government revenue is the sum of tax revenue and natural capacity fees, the latter predicated to natural capacity programs and material intensity subsidies and incentives in equal amounts on the budget side.

Figure 16.10 shows the results. The outsourcing of tax-funded social and income security and healthcare in year 1 to a separate basic living program

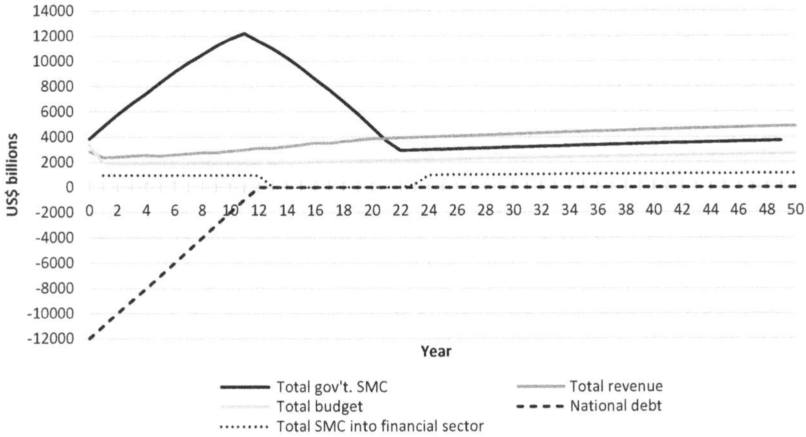

Fig. 16.10 Illustrative projection for the US, in constant 2013 dollars, showing annual sovereign money creation, total government revenue and total budget, along with a paydown of the national debt and flows of new sovereign capital into the financial sector for commercial lending and investment. (Source: Author's creation, based on assumptions and data described in the text. Note: The projection incorporates a variety of measures, including tax-shifting from income onto material intensity, a basic income scheme, natural capacity fees levied on financial trades, and investments in material efficiency and natural capacity. Assumptions and method explained in text and in the Appendix)

drastically reduces the government budget. However, $1 trillion in debt-reduction and about $430 billion for the basic living program bring the total sovereign money requirement to nearly $3.9 trillion. Tax revenue also drops, although over the first ten years it regains that lost ground.

The debt-service requirement drives the year-on-year sovereign money requirement up over the first ten years to over $12 trillion. The financial sector sees an inflow of $1 trillion per year to soften the effect of eliminating fractional reserve banking. Once the national debt is paid off, the sovereign money requirement begins to fall. Revenues by this time exceed the budget, enabling this decline. Once the year-on-year deficit falls to the target value of $100 billion, excess sovereign money becomes available to the financial sector. However, there is a period of ten years between the debt being paid off and the deficit falling to the target value in which no money is paid into the financial sector. In reality, this gap could induce a credit crunch, which would be avoided by allowing some sovereign money flowing through the government budget process to flow out into the financial sector.

Once the deficit is reduced, the long-term projection becomes stable.

These results are, of course, sensitive to the initial assumptions, and surely would be improved through more detailed, realistic analysis. However, in providing an overview of what is possible, the analysis suggests that the combination of measures described above could be very beneficial for the economy in several ways. First, it shows the effect of government taking control of the public currency. Flows of new money into the financial sector take place after, not before, the requirements for government investment, debt-service and deficit-maintenance have been met. In the model, this re-prioritization ushers in an initial period of relative upheaval, as the graph shows. However, as it also shows, the same re-prioritization results in a more stable system in the long run. Second, the government can fund its operations debt-free, without risking inflation, provided the national monetary authority has control over the amount of money flowing around. This means that, third, the government finally has the necessary ability to invest in measures to encourage material discipline within the economy and to eliminate any prior, perverse incentives against it. Tax-shifting from incomes onto material intensity, investing in new technologies, improving the nation's infrastructure, and regenerating its natural capacity are examples of these kinds of measures. The key to success is that the instruments in question would work, and could only work, in concert.

International Implications

This analysis, using US data, could readily be replicated for most other higher-income countries, including the OECD ones. All these countries have government programs and institutions similar to those in the US. If the OECD countries alone were to implement the kinds of changes outlined above, then a substantial majority of the global economy would be placed onto a path toward greater alignment with natural systems and higher internal participation in this objective.

An advantage of several or many higher-income countries embarking on this path at the same time is a reduced likelihood that any one national economy could gain any short-term economic advantage over others, or that the world's transnational corporations might be tempted to relocate their operations to more 'accommodating' markets.

In regions with weak institutions or widespread absolute poverty, these kinds of measures would not have the desired effect. In countries where little tax is successfully collected, one cannot expect a tax-shifting program from income onto material intensity to be successful. If living costs increased due to taxation on material intensity in these countries, then a greater number of people would be pushed into absolute poverty.

One potential remedy for lower-income regions is an international elective redistribution program. Another is basic living funded by artificial capacity fees on corporations doing business in these countries. Foreign companies would be charged a higher rate than domestic companies, thereby encouraging investment in domestic businesses, while bringing in as much revenue as possible from outside. These measures could be established within lower-income countries having at least a modicum of institutional integrity.

Some of the revenue collected from natural capacity fees in wealthier countries could be used for the maintenance and enhancement of natural capacity within poorer regions unable to do so by themselves. A long-term advantage these regions would have is that, in starting from a relatively low per-capita material intensity, they may be able to more quickly and effectively decouple economic growth from material intensity. Analogous to the way some developing regions have bypassed telephone wires altogether for mobile technologies, and as some of these regions now are bypassing electric grids for local renewable generation and storage, so they may in the future avail themselves of other materially efficient technologies as capital allows. They may be able to innovate quicker than established economies, particularly as credible domestic institutions became established and absolute poverty were overcome.

In the longer run, this kind of co-ordination would become a stepping stone to establishing a transnational governance structure of some kind, or later perhaps a global one. The need would arise for global co-ordination of money flows and for political governance, not to concentrate power at the global level but actually the opposite: to distribute it consistently to smaller jurisdictional units over which people have relatively more influence.

The innovations that could come out of this institutional evolution cannot be anticipated. The instruments introduced here, along with their potential interactions, represent only what can be foreseen over the next few decades at the most. They might not by themselves bring the economy into alignment with natural processes, nor bring a prosperous, meaningful life to everyone, but they would represent initial steps along that path.

Appendix: Assumptions and Data for Economic Projection

Baseline Data and Period Covered

US budget data for the scenario analysis come from 2013. This is taken as year zero. The US population is projected forward for 50 years using the United Nations' medium-fertility projection toward a long-term plateau of 500 million. The standard logistic growth function is used.

The economic instruments and income projections are run over 20 years. The government budget projection is extended out to year 50 to illustrate the long-term consequences of the implementation of the instruments.

Economic Instruments

Basic Living Program

The limitation of the basic living program to four people per family, along with any residency requirements, probably would reduce the number of eligible individuals down from the current US population of around 325 million to around 300 million. From this starting point, the US population is projected forward according to the United Nations' medium-fertility scenario toward a theoretical long-term plateau of 500 million. This population projection determines the total funding requirement for the basic living program.

Income and Payroll Taxes

The model does not incorporate forward projections of changes to pre-tax incomes. These vary by income or asset bracket, and are uncertain. For simplicity, pre-tax incomes are assumed to remain constant in 2013 dollars.

For individual taxes, the average gross taxable income per return is calculated as the total adjusted gross income by income bracket divided by the number of returns in that bracket. The baseline average gross income per return is then used to calculate an average tax per return under the new tax rates in the scenario. This tax is a weighted sum of the taxes payable within each bracket. Dividing this amount into the average gross income per return gives an effective, or aggregate, tax rate by income bracket. The total tax revenue within each income bracket is the average tax per return multiplied by the number of returns. The number of returns is projected forward from the 2013 baseline by US population.

As in the tax-shifting analysis in Chap. 14, this analysis is limited to filers married filing jointly and single filers, these comprising 90 percent of all individual filers.

For corporate income taxes, CBO data are organized by the value of total assets. Within each asset bracket, the CBO lists net income and income subject to tax as a total from all returns, the latter incorporating the net operating loss deduction and special deductions. The CBO reports total income tax before credits and after credits, of which there are many.

From these aggregate data, the average income subject to tax within each asset bracket is calculated as the total income subject to tax divided by the number of returns. The statutory tax rate before credits, by asset bracket, is the total income tax before credits divided by the total income subject to tax. These rates serve as the basis for comparing against the effective tax rates under the proposed scheme because any new rates established under a new scheme would be defined as being before credits.

The new rates are applied to the asset brackets to estimate the average tax per return on the average income subject to tax. Asset brackets where the average income subject to tax is greater than ten times the median are subject to the higher marginal rate of tax for the portion of income above that threshold. The average tax per return by asset bracket is then divided into the average net income to calculate the effective tax rate under the new scheme. The average tax per return is also used to estimate the total income tax revenue by multiplying by the number of returns in each asset bracket. The scenario does not incorporate a forward projection of the number of returns because, for corporations, it cannot be expected to increase with population. For simplicity, it is held constant.

As the main text discusses, the establishment of a basic living program funded through artificial capacity fees would replace tax-funded social and income security and healthcare. This means that employees and employers would no longer be subject to payroll taxes. The scenario estimates the potential effects of this change upon individual incomes. It is assumed for simplicity that the change does not affect corporate incomes because the employer portion of payroll taxes usually is taken as a withholding from employee salaries. In other words, the employee pays the entire payroll tax either through their employer or directly to the government.

Payroll taxes in the US comprise Social Security taxes and Medicare taxes. The scenario uses rates in effect during the 2010s, which were 12.4

percent on the first roughly $120,000 of employee earnings for Social Security and 2.9 percent on all employee earnings. Not all income, however, is subject to payroll tax: only wages. Other forms of income, such as dividends and rent, are not subject to payroll taxes. The average gross income per return by income bracket, used for the scenario, thus must be adjusted to reflect only wage income. This adjustment assumes that high earners generally will have diverse sources of income not subject to payroll tax, whereas filers earning the median income or less generally will rely exclusively on wage income. An illustrative schedule of adjustments is applied to the average gross income per return by income bracket, derived from the CBO baseline data, so that filers slightly above the median report 90 percent or more of their income from wages, whereas filers reporting millions of dollars of income per year derive only about 20–30 percent. The payroll tax rates then are applied to this income subject to payroll tax to produce per-capita estimates of payroll taxes paid by income bracket. The schedule of adjustments is calibrated to produce a total of payroll taxes from all filers reasonably close to the actual reported payroll tax revenue of $948 billion for the baseline year of 2013. This ensures the adjustments at least are in the right ballpark.

The average per-capita payroll tax amounts by income bracket are applied to the income portion of the scenario as increases to the gross taxable income by bracket. With the establishment of a basic living program and the elimination of payroll taxes, it is assumed that people's gross incomes, on average, will increase accordingly. This assumption ignores the possibility, for convenience, that employers might not pass along all of the savings to their employees. This new gross taxable income then forms the basis of the projections to net income after the new income taxes, basic income, and redistribution dividends have been applied.

Material Intensity Tax

The material intensity tax is increased according to the standard discrete logistic growth function, $N_{t+1} = rN_t((K - N_t)/K)$, where r is the annual rate of increase, N is the quantity increasing, and K is a maximum value. The rate of increase for the tax on extraction is 25 percent per year, with a ceiling of $150 per tonne. The rate of increase for the tax on effluents is 15 percent per year, with a ceiling of $200 per tonne. The effluents tax begins from a higher starting value in the scenario, hence the lower growth rate. However, it has a higher ceiling because of the economic rationale for taxing effluents higher than extraction discussed in Chap. 14.

Table 16.2 Estimated growth rates and ceilings for forward projection of tax revenues in the economic projection scenario, by category

Tax category	*Starting value ($ bil.)*	*Annual growth rate (r) (%)*	*Ceiling (K) ($ bil.)*
Estate and gift taxes	19	0	29
Excise duties	84	0	130
Customs duties	32	1.8	50
Miscellaneous receipts	103	4.7	160

Source: Author's creation, using data from US CBO (2017)

The rates and ceilings obviously are illustrative, although they are scaled to produce an increase in average consumption costs somewhat less than the per-capita basic income. In a practical setting, the purpose would be to render consumers indifferent to the changes, to the greatest possible extent, although actual consumption costs would vary considerably from one consumer to another.

Other Taxes
Other kinds of taxes are carried through the analysis largely unchanged. However, a reasonable forward projection of revenue in these taxes is needed. For this, CBO budget data from 1967 to 2016 are used.[6] Because the CBO data are in actual dollars, annual rates of inflation are used to convert them to constant 2013 dollars. The forward projection is logistic. The annual growth rate in each budget category is calculated from the average annual growth rate over the prior 50 years. The ceilings for these tax revenues (K in the logistic function) is estimated by proportion from the assumed long-term, stable population of the US for the basic living program projection. The values are given in Table 16.2.

US Budget Projection

Budget Categories
US federal budget outlays are divided into 'discretionary' and 'mandatory' categories. Discretionary outlays, which in 2013 totaled around $1.2 trillion, are divided roughly evenly between defense and non-defense

[6] Data supplements to US CBO (2017): www.cbo.gov/publication/52801.

expenses. The non-defense outlays encompass a myriad of programs across numerous departments. Science, transportation, national parks and federal law enforcement are a few examples. The mandatory outlays include social security and income security (such as unemployment insurance); major healthcare programs, including Medicare, Medicaid and the Children's Health Insurance Program; federal civil and military retirement programs; and veterans' programs. The CBO includes a category of 'other' mandatory programs, which accounts for a fairly small proportion of the total. The category 'offsetting receipts' comes from employer contributions to social and income security and healthcare. In 2013, mandatory programs cost about $2.3 trillion, about half of which went to social and income security. Major health care programs cost an additional $768 billion. The other programs were relatively small by comparison. Offsetting receipts in 2013 were $305 billion. The total net mandatory outlay in 2013 was just over $2 trillion. These are the year-0 baseline values for the budget scenario.

Budget Projection Method

In year 1, social and income security and the major healthcare programs are replaced with a basic living program funded through artificial capacity fees. The budgets for these programs, along with offsetting receipts, go to zero. Federal military and civil retirement, however, is retained. The basic living program is not a substitute for it but a complement. In the private sector, employees would not be precluded from establishing retirement accounts over and above their basic living dividend. The same benefit should accrue to federal employees. So, in the scenario, just as in today's world, your income taxes pay the people who run the government.

The total US budget in year 0 (2013) was $3.37 trillion, $1.2 trillion of which came from discretionary programs and the remainder from mandatory ones. The government's total revenue, which came from taxes, was $2.86 trillion. The balance of $510 billion was a deficit, which was added to the national debt of just under $12 trillion.

In year 1 in the scenario, the government has the ability to create sovereign money. It also has a commitment to reduce the national debt by $1 trillion per year until it is paid off. The payment of this debt would remove once and for all any obligation the government had to the commercial banking sector. From that point on, it would have sole control over the public currency. The $1 trillion in debt reduction is paid into the financial sector, a cash infusion which would compensate for the termination of fractional reserve banking. Sovereign money creation serves three purposes: funding the balance of the budget requirement, debt reduction, and funding the balance of the basic living program, above and beyond the revenue from artificial capacity fees.

In year 1, the government's budget has dropped from $3.37 trillion to about $1.9 trillion. This amount includes natural capacity investments and material efficiency subsidies and incentives, funded by natural capacity fees. The SMC requirement for the budget is this amount plus the deficit of $510 billion from the previous year. The SMC requirement for debt reduction is $1 trillion and that for the balance of the basic living program is $431 billion, bringing the total SMC requirement to $3.86 trillion.

On the revenue side, tax revenue in year 1 drops from $2.86 trillion to $2.23 trillion. Revenue from natural capacity fees is $155 billion, bringing total revenue back up to $2.38 trillion. Because natural capacity fees are predicated, however, they are essentially a pass-through. The net government revenue then is the total SMC requirement minus the total revenue, which is a deficit of $1.48 trillion, which is fed into the SMC requirement for year 2. Meanwhile, the national debt is reduced from just under $12 trillion to just under $11 trillion.

The budget iterates in this way out to year 20. This is the point at which the income tax projection stabilizes at the values for years 17–20 in Chap. 14: in other words, income taxes do not change after year 20. The same goes for natural capacity fees. From years 21 to 50, these revenues are projected forward using a logistic function whose growth rate is the average of the year-on-year growth rates from years 1 to 20, and whose ceiling is an arbitrary value that attempts to balance the effects of long-term population growth with the effects of the new régime upon business and finance. The long-term ceiling for annual tax revenue is set at $5 trillion, slightly less than twice the starting value. The long-term ceiling for annual revenue from natural capacity fees is set at $1 trillion, about seven times the starting value.

The basic living program is projected all the way out the year 50 in line with projected population growth. The mandatory budget categories that are retained, covering federal retirement and other programs, are projected out using the average year-on-year growth rates from years 1 to 20 and ceiling amounts in proportion to the projected US population by year 50. For convenience, the discretionary programs are projected out using the same method, although it is acknowledged that these will be less tied to population growth than the mandatory ones. Net interest on debt is

Table 16.3 Starting (year 0) and projected values for the government budget projection

Government outlays (US$ bil.)											
	Discretionary	*Mandatory*							*Predicated from NC fees*		*Total*
	All discretionary	*Social & income security*	*Major healthcare programs*	*Federal civil & military retirement*	*Veterans' programs*	*Other programs*	*Offsetting receipts*	*Net interest on debt*	*Natural capacity investments*	*Material efficiency subsidies & incentives*	
r	0.8%	N/A	N/A	3.0%	2.0%	1.0%	N/A	1.8%	Apportionment		
K	1878	N/A	N/A	238	126	165	N/A	N/A	0.5	0.5	
Year											
0	1202	1147	768	153	80	105	−305	221			3372
1	1206	0	0	154	81	106	0	221	77	77	1922
10	1238	0	0	168	86	109	0	55	128	128	1913
20	1274	0	0	182	91	113	0	0	236	236	2131
30	1308	0	0	194	96	117	0	0	310	310	2335
40	1341	0	0	204	100	120	0	0	375	375	2516
50	1373	0	0	212	104	123	0	0	424	424	2660

Government income (US$ bil.)								
	Sovereign money creation (SMC) for government programs ($ bil.)				*Government revenues ($ bil.)*			
	For government budget (net of last year's revenue)	*For debt reduction*	*For basic living program*	*Total SMC*		*Tax revenue*	*NC fees*	*Total revenue*
					r	2.2%	6.0%	
					K	5000	1000	
Year								
0						2862		2862
1	2432	1000	431	3863		2229	155	2384
10	9685	1000	525	11209		2618	256	2875

(*continued*)

Table 16.3 (continued)

Government income (US$ bil.)							
	Sovereign money creation (SMC) for government programs ($ bil.)				*Government revenues ($ bil.)*		
	For government budget (net of last year's revenue)	*For debt reduction*	*For basic living program*	*Total SMC*	*Tax revenue*	*NC fees*	*Total revenue*
20	5213	0	634	5847	3390	471	3862
30	2435	0	743	3177	3625	620	4245
40	2616	0	853	3469	3837	751	4587
50	2760	0	967	3727	4025	848	4873

Balances and macroeconomic sovereign money creation (SMC)				
Year	*Balances*		*Macroeconomic SMC ($ bil.)*	
	Net government revenue (deficit or surplus)	*National debt*	*Excess SMC from government operations*	*Total SMC into financial sector*
0	−510	−11983		
1	−1479	−10983	0	1000
10	−8335	−1983	0	1000
20	−1985	0	0	0
30	−100	0	1068	1061
40	−100	0	1118	1114
50	−100	0	1146	1144

Source: Author's creation

Notes: Discretionary and mandatory program projections use logistic functions employing long-term growth rates (r) in program budgets from CBO data. Ceiling (K) established by proportion from US population projection; in year 1, social and income security and healthcare programs are replaced with a basic living program. Offsetting receipts from payroll taxes, which were used to fund these programs, also are eliminated; Natural capacity investments and material efficiency subsidies and incentives are funded by natural capacity (NC) fees; government budget that is funded through sovereign money creation is the total budget in any given year minus net revenues from the prior year; Debt reduction through SMC is $1000 billion annually until debt is paid off (year 12 in the scenario); SMC funding of a basic living program makes up any shortfall from artificial capacity fees; tax revenue is the total from the tax projection described in the text for first 20 years, then projected logistically from years 21 to 50 using an average growth rate over the first 20 years (r) and an assumed ceiling (K); revenue from natural capacity fees is the total from the natural capacity fee projection over the first 20 years, then projected logistically from years 21 to 50 using average growth rate over the first 20 years (r) and an assumed ceiling (K); net government revenue is the total revenue minus the total budget. The target deficit is −$100 billion; national debt is the prior year's debt minus the current year's SMC for debt reduction; if the revenue exceeds the budget, then any excess is carried over into the following year and fed into the financial sector as investment.

calculated at a rate of 1.8 percent, which is the 2013 (year 0) interest paid divided by the 2013 debt.

As Chap. 16 explains, the sovereign money creation requirement consists of a portion serving the government budget, net of the prior year's revenue; a portion for debt reduction, until the national debt is paid off; and a portion to fund the balance of the basic living program, as needed. The government's net revenue then is the total revenue minus the total SMC requirement, up to a maximum of −$100 billion. In other words, the budget policy is to aim to operate at a modest deficit from one year to the next. If the SMC requirement exceeds total revenue, then the excess is carried over into the SMC requirement for the following year. If revenue exceeds the SMC requirement, then there is an excess of funds from government operations. The excess from one year goes into the financial sector as investment the following year. In this situation, which prevails in the later years of the projection, after the national debt has been paid off, SMC issued in one year is being mopped up by taxation.

Table 16.3 summarizes the starting values and projected values for the budget projection every ten years up to year 50, along with associated assumptions.

References

Acworth, W. (2017). *2016 Annual Volume Survey*. London, Singapore, Washington, DC: Futures Industry Association. Retrieved from www.marketvoicemag.org.

Berman, M. (1981). *The Re-Enchantment of the World*. Cornell University Press.

Kelly, M. (2012). *Owning Our Future: The Emerging Ownership Revolution*. San Francisco, CA: Berrett-Koehler.

US CBO. (2017). *An Update to the Budget and Economic Outlook: 2017 to 2027*. Washington, DC: United States Congressional Budget Office.

CHAPTER 17

A Manifesto for Market Planetarianism

The bulk of this volume has addressed, in a largely apolitical way, the question of the economy's relationship with nature and its implications for economic policy. This section departs from that narrative to lay out a political manifesto in support of material discipline and widespread prosperity, describing actions that could be taken today, early in the twenty-first century, toward those ends. A working label for this manifesto is market planetarianism. It is both a social philosophy and an economic doctrine.

A manifesto, writes Yanis Varoufakis, is an emotional plea. "It must speak to our hearts like a poem while infecting the mind with images and ideas that are dazzlingly new." One must be awakened to the "possibilities with which our current reality is pregnant", and be left asking why and how we have been complicit in the state of things as they have become. A manifesto must carry urgency to inspire the reader to make changes to their own lives and to the world around them.[1] Yet a manifesto is equally a set of prescriptions, argued rationally and presented coherently. To succeed, a manifesto should lift our eyes to the big picture, synthesizing apparently disparate ideas and concepts. For this, it needs to show clearly how those ideas and concepts relate to one another. The prescriptive part of a manifesto, like the emotional part, should lift the heart. "Our feelings always deal with the totality of any system", wrote the architect Christopher Alexander.[2] Seeing the whole allows the familiar to be reinterpreted in a

[1] Varoufakis (2018).
[2] Alexander (1979).

F. Murison Smith, *A Planetary Economy*,
https://doi.org/10.1007/978-3-030-49296-0_17

new light. A new understanding is something that 'feels right' to the receptive reader. Such an understanding drives and energizes change.

Market planetarianism is a doctrine of harnessing the power of markets toward one or more economic ends serving the human collective on Earth. Implications arising from specific market solutions, if they were significant enough, might from time to time permeate up into society's normative space, prompting the collective to shift the boundaries of the market, such as through legislation. In this way, society would evolve toward the normative goal enshrined in a prime directive; or, in a crisis, perhaps revise the prime directive itself. Market planetarianism, or a similar doctrine, is necessary for a society facing existential threats arising from the scale of its activities. At the societal level, it is outward-looking, rather than introspective, unlike traditional collectivist or capitalist social theories. Humanity, early in the twenty-first century, is careening into a new world, unequipped with a coherent social philosophy or a set of prescriptions to deal with it. Market planetarianism offers one.

The manifesto is organized loosely along the lines of the rest of the book, condensing the prescriptions of the earlier chapters into a series of declarations followed by a set of propositions for the establishment of a new social and political system for long-term stability on a crowded planet.

Although the emphasis is global, the prescriptive part of the manifesto, like the foregoing analysis, applies particularly to OECD countries. It does not apply very well to regions where absolute poverty remains widespread. By focusing on economically wealthy regions, it proposes mechanisms by which they may help to lift disadvantaged regions out of absolute poverty, establishing the institutional and instrumental foundations for those regions eventually to participate in material discipline. Cultural differences also should be acknowledged. It would be tone-deaf to propose that the entire world should become some form of market democracy. The ways markets drive material discipline, while also ensuring widespread prosperity, would vary from one culture to another. The significance of their role also would vary. Neither market nor democracy are necessary conditions for material discipline, only sufficient ones. They are not the only conceivable agents of change toward a stable, prosperous economy but they are probably those most powerful.

Declarations

On a Crowded Planet, Everybody Is in It Together

Well-worn and self-evident though it may seem, the declaration that 'everybody is in it together' represents a conundrum for early-twenty-first century society. If it is really true, then how is it that society remains dominated by rampant individualism? We may be living in a transition period between Cartesian modernity—the era of individualism inspired by Descartes, Hobbes and Locke—and a future planetarianism. The age of European exploration and expansion that was enabled by an ethos of dominion over nature exported to the world the Pioneer spirit, a culture of self-sufficiency, acquisitiveness and fierce protection of individual rights, qualities that have persisted long after the disappearance of anywhere to pioneer. On a crowded planet, they begin to look like anachronisms.

A future planetarianism will not so much supersede these qualities as subsume them within a broader awareness of the unity of the world and of the primacy this unity takes over individual maximization. If society is to remain civilized, there will be no alternative to planetarianism in some form. Under a planetarian ethos, self-sufficiency becomes group sufficiency; acquisitiveness is tempered by sufficiency for all; and individual rights become bounded by collective imperatives. In other words, you can do what you like as long as it does not compromise society's alignment with nature or any pre-existing moral constraints. If it did then it would compromise society: in mainstream economic parlance, your actions would represent an externality upon everybody else.

This is not 'saving the planet' thinking, it is 'saving civilization' thinking. It becomes relevant once human society becomes materially significant relative to nature. Even though the scale of human society was already significant in Descartes' day, neither he nor any of his Enlightenment contemporaries saw this. Four hundred years later, there is no avoiding it.

On a crowded planet there are no 'haves' or 'have-nots'. If everybody is in it together then everybody *has* to be in it together for a planetarian ethos to emerge. Put another way, if everybody is given the means to change, and substantially all agree on what that change should be, then the change in question will come about.

Widespread Prosperity Begets Material Discipline; Both Are Needed for Economic and Social Stability

The idea of everybody being in it together is an idea of everybody being able to live above a sufficient minimum. Until that condition is satisfied, everybody will not be in it together, and society as a total system will continue to fail in addressing the consequences of its crowding. Despite decades of scientific advice to policymakers that the economy should move toward alignment with natural processes, it has not done so because its existing internal structures and processes perpetuated social problems preventing long-term social cooperation. Trying to address the interconnected problems of climate change, habitat loss, biodiversity decline, sea-level rise and water security before addressing the pervasive, systemic inequalities within the economy is putting the cart before the horse. People will begin to take the long view only once their near-term security is guaranteed. Only then would it become possible to entertain the idea of a 'prime directive' or some shared *telos* to help guide and direct individual actions and decisions. If people began to act upon a long-term guiding principle of some kind then this principle eventually would manifest itself in the behavior and decisions of legislators and policymakers, who are, after all, of the people themselves. The principle thus would become established in law, enforced through regulation, and enabled through the action of economic instruments upon markets. Over time, the economy and society would evolve toward a state consistent with that principle.

Many writers on these issues refer to a culture of 'sufficiency', in which there is no such person as a 'have-not' and therefore, by association, no such person as a 'have'. Sufficiency as a culture means that if higher earners must give up something then they would do so gladly because, if nothing else, their great-grandchildren will not be spared the consequences of their inaction or irresponsibility any more than anyone else.

Historically, much of the direct redistribution within the economy was voluntary, in the form of charitable contributions and philanthropy. To the extent the rich have been taxed more than the poor, those taxes have gone into government coffers to fund welfare programs whose success has been mixed, at best. There has never been any *institutionalized* direct redistribution. The present volume has suggested a few forms it could take, whether on an individual basis, in the form of elective redistribution to replace part of the income tax on the wealthy, or on an institutional basis, in the form of a basic living program funded by artificial capacity fees on businesses.

Framed as a question of sufficiency, the distribution question becomes one of establishing a firm lower bound to prosperity without arbitrarily limiting the upper bound. This is the distributional philosophy not of a socialist or a capitalist economics but of a market-planetarian one. The collective impetus is a characteristic clearly shared with socialism, yet the promotion of the market as an engine of economic change, and the absence of a firm upper bound on prosperity are capitalist in origin. The distinct feature of this economics is the directive imposed upon the market in the interests of long-term economic stability and widespread prosperity.

Taking widespread prosperity as a prerequisite for economic alignment with natural processes does not mean waiting for distributional measures to have the desired effect before taking up the challenge of material discipline. Both should be pursued in parallel. Yet any instruments established to steer economic activity toward greater material efficiency will only work if they are accompanied by measures to ensure full social participation in those measures, namely, to ensure individual sufficiency.

Governance Must Be Inclusionary

Widespread prosperity and participation in the fulfilment of a prime directive extend not only to economic activities but also to political ones. Prosperity, as imagined here, can be measured many ways, one of them being financial; others being health, education, family, community and governance. An ability to participate directly in governance provides a sense of belonging to society and a sense of purpose. One becomes invested in decisions, their outcomes and their broader intent.

Historically, with few exceptions, civilizations have been marked by a disconnection between citizens and decision-making bodies. So-called democracies resolve that disconnection only on election days, occurring at most once or twice a year. At all other times, elected representatives act on their own. Granted, citizens in most democracies have the ability to provide public comment on deliberations, a process most commonly seen in local government and in the regulatory process. Decision-makers in most situations treat it sincerely and they take it seriously: it is more than mere window-dressing. Yet when a decision is actually made, it is still only the elected representatives who make it.

This arrangement exists out of historical convenience. Back in the days when it was arduous to move people or information, the delegation of decision-making responsibility to elected representatives represented an efficient solution. It also served as a protection against the ill-considered judgement of an uneducated electorate, as the framers of the US Constitution articulated. Today, with the global availability of information and widespread literacy, it is an anachronism and a hindrance to forward-thinking governance. It is exclusionary. It goes against the reality that everybody is in this together.

If the problems include everybody then the solutions should include everybody. The physical technologies already exist; it is a matter of establishing the accompanying social technologies. Options include sortition—the selection of individuals for short terms within legislatures, along the lines of jury duty—and cross-participation, in which jurisdictions place representatives in one another's chambers, analogous to ambassadorships in the diplomatic arena, but with actual voting ability.

Full planetarian governance, though, must be hierarchical, from the individual to the global, involving the local, regional and national in between. As components of a larger hierarchical governance structure, sortition and cross-representation can be tried and tested on small scales first. The concept and structure of a global governance council outlined in Chap. 5 represents one way in which hierarchical, inclusionary governance may unfold: the individual, community, city, state, nation, region and planet all participating to address issues at all levels, facilitated by communications technologies unimaginable when democratic structures first were established.

Public Money Belongs to the People

Money is a daily necessity for a civilized society to function, yet hardly anybody ever claims to have enough of it. Money can take many forms, not only the public currency with which we are so familiar but also private, local and virtual currencies, which fulfill many specialized functions. For collective governance to be successful, however, a currency needs to exist over which the collective interest has full control. This is so that investments can be made in initiatives supporting a prime directive, and taxes and fees can be collected on activities detrimental to it, thereby also closing the loop of money flow.

By default, this currency is the public currency, the national legal tender. However, in today's economy, the public currency is not fully under public control. It has been co-opted for private, commercial finance, which has been granted virtually unregulated freedom by governments to create new debt denominated in the public currency. Not only has the public currency become privatized but worse, governments themselves are placed in the position of having to finance their operations by borrowing their own currency. Currently, there exists no public right to money.[3] Public money should be a public right, not a private commodity. It rightfully belongs to the people, not to the banks.

Being able to issue new money without having to borrow it, the government would be able to provide all kinds of essential services for free, without having to raise income taxes. These services would replace outdated, ineffective and undignified wage-based welfare programs.

There is nothing to stop commercial banks from creating their own parallel currencies for conducting their business. "In the same way that an elastic supply of credit and debt enabled the production and accumulation of private wealth", writes Mary Mellor, "[so] the elastic supply of public [expenditure] could enable the expansion of public provision and public wealth."[4] The establishment of a national monetary authority of some kind as the sole issuer of the public currency, and the institutional separation of the depository and lending functions within the financial sector, would accomplish this end. Commercial finance still would do business in the public currency because the flow of public currency through the economy would incorporate private investment and lending, as it does today. The main difference would be that banks would not be able to create new public currency through creative accounting, such as making loans against their deposits. Commercial lending would be made against existing reserves. In this way, a national monetary authority would have full control over the flow of the public currency within the economy, quantitatively easing or tightening supply, balancing the government investment against macroeconomic requirements as needed.

Within the context of inclusionary governance, this system would give the people control over their money, thereby placing commercial finance at the behest of the public interest rather than the other way around.

[3] Mellor (2016, p. 11).
[4] Mellor (2016, p. 186).

A Basic Living Standard Is a Basic Human Right

Given the pervasiveness of money within the modern economy, it is perhaps surprising that basic income has not previously been asserted as a basic human right. Although many authors proposed variants of the idea in the past, none of them accorded it the same status as food, shelter and individual security, these being three basic human rights enshrined in the United Nations Universal Declaration of Human Rights.[5] Yet money is the medium through which the vast majority of people struggle to live their daily lives. Surely it stands alongside food, shelter and security as a basic need?

As Emma Seeri describes in a survey of more than 100 countries, "free public health and education services are a strong weapon in the fight against economic inequality. They mitigate the impact of skewed income distribution, and redistribute by putting 'virtual income' into the pockets of the poorest women and men."[6] Seeri's study is a strong counterargument to the neoliberal doctrine that the privatization and monetization of public services would render them more efficient and productive. That proposition has never been convincingly demonstrated. This is not to say necessarily that all public healthcare and education should be in-housed by government but that, if some of it were outsourced, then the public sector should always retain control over its funding and would exercise keen oversight. Another problem with privatizing public services is that it renders them democratically unaccountable. On the other hand, "by taking things off the market", writes Ben Tarnoff, "government can democratize access to the resources that people rely on to lead reasonably dignified lives".[7] As numerous studies have shown, societies exhibiting high personal dignity are relatively peaceful, prosperous and stable, not particularly prone to social tension or unrest. Such conditions are an essential prerequisite to participation in the grander challenge of aligning the economy with nature.

By adding a basic living standard to the Charter on Human Rights—incorporating income, health and dependent care and education—the countries of the UN would formalize a commitment to implement basic living programs within their jurisdictions. Artificial capacity fees and sovereign money creation would avoid the need to fund the programs through

[5] United Nations (1948).
[6] Seeri (2014, p. 1).
[7] Tarnoff (2017).

taxes. Elective redistribution from taxpayers in wealthier countries would represent an additional potential flow of funds to people in poorer regions. UN oversight would combat any institutional weaknesses on the ground to help ensure that the funds went to their intended recipients. Absolute poverty could be all but eliminated within a generation or two except, perhaps, in a few remaining war-torn pockets of the world.

Tax Social Bads, Not Social Goods

Government must raise revenue for civilization to persist. Yet taxation of human productive labor is perverse and counterproductive. It actually gets in the way of solutions to the problem of aligning a large economy with natural processes, because it perpetuates the imprisonment of wage labor. Governments should raise revenue by taxing material intensity. The lower the economy's material intensity becomes, the higher the tax rate on it. A material intensity tax is one means by which sovereign money issued into the economy would be recovered.

Freed, for the most part, from taxation on wage labor, and supported by a basic living program, people would be able to afford materially efficient options in the market. They would have time to realize their potentials, contribute creatively to society, and enjoy a higher level of prosperity, however measured. Communities would become more cohesive, crime would diminish, people would live healthier lives, and civic participation would flower. All these things are possible beginning today.

Guide the Markets and Let Them Do Their Work

As a general principle, even on a relatively empty planet, no market should be allowed to degrade natural capacity. Yet the principle applies with extra force on a crowded one. This means that society must make decisions about what kinds of markets to allow, or disallow, and how to tax them. It implies a conscious decision relative to a guiding principle of some kind. A society that allows any and all markets to operate indiscriminately is one that allows itself to be controlled by the hegemony of the market. This is the tail wagging the dog. It is approximately the state of the present global economy.

Society cannot allow itself to be pulled this way and that by market whim. At the dawn of the industrial revolution, nobody bothered to ask whether a market for coal should exist. Coal was there for the taking and it could meet the emerging demand for industrial production in a way that the existing energy sources of the day—water, wind and wood—could not. Nobody bothered to ask this normative question because the consequences of burning coal seemed so remote at the time. Now that the consequences of burning fossil fuels are no longer remote but very much in the news, almost daily, the various markets for fossil fuels have become so much part of the modern economy that asking whether they should or should not exist might seem fruitless. Yet the dog has to wag its tail. Just because a market already exists is no justification whatsoever for its continued existence in the future. Even if a new market could be created for an as-yet untapped exhaustible resource, that would be no reason simply to go in and tap it. Society has to make forward-thinking decisions to encourage new markets and to discourage detrimental ones. If the guiding principle is material discipline, then it implies investing in non-polluting energy sources and taxing the polluting ones. This is how collective interest takes back control over the global economy from commercial interests and begins to employ markets to serve that long-term goal.

To the extent that these structural changes would create economic upheaval, the other provisions outlined above would significantly diminish it. People's lives are directly affected by the replacement of old markets with new ones. If their prosperity is guaranteed in advance, such as through basic living, then the necessary structural changes to the economy will become easier to accomplish.

As an interconnected system of markets, the shape of the global economy represents an embodiment of what society wants markets to accomplish. Historically this that been whatever markets wanted to accomplish. As a consequence, modern society has come to reflect the myopia of the unguided market. On a crowded planet, it is not sufficient that markets merely create wealth. They have to create long-term economic stability while also creating wealth. This necessitates guiding markets according to a prime directive of some kind. Markets themselves are the engine of this change and they are the quickest, most effective way to get there.

Proposals

Following from the declarations, the manifesto offers a set of proposals for institutions, governance, policies and instruments. The overarching theme is long-term social stability through alignment with nature. Other authors have made similar suggestions in the past. To quote Goldsmith et al., "if we plan remedial action with our eyes on political rather than ecological reality, then ... we shall muddle our way to extinction."[8] Morris Berman wrote that "the preservation of the planet may be the best guideline for all politics".[9] He did not make explicit whether he was referring to society or nature, or both; however, the sentiment is apt for today's world.

Table 17.1 summarizes the propositions, identifying dependencies and other connections among them.

The legislation required for most of these proposals is national, as prior chapters have discussed. Some of the governance changes would percolate up from the local level to the national level and beyond. Even though the sequence of proposals in Table 17.1 ends with a prime directive, this milestone really is part of a cyclical process, which then loops back to guide and shape future changes, yet to be anticipated. The proposals preceding the prime directive anticipate it, of course. Legislation to enact them would be driven by a shared desire to restructure the economy with material discipline and widespread prosperity in mind. The establishment of a prime directive by a global governance council simply would formalize a guiding principle already recognized and followed in the changes that led up to the creation of the council in the first place.

As the foregoing analysis suggests, the way government does business can feasibly be altered to begin guiding the economy, and society at large, on a path toward alignment with natural processes and widespread prosperity. It can be done without sending the government financially insolvent or destabilizing the capital markets. It will, however, require some major structural changes, such as the abandonment of the 'tax and spend' paradigm in favor of alternate vehicles for raising revenue. It will require the abandonment of the wage-based welfare paradigm for a non-wage-based one, in which an equal basis of shared prosperity is enjoyed by everyone, regardless of any other income. It will require the government to boldly take back control of the national currency, while continuing to provide liquidity to the capital markets. It will require action to shift taxes from social goods onto social bads, accepting that, although this will increase the cost of consumption, it also will drive market-based solutions to reduce it. All of these changes are feasible today, provided that legislators begin to think differently about what government does, or is supposed to do, and how it does it; and that they act across party lines.

[8] Goldsmith et al. (1972, p. 15).
[9] Berman (1981, p. 295).

Table 17.1 Proposals for a market-planetarian economy embodying capacitism as its guiding philosophy, expressed through a suite of market-planetarian institutions (I), governance structures (G), policies (P) and instruments (N)

Proposal	*Type*	*Depends upon …*	*Purpose*	*Effect*	*Enables or encourages …*
Material accounting standards	G, P		Measure businesses' material throughput and intensity	Identify efficiencies	Material intensity tax
Planetary trust	I		Independent, publicly accountable institution for returning commons rent to all	Capacity securitization; distribution of prosperity	Common capacity fees and natural capacity share ownership.
Natural capacity fees on financial trades	P, N		Material efficiency, reduced financial speculation	Material efficiency, reduced financial speculation	Material efficiency subsidies and incentives
Material efficiency subsidies and incentives	P, N		Encourage materially efficient production	Reduce costs to establish new technologies	Material intensity tax
Phase-out of subsidies and incentives on materially inefficient production	P, N		Discourage materially inefficient production	Reveals true cost of materially inefficient practices	Material intensity tax
Public financial rating agency	I		Public governance of financial ratings	Independent financial ratings according to criteria in public interest	Other monetary reforms; prime directive
Separation of bank deposits and lending	I, G, P		Elimination of fractional reserve banking	Long-term financial stability and debt reduction	Sovereign money creation

(*continued*)

Table 17.1 (continued)

Proposal	*Type*	*Depends upon …*	*Purpose*	*Effect*	*Enables or encourages …*
Sovereign money creation (SMC)	I, G, N	Separation of deposits and lending	Public control of public money	Nationalization of national currency	Basic living program
Artificial capacity fees on businesses	P, N		Monetize common assets and fund basic living	Non-tax government revenue	Basic living program
Basic living program	P, N	Artificial capacity fees and/or SMC	Basic income, free health and dependent care and public education	Widespread prosperity	
Elimination of social and income security	I, P	Basic living program	No longer needed	Eliminate wage-based welfare, reduce government budget	
Elimination of capital gains tax	P, N	Other reforms listed here	No longer needed because capital gains are now regenerative (see Note)	Full reward of capital appreciation	
Material intensity tax	P, N	Reducing income tax	Encourage material efficiency	Increased material efficiency	Reducing income tax
Reductions in income tax	P, N	Material intensity tax	Tax-shifting from income onto material intensity	Consumer control over consumption	Material efficiency
Elective redistribution	I, P, N	Material intensity tax	Direct redistribution of income or assistance from high earners to low earners	Narrowing of the income or welfare gap; control over how resources are spent	Success of governance reforms

(*continued*)

Table 17.1 (continued)

Proposal	*Type*	*Depends upon ...*	*Purpose*	*Effect*	*Enables or encourages ...*
Property titling and tax reform	P, N		Title to property necessitates maintenance of natural capacity, whose state determines level of property tax	Encourages improvement of natural capacity by property owners	
Sortition (local first, then regional)	I, G		Inclusion of citizens in policymaking	Citizen investment in governance	Cross-representation
Cross-representation	I, G	Sortition	Jurisdictional interdependence	Broadens stakeholder base	Global governance council
Global governance council	I, G	Earlier institutional and governance reforms	Hierarchical governance for global society and economy	Streamline global cooperation	Establishment of prime directive
Global prime directive	G, P	Institutional and governance reforms	Material discipline and widespread prosperity, globally	Guiding principle for policymaking and economic activity	

Source: Author's creation

Note: With other measures in place, capital appreciation is no longer exploitative but becomes regenerative, that is, it results from maintenance or improvement of natural capacity. To tax it would be economically perverse. Capital gains also would become absolutely smaller than in today's leveraged financial markets

Although some of these proposals may appeal to one party, some to another, only in concert can they have the desired effect.

There is nothing to lose in pursuing proposals of this kind right away, and everything to gain. Failing to pursue them will introduce a grave long-term risk to the global economy and, by extension, to human society.

References

Alexander, C. (1979). *The Timeless Way of Building*. New York: Oxford University Press.

Berman, M. (1981). *The Re-Enchantment of the World*. London: Cornell University Press.

Goldsmith, E., et al. (1972). *Blueprint for Survival.* New York: Signet, New American Library, Inc.

Mellor, M. (2016). *Debt or Democracy: Public Money for Sustainability and Social Justice.* London: Pluto Press.

Seeri, E. (2014). *Working for the Many: Public Services Fight Inequality.* Oxfam International Briefing Paper 182, Oxford, UK.

Tarnoff, B. (2017, June 21). How Privatization Could Spell the End of Democracy. *The Guardian.*

United Nations. (1948). *Universal Declaration of Human Rights.* New York: United Nations.

Varoufakis, Y. (2018, April 20). Marx Predicted Our Present Crisis – And Points the Way Out. *The Guardian.*

CHAPTER 18

A Planetarian Society

From Globalism to Planetarianism

What do the economy and society want to be when they grow up? Numerous writers, from the ancient Greeks to Thomas More to Karl Marx and many who followed, argued that the particular society in which they lived was at an incomplete stage of development. They proposed a future state of some kind that would represent what they considered a 'complete' or 'ideal' state of development. More called his ideal 'Utopia', as a veiled critique of monarchy. Marx called his ideal socialism, as an alternative to capitalism. A lesser-known visionary, Peter Berg, gave the name 'planetarianism' to his ideal.

Unlike the others, Berg did not elaborate his ideal in great detail, which is perhaps why it was largely forgotten. His vision was of a society of local and 'bioregional' communities united in a kind of collective biophysical consciousness. It was in this sense the outline of a paradigm—a system of norms, ethics and practices characterizing a society yet to exist—rather than a collection of economic prescriptions.

What distinguishes these and other 'ideal' societies is their author's descriptions of people's prevailing norms and attitudes: the systems of ethics in operation and the ways people would perceive the world. As Erich Fromm observed, to change from the current society to some idealized future state would necessitate not only a radical transformation of the

F. Murison Smith, *A Planetary Economy*,
https://doi.org/10.1007/978-3-030-49296-0_18

socioeconomic system but also, more fundamentally, a profound change in human character.[1]

The difference between the visions of Berg, Fromm et al. and those of earlier authors is that the Bergs and Fromms of the world, writing during the 1970s and 1980s, could foresee the current, global trajectory of society and its economy. This trajectory simply was not on Thomas More's radar early in the sixteenth century. Even Marx could never have anticipated that, within a hundred years of the publication of *Das Kapital*, the economy would grow so much, so fast, as to materially affect natural processes on a global scale The likes of Berg, Fromm, Berman and others tackled globalism head-on because it was already here. They, like their forebears, acknowledged the dysfunctions of modernity; but the modernity they discussed was not limited to a handful of industrial or imperial nation-states. It was the modernity of a rapidly globalizing economy, dragging as it did so many diverse societies into the age of the global market. Economic globalization, within the span of a few decades, imposed upon societies around the world a whole new paradigm, a whole new set of values and ways of thinking about the world, some of which must have seemed completely alien. But instead of replacing the old ways of thinking and doing things, modernity has been absorbed into a kind of cultural coexistence, governing the realms of commerce and industry, but not altogether replacing the cultures of people's everyday lives.

Globalism is both a phenomenon and a doctrine. As a phenomenon, it is the interconnection of diverse, far-flung societies by the technologies of communication and commerce. Where the phenomenon has been interpreted as a good thing, it is also a doctrine. It is the doctrine of 'economic development' practiced since the Second World War by the World Bank, the United Nations and other supranational organizations. It is the belief that opening poor, backward nations to industrialization, banking and commerce would lift them out of abject poverty, bring them more fully into the money economy, and integrate them into the community of nations. It has been assumed to a large extent that this 'community of nations' is a community of market economies embodying the tenets of capitalism. The centrally planned economies resisted this program for a few decades yet, by the 1990s, there was no stopping market globalism.

In lifting people out of absolute poverty, it has been largely successful. Along the way, though, it has enriched a relatively small number of people

[1] Dodd (2014, pp. 337–338), citing Fromm (1976).

at the top in ways that would have been unimaginable a hundred years ago. Also, globalism is not the only means through which millions of people's lives could have been improved. Its greatest shortcoming is that it failed to recognize and adjust for the planetary systems enabling it. Globalism, in its current form, is a recipe for biophysical, economic and social instability. It needs to evolve into something more holistic. Peter Berg coined as good a term as any for that something: planetarianism. This final chapter discusses some of the ways the globalist paradigm might evolve into a planetarian one, along the way putting a little meat onto the bone Berg threw.

Perceptions, Norms and Ethics

Falling off the Edge of History

In the day-to-day business of modern life, it is easy to forget, and equally difficult to comprehend, the duration of all that came before: the 'deep time', comprising hundreds of millions of years of multicellular life, and four billion years of unicellular life and of the pre-living Earth before that. Five thousand years of human civilization is only one ten-thousandth of the time since the dinosaurs went extinct. That period, in turn, is about 1.1 percent of the geological history of Earth. Human civilization represents about half of the sliver of geological time since the last ice age known as the Holocene. Only within this little sliver has human activity begun to have effects upon natural systems measurable at the global level.

The prior history of Earth's natural systems is so long that if a time-lapse movie of Earth were made using snapshots taken every thousand years then, at a standard 25 frames per second, a million years would take 40 seconds, and the history of life since the extinction of the dinosaurs would take about an hour. Right at the end of this hour, in the very last second of the movie, a 'thing' would appear on Earth, all of a sudden, as if out of nowhere: a network of artificial structures lain over the world's ecosystems, and thousands of artificial objects orbiting the planet. Its appearance would happen so fast, in the space of one-fifth of a second, that the eye could barely catch it. The movie then would abruptly stop at this new image. If the movie also displayed the chemical composition of the atmosphere and oceans, species diversity, and other metrics over time, then their values also would abruptly change within the last second of the movie.

Some writers use a metaphor of Earth as a living organism becoming infected with a human 'parasite'. A better analogue, conveying the extreme suddenness of this event, is a chemical state change, like the familiar 'iodine clock' experiment from school chemistry class, where the clear solution suddenly turns blue. That change is due to a kind of competition among chemical species, ultimately won by a blue triiodide starch complex. In systems terms, this is the meaning scientists seek to convey when they talk about systems 'flipping into new states'. Scientists, and many of the rest of us, worry whether people will be able to adapt to new states into which natural systems may flip, or whether such new states could cause society itself to flip into a new state. Within the context of the time-lapse movie, nature has just undergone a sudden state change. It is unknown how this state change will play out. We are falling off the edge of history: what lies in the future is highly unpredictable. Earth has never experienced anything like this before.

What can be predicted is that the sudden change brought on by the emergence of human civilization, and recently by the expansion of the modern economy, will leave a lasting imprint on the geological history of Earth. This imprint will persist for hundreds of millions of years.[2] Not only will the geological record reveal abrupt changes in the chemistry of Earth systems—atmosphere, oceans and soils—but also it will reveal a dazzling array of novel chemical species never before found in nature, such as plastics, aerosols and pesticides. The record also will feature a very sharp line representing a major extinction event, perhaps not on the scale of the great mass extinctions of the geological past, but much more sudden. Aliens visiting a few million years hence may or may not find people on Earth but they would find this record as a testament to a prior society whose physical technologies perhaps outstripped its social ones.

Some predict the collapse of civilization.[3] Others are optimistic. There is room for both perspectives. To the best of our knowledge, we live today in a unique time on Earth, both witnessing and participating in a new event. However, it might not be the last of its kind; it could be the first of many. Even if the current event were to cause the collapse of modern civilization—because, for instance, we might not know what we are doing—it is unlikely that the human species would die out. The persistence of a remnant human population would leave open the possibility for

[2] Brannen (2017, p. 166).

[3] Indeed, as Lovins et al. (2018) describe, a burgeoning literature exists on this.

new civilizations to emerge. Even if humanity did not succeed in aligning its activities with natural processes the first time around, it might still get things right on subsequent tries. Future civilizations would have to contend with the mistakes of past ones, of course, as the effects of those mistakes would reverberate around Earth's natural systems for centuries or millennia. One would hope, though, that people's ability to adapt and to learn eventually might overcome those obstacles. Only an utterly catastrophic change to Earth systems would wipe out humanity forever and, if it did, a majority of other species probably would be lost as well, if earlier mass extinctions are any guide. Such a catastrophe would represent a sixth great mass extinction, from which life would take millions of years to recover.

Our modern economy has not yet grown so devastating as to have that kind of effect upon natural systems. Current increases in carbon emissions, for example, are orders of magnitude smaller than those that emanated from the Siberian Traps to cause the end-Permian extinction, the greatest in geological history. And the awakening of social awareness about the economy's scale and its exogenous effects suggests that corrections could be made long before such scales would be attained. The issue is not the survival of the human species, or of life on Earth, so much as the survival, stabilization and persistence of human society.

So the perceptual or attitudinal shift that needs to happen is to look at the whole Earth, as if from orbit, and to acknowledge human society as a subsystem of it. This acknowledgment will accomplish two things. First, it will remove social myopia. Viewing the whole system from outside looking in will pull the mind up to large spatial scales and large temporal scales. It will require people to take the long view of decades to centuries rather than of days to years. By considering the whole system, the normative orientation will become intrinsically more collectivist and less individualist. This is not to dismiss the relevance or value of individual endeavor, only to place it within a wider collective context, which recognizes certain limits to individual actions. The effect at the individual level would be seen in behaviors which, as a matter of habit, would consider individual decisions and actions within the context of a collective norm of some kind, such as material discipline. This brings us to the second effect, which is that perceived units of value within society no longer would be purely social in nature, such as money, time or relationships. Looking in at the whole system from outside, it is immediately clear that the primary, exogenous circulation of value is material. Some of this circulation is abiotic in nature and some of it is biotic, the biotic part being made possible by the conversion of solar energy through photosynthesis. If one looks only for social measures of value, all one will find is social measures of value. To do so is to miss the underlying circulation of value that makes all the social value possible. This circulation of material value is what must be conserved.

Growth and Prosperity

One of the major perceptual and normative changes that will be needed to enhance nature's ability to support the economy is from 'growth' to 'prosperity'. Economic growth without any compensating increases in material efficiency increases the material intensity of the economy. Growth is considered necessary because the world's population is growing. It has been considered desirable within the economic orthodoxy because it is assumed to be the 'rising tide that lifts all boats'. Empirical research, however, consistently refutes this assumption, instead showing that, when left to its own devices, growth is a rising tide that lifts only some boats—a few of them very high indeed—leaving many others stuck underwater on the mud.

In a world one hundred to two hundred years hence, in which the population is expected to approach a steady maximum, the case for the material growth of the economy should weaken. There may remain, however, a pronounced distributional problem, that of lifting all the boats off the mud flats, this in turn creating an upward pressure on material intensity. It could become a zero-sum game, because if increases in material intensity sufficiently degraded natural capacity then that could become the very factor preventing the establishment of widespread prosperity. Social unrest could result. The key to success is through material discipline, similar to a doctrine of fiscal austerity but applied to material intensity.

It would represent a major and fundamental shift in norms and attitudes to move from a doctrine of growth to one of widespread prosperity. As earlier chapters have discussed, it would necessitate a recognition of the global collective as an actual and legitimate regulator of individual self-maximization. Individual choices would be nudged by economic instruments designed for the long-term interests of the collective. Over time, these new choices would become the norm, habituated into people's attitudes. To get the ball rolling, however, will require a sufficiently broad acceptance of the need for this change so as to apply the necessary political pressure. Such a broad acceptance may be possible if the case can be made that the mudbound boats can be floated without sinking those already afloat.

Because economic growth conventionally has been measured in money terms, the culture of growth is one of financial accumulation. Essentially, it is a culture of surplus. Growth culture is made possible by economic surplus, which ultimately is supported by natural capacity. Money, in its various forms, is a product of this surplus. In hunter-gatherer societies, there was no surplus, by definition, so there was no money. Everybody lived a subsistence existence. Surplus became a characteristic of early agricultural and civilian societies because agriculture enabled an accumulation and hoarding of the products of nature. However, differences in surplus among individuals, which may have arisen for a variety of reasons, created wealth inequality. As a consequence, many people spent their whole time struggling to get by, while a few, who were exceptionally able, powerful or lucky, escaped subsistence and spent at least part of their time on higher things, such as art, music, writing or science.

Early money tokens facilitated an exchange among heterogeneous products of surplus, such as crops and animal products. Inevitably, however, money itself became hoarded and thus became a commodity in its own right, albeit a special one that could represent other commodities—analogous to a blank tile in a Scrabble game or the Joker in a pack of cards. In societies through which money circulated, subsistence needs thus became expressed through money: to do so was more convenient.

Looking from today's economy into the future, maintaining economic surplus will necessitate increasing material efficiency. Yet the total financial wealth of the economy already is sufficient to support everyone's financial subsistence. The earlier analysis on economic controls, and those of other authors, show this. In today's economy and the economy of the future, everyone should be able to spend significant time on non-wage pursuits. To do so would represent a maturation of civilized society.

The reason why not everyone's financial subsistence is met—aside from the small and diminishing number of people living wholly outside the money economy—is due to scarcity. Scarcity is an artifice created by distributional inequality. Since the 1970s, economic growth in emerging nations has delivered real and significant increases in income to large proportions of the population, while also enriching the top few percent. In developed nations, the top few percent have seen real incomes multiply, whereas large swathes of the population have experienced stagnation or decline.

To the seventeenth-century English philosopher Thomas Hobbes, competition, mistrust and glory were universal characteristics of society. It was every man for himself. "Hobbes's war of all against all was in the first instance a struggle over a fixed economic pie", writes Stephen Marglin. The struggle for power therefore was an economic struggle. Hobbes's successors, beginning with Locke, tried to circumvent this struggle by proposing that policymaking should increase the pie overall. Then everybody could have a bigger piece. Here was the birth of trickle-down economics. "Growth", writes Marglin, "has been chasing Scarcity ever since." Scarcity "[was] able to remain comfortably ahead of its pursuer [only once] the engine of production [became] sufficiently well developed to deliver steady growth". The "genie of Scarcity", he writes, could be let out of the bottle only once the conditions were in place for rivalry, *à la* Hobbes, to be channeled into the economy.[4]

But what if those conditions were to disappear? Would the economy continue to act as a conduit for rivalry? How else could rivalry be played out? Could rivalry ultimately result in social breakdown?

What Marglin referred to is the *relative* scarcity of resources within the economy. At any point in time, the total wealth of the economy is distributed unevenly, sometimes more so than others. The capitalist system, writes Nigel Dodd, tends to encourage the hoarding of wealth.[5] Socialist systems tend to do so as well, for different reasons. The hoarding of wealth creates scarcity; however, this scarcity is relative. It is what compels people to keep up with the Joneses. *Absolute* scarcity, on the other hand, occurs when an insufficient supply of a good or service is experienced. Think of bread lines in Soviet Russia or famines in Ethiopia. The growth that keeps relative scarcity at bay is an absolute increase over time in the volume of material passing through the economy. Consequently, some thinkers have described the economic process in terms of waste rather than scarcity,[6] but it comes to the same thing. The hoarding of wealth enables overconsumption, or conspicuous consumption, which generates relatively more waste than average consumption or subsistence consumption. Invoking Amotz Zahavi's handicap principle of sexual selection among animals,[7] the

[4] Marglin (2008, pp. 214–216).
[5] Dodd (2014, p. 166).
[6] Dodd (2014, ch. 5).
[7] Zahavi (1975).

attitude within civilized societies became, 'look how much money I can waste, and get away with!'

The growth keeping relative scarcity at bay forestalls social unrest, but not for ever. The conundrum facing a materially large and growing economy is that this growth, if unchecked, eventually would result in absolute scarcity—specifically, of natural capacity—and therefore by extension also of economic wealth. The pursued would turn on the pursuer.

The problem with absolute scarcity of natural capacity is a perceptual one. If it is not directly experienced, then it remains abstract. Then it is not confronted. However, another aspect of the prevailing attitude toward growth is directly experienced by many people, and has been for decades. This is financial volatility.

'Growth' is both a characteristic of the modern economy and a doctrine. For many of its proponents, 'growth' is a system of belief. Mass adherence to the growth credo produces an observed aggregate effect of economic growth. The credo recognizes no limits. There is, for instance, no overarching norm imposed upon the modern economy encouraging growth in moderation, even if cool heads in macroeconomic policy might advocate for it. The way the prevailing incentive structures are set up, more growth produces greater rewards. With no immediate discouragement of *over*-growth in personal wealth, there arises a tendency for overgrowth in the aggregate. This tendency is played out in full force in those sectors of the economy subject to few material limits to value creation, such as in financial services. Peter Warburton described the situation as "anarchy, masquerading as an agent of national prosperity and personal freedom".[8]

Anarchy it is, because it is largely ungoverned. Anarchy in the financial markets is what ensues when over-growth propels the financial sector into a meltdown. All the financial crises of the past hundred years came after what are euphemistically referred to as 'valuation bubbles'. Even under normal circumstances, stock markets tend to be much more volatile than underlying economic indicators. Volatility is a product of growth culture. It produces uncertainty. And uncertainty produces myopia. What point is there in taking the long view when the near future is so uncertain or risky? Better to make the most of what is available now. Myopia feeds back into growth culture as a drive for near-term profits. Make the money and get out before the next crash.

[8] Warburton (1999), cited in Dodd (2014, p. 199).

At the heart of the misperception about scarcity within the modern economy rests a culture of exploitation. There is no room for a concept of absolute scarcity of natural capacity when a drive exists only for exploitation. Instead, natural capacity is ignored or at best presumed to be endlessly abundant and therefore irrelevant to maximizing personal wealth. Yet a culture of exploitation also produces *economic* instability, in and of itself. This instability, in effect, serves as a warning sign of an impending Hobbesian nightmare: a collapse of the modern economic order into social upheaval, exacerbated by a prolonged diminution of natural capacity to support the economy.

What if the absolute scarcity of natural capacity forced us to abandon the doctrine of growth for one of material discipline? How would attitudes change regarding the individual and the collective? Are people really as competitive as Hobbes supposed? And how would the pie be shared?

A doctrine of material discipline represents a strategy for long-term economic stability on a crowded planet. Material growth of the economy would be superseded by a progressively increasing material efficiency of the economy, thereby reducing the economy's reliance on natural capacity. As argued earlier, the enormous headroom that exists today to increase material efficiency represents a huge economic opportunity. Economic growth would become progressively decoupled from the economy's material growth. In this scenario, recirculative industries would grow while extractive and wasteful ones would decline.

One might ask how, given a fixed circulation of money, the gross global product of the world's economy would grow. The answer is that, adjusted for population growth and price inflation, it might grow relatively little. But it would grow through increased spending: that is, the un-hoarding of money on the part of corporations and government, such as through sovereign money creation and a basic living program. These entities would invest in innovations driving increased material efficiency. Those innovations in turn would result in technologies that reduce the operating costs of businesses, government and households, even while tax burdens were shifted from income onto material intensity. The result would be increases in real *net* income both at the individual level and in the aggregate.

When one views the global economy and nature as connected systems, 'economic growth' as a concept, a doctrine and a culture, is revealed as too narrow for a crowded planet. The culture of growth is too individualistic and the doctrine of growth is too nation-centric. If we assume for a moment that the amount of money in circulation is fixed, then the GDP growth of one nation necessarily implies the GDP decline of another, just as the income growth of one individual implies the impoverishment of another. This is Hobbes's struggle over a fixed economic pie. That the actual gross global product has grown over time is a function of increasing productivity enabled by improving technologies.

So in a world where the economy becomes materially more efficient, 'economic growth' no longer would refer to the increasing material intensity of the economy—except insofar as it may continue to grow materially for a while until efficiencies take hold—but rather to its increasing productivity. It also refers to societal maturation, evidenced through the broadening of prosperity—the floating of all boats—through the establishment of a minimum living standard. It is growth as both a quantitative distribution and a qualitative development. The broadening of prosperity will power up the engine of innovation.

A culture of broad prosperity will be a culture where a planetary, or whole-system, perspective is the norm. Just as in a growth culture, where growth begets more growth, so, in a prosperity culture, prosperity will beget more prosperity. Provided substantially everybody lived well above financial subsistence, and that the surplus allowing this placed minimal burden on natural capacity, the economy ought to remain stable for the long term.

Planetarian Society: A Vision

The discussion in Chap. 3 of norms for a future economy is a launchpad here for a broader cultural and social vision of a planetarian society, encompassing people's perceptions, attitudes and ethics. To quote Imam Khalid Latif and Linda Sarsour: "Our goal is not to prepare our children for this world, our goal is to prepare this world for our children."[9]

A World of Opportunity, a World of Tranquility

Karl Marx, John Maynard Keynes and others foresaw in the far future a coming 'age of abundance', in which rivalry would cease to be a principal motivator of people's economic participation, and relative wants would fade away.[10] In thinking about economic 'abundance', though, they had

[9] Latif and Sarsour (2016).
[10] Marglin (2008, pp. 209–222).

not anticipated the risk of absolute scarcity arising through the material growth of the global economy. An age of abundance, if it is to come, must be predicated upon material discipline. In fact, the pursuit of material discipline will enable this abundance, although it must be accompanied by the elimination of absolute wants. Contrary to Marx and Keynes, relative wants would not be eliminated in this future world, yet their strength—the degree to which they created economic demand—would be weaker than in today's growth economy. This is for two reasons. One, because the satisfaction of relative wants en masse would have to wring out of the economy additional increments of material efficiency, this within an economy already recirculating most of its material. The economic cost of achieving that increment in most situations might outweigh the benefit. Second, people's value systems likely would have changed. The elimination of absolute poverty implies a minimum living standard. With such a floor in place, people's valuation of time against income would shift away from income. The urgency to earn income above the floor would be relatively weaker: therefore, so would rivalry.

This is not to say that a future 'age of abundance' would be an age of indolence. The costs of basic living must be funded from somewhere. Assuming the bulk of it came from artificial capacity fees, it would rely upon the financial viability of businesses which, in turn, would rest upon value delivered by their employees. This scenario implies a market economy within which, just as today, firms would compete for customers, this competition driving innovation.

It would be a world of opportunity, though, precisely because opportunities would be truly diverse, not merely or even primarily financial. Financial value would coexist with other forms of value, such as time value, these coming into greater play in people's lives by virtue of the basic living dividend. The 'abundance' of this world would be multivariate in nature, dampening any tendency toward limitless accumulation. If wants are unbounded, as economics traditionally has taught, argues Marglin, then abundance remains an impossibility.[11] But wants are not quantitatively unbounded, only qualitatively so. They could take an endless variety of forms, while being materially limited.

The seeking of individual fulfilment is much broader than the 'leisure' concept of traditional economics. In this future economy, a distinction between 'work' and 'leisure' might be superfluous. The assurance of a

[11] Marglin (2008, p. 221).

basis of well-being and the bounding of material wants would enable people to explore individual potentials, experiment with their lives, try new things and fail in them. In fact, a key purpose of establishing a basic living program would be to set an expectation that earning a money income does not necessarily have to be one's prime objective. Individuals of an entrepreneurial bent would seek activities having a potential for financial remuneration, as they do today. Yet this world would provide equal support to people who were not entrepreneurial. People with artistic, literary or scientific talents, for instance, would have the potential to contribute to society in their own particular ways, without the implicit discrimination associated with an absence of financial return from their activities.

It would be a world of opportunity in so many ways that the current economy is not. It implies a multitude of currencies of value circulating within society, some of them financial, many of them cultural. Indeed, the evolution from the present economy to this future one would entail a significant cultural shift away from a hegemony of money toward a more equal coexistence of a profusion of cultural currencies. Money, writes Dodd, is seen by some commentators as "an alien *and alienating* presence in social life, bringing people together simply by rendering them as tools for each other" (italics his). It "turns social processes into objective things".[12] Yet by providing all people an equal financial basis, the economy would remove from society much of the alienating presence of money, thereby allowing people to establish many more kinds of non-money relationships than are possible today.

As a consequence, less time spent would be spent doing paid work than today. Paid work itself would take many forms. The idea of an employee on a salary, with medical and retirement benefits, likely would be consigned to history, an anachronism of the earlier growth economy. As earlier chapters have discussed, taxes on earned income would be minimal, with most of the tax burden being placed upon material intensity. Corporations would not have to contribute employee benefits above and beyond payment for work performed, nor would they be expected to. Payments might not even be wholly financial; they might take the form of services rendered in kind, thereby bypassing money altogether. All these beneficial changes would arise from taking money out of people's survival equation.

[12] Dodd (2014, p. 273, 274).

Work in this world of opportunity would become less drudgery and more enjoyment. People would have more time to do the work they feel passionate about than work they need to do to support themselves or their families. Much more time would be spent engaged in activities the individual would find meaningful than in today's economy lacking a welfare floor. Whereas those people who today experience their work as 'play' are the lucky few, in this future world it would commonplace. "Every ordinary man has tasted the paradise of play in his own childhood", wrote Norman Brown. There is an "immortal instinct for play ... merely repressed by the habits of work". To Brown, "the foundation on which the man of the future will be built is already there;" it would not have to be created from nothing.[13]

Today, many people work twelve-hour days in two or three jobs just to make ends meet. In a future world of opportunity, they might spend twelve hours a day in two or three major activities, but would not have to rely on all of them to make ends meet. Perhaps only one of them need be paid, occupying a few hours a day or two to three days a week, the others contributing to individual fulfilment or to cultural enrichment. As a consequence, the culture of this future world would not be as heavily shaped by money as our current culture of modernity. Flows of financial money would be complemented to a much greater extent by flows of other kinds of non-money value.

An outside-looking-in perspective on economic activity would interpret financial money as one specific kind of value-flow among many, occupying its own cultural spaces. By the end of the twenty-first century, the rather narrow culture of money that has dominated the world's societies could be fading, and a future world of broader opportunity beginning to appear on the horizon.

By ensuring an essentially unbounded array of opportunities, such a world would become one of tranquility. Financial and economic volatility would be significantly reduced. The future would appear much more predictable than it does today. With economic uncertainty diminished, planning for the long term would become much more culturally ingrained. Ironically, the time when the long view is most needed—today—is a time of great volatility and uncertainty. Yet there are ways to sow the seeds of that long-term thinking. Policies to stabilize people's personal futures will ensure that they do not have to try so hard to make ends meet. With

[13] Brown (1959, p. 36).

individual short-term survival assured, today's culture of self-maximization would gradually give way to one of self-realization. With more time on their hands, people would engage less in exploitation and more in preservation. Communities would re-grow. A progressive de-commodification of the individual would lead over time to a culture that values stability over growth. The economy as a whole consequently would become much less dependent upon debt, therefore intrinsically less exploitative. The growth imperative would cool. Because people's livelihoods no longer would depend primarily on the debt engine, growth could tail off without causing widespread hardship. People's lives, in fact, would be richer and more balanced anyway. In the aggregate, less debt and lower growth would diminish financial and economic volatility, thereby reinforcing the cycle. The economy would become relatively much more stable, predictable and tranquil.

It would be tranquil also in a distributional sense. Economic disaffection would become largely a thing of the past. The majority of social conflicts can be traced to resentment of the 'haves' by the 'have-nots'. The existence of a welfare floor would eliminate the problem of 'not having' in an absolute sense. An income tax system that encouraged elective redistribution on the part of the 'haves' would further alleviate the problem of 'not having' in a relative sense. By applying mechanisms to narrow the income gap, the distinction between 'having' and 'not having' might fade over time.

Everyone would be subject to the same requirements for material discipline, implemented through the kinds of economic instruments outlined earlier. In today's world, only the affluent can afford material efficiency. Energy-efficient vehicles, solar panels and home energy upgrades are relatively expensive, and are purchased as much for status as for their financial payback. At the same time, widely disparate attitudes toward consumption within today's economy create a social tension. A future world in which status were decoupled from material throughput could accommodate conspicuous consumption, provided such consumption were materially efficient.

This is not to say that conflict would be eliminated. Conflicts arise over issues other than economic disparity, such as access to land of historical or cultural significance, access to natural resources, or access to equal opportunities or equal justice. They represent disparities in access to flows of value that are not monetary in nature. However, by resolving the issue of economic disaffection, the door opens to resolving other kinds of disaffection as well.

A future world of opportunity and tranquility might seem like a place where nothing ever happens, a social analogue to the standard economic equilibrium. Yet a world of tranquility would not be static. It would be a dynamic world: largely peaceful but evolving nonetheless.

Tranquility would not be limited to the social sphere. It would extend also to the natural sphere. The provision of a basic living program in and of itself would reduce the economy's material intensity because it would sever the connection between subsistence and debt, thereby diminishing the imperative for growth. The emergence of a culture of greater farsightedness, arising as it would from people's worrying less about where the next paycheck was coming from, would result in all kinds of normative changes reflecting an appreciation and understanding of our place within the systems of Earth. Material discipline thus would become relatively more valued. Over time, the economy's reduced intensity upon natural systems would allow natural capacity to regenerate. Natural cycles eventually would become relatively more stable and predictable. Natural systems would become more robust and resilient to perturbations. Some of these improvements might take hundreds or even thousands of years to complete. In the early decades, the effects of past excesses might continue to dominate. However, if it became common practice to watch for early signs of improvement in natural systems—a vigilance sorely lacking over the past two centuries—then those early signs would indeed be found, and they would multiply over time.

In the long run, one would hope that human culture became modeled in some way upon nature and, as a consequence, became seen as not separate from it but part of a continuum with it. As biological agents, most organisms are indeed self-maximizing: the theory of natural selection predicted this and evidence supports the prediction.[14] However, individual self-maximization bumps up against ecological limits, such as of food, water, space or reproduction, so that, in the aggregate, an ecosystem is a system of sufficiency—just enough to persist—and of efficiency: making the most of available resources. In a mature and stable state, it becomes a profoundly rich and diverse system, an analogue for a human social world of opportunity and tranquility.

[14] For example, Malhotra and Thorpe (1991) and Jensen et al. (2007).

The Storm Before the Calm

Yet there is no illusion about how much of a struggle it may take to get there. Today, there exist barriers of perceptions, mindsets, attitudes and norms. Pushing through these barriers will entail an evolution in ethics.

Some worry about a future dystopia brought on by economic collapse. Yet a great deal of evidence presented here and elsewhere suggests that the dystopia exists today, a dystopia of modernity. It is a dystopia of rampant excess, poverty, inequality, mistrust, conflict and war. "The established disorder of our present era is not necessary", asserts the philosopher Donna Haraway.[15] In the long run, this is true; but maybe in the near term, it *is* necessary, as a way to awaken people to the need for an orderly, tranquil future. The world imagined above is not a 'utopia'; it is a *syntopia*, the antonym of 'dystopia'. It is a 'together place', as distinct from an 'apart place' or a 'no place'. The task facing the people of the world is very simple: to stop considering themselves apart from one another and to work together. Clichéd as this may sound, the reality of the situation requires it.

The coming-together of the world boils down to a syncretion in ways people look at the world and their place in it. This is not to say everyone should agree on everything, but there are some things everyone should agree about if the economy is to come into alignment with nature. One of these is what one could call a Copernican revolution of the individual: a reconstruction of the social cosmos in which the individual no longer sits at the center but exists in orbit around another center: the system of Earth, and our society upon it. Another, related, requirement is an abrogation of 'otherness'. A continued recognition of 'the other' within society is what perpetuates tribalism and keeps nation-states alive. Perceiving other people as 'alien' is a mindset that blocks the way toward a syntopic worldview. Nature cares not one jot how any particular group of people may view another. In the face of an economic coevolution with nature, there is only a single group of people. Their actions have consequences affecting all and sparing none. A society, on the other hand, unburdened from a prevailing 'otherness' among its members would be a society unified: not all the same, but all of a similar mind. It is clear that the de-individuation of society is not only a recognition of something more important than 'me' but also a recognition of something more important than any differences between me and you. The two go hand in hand.

[15] Weigel (2019).

Heedless self-actualization or self-maximization leads to the unintended collective maximization of effects upon natural systems. For an intentional collective *minimization* to be successful, self-maximization must become constrained in certain ways. Self-maximization must be bounded from the outside by laws or economic instruments, but equally it may be bounded by the ethics of the individual, through a consciousness of the collective, a consciousness largely missing from modern, Western society. Today's economy and its associated liberal democracy prides itself on being a meritocracy, at least in principle. Rewards, supposedly, come to those who succeed. However, as David Brooks writes, our current version of a meritocracy is self-centered and self-maximizing rather than being grounded in an awareness or recognition of social institutions and communities. Brooks agrees that all people should have equal opportunities; yet what is lacking is a 'common telos': that is, a purpose toward which the development of communities and institutions, and the fostering of diversity, would be directed. Diverse institutions and communities reflect an individual identity defined less by social groupings and more by commonalities. The recognition of commonalities is a conduit to the recognition of the collective. A recognition of the collective opens a pathway to a common telos. Diverse communities and institutions provide the range of perspectives necessary to support a common telos.[16]

While progress has been made in that direction, there is still a long way to go. The twenty-first century could be a stormy one, not only climatologically. A hundred years ago, much of the world was embroiled in an ideological struggle between capitalism and communism. After the stock market crash of 1929, the excesses of unchecked capitalism were reined in by the governments of the market economies, ushering in a Keynesian era of oversight and control. This era coincided with the postwar establishment of the United Nations and the emergence of newly independent states formerly under colonial control. The era of relatively strong government did not last long, however. Postwar economic growth and incipient political globalization brought increasing economic globalization. Corporate multinationals grew in power to a point where they could play national governments off against one another to their own benefit. Financial markets became globally connected. Emerging economies industrialized, while many existing industrial economies began to de-industrialize.

[16] Brooks (2018).

It is not so much that national governments enabled economic globalization but rather that they were powerless to stop it. Some, such as the British and US governments of the 1980s, adopted the attitude that if you can't beat them then you may as well join them and make the most of it. Government support to households was scaled back, leaving people to make what they could of the new 'free' market economy. Manufacturing jobs went to countries with low labor costs and little regulatory oversight, to be replaced in the de-industrializing countries with jobs selling things. The establishment of transnational trading areas lowered barriers to international trade, ostensibly in the interest of stimulating domestic economic growth; however, they served also to accelerate the economic globalization already under way.

Little wonder that, early in the twenty-first century, there has been a backlash, particularly within the very same OECD countries who built such great wealth through industrialization, and who established Keynesian policies to spread it. People whose livelihoods were ripped away by globalization wanted their countries back. They wanted their old jobs back, their old lives back, their old security back. And they did not want 'foreigners' coming in competing for what little economic security was left. Their anger was misplaced. If their governments had invested in innovation, they would have stable jobs in new industries, industries with the prospect to establish a new economy for the twenty-first century and beyond. Instead, their elected representatives were busy genuflecting themselves at the altar of market liberalism. The 'foreigners', for their part, were trying to escape far worse situations. Communications technologies that had been unavailable during the early days of globalization enabled them to see how good things could be in the wealthy countries. So they packed their bags.

In the grand scheme of things, it should not come as much of a surprise that globalization, economically, politically and culturally, should result in a pendulum swing toward xenophobia, nationalism or nativism. Voters in many OECD countries have blamed political elites for abandoning them to globalization, taking out their frustrations at the ballot-box by electing populists and nationalists. This, they assumed, was all they could do. Yet their anger is misplaced. De-globalizing the economy is not an option. It is a phenomenon far beyond the control of any single national government. Pulling up the drawbridge will not make anybody on the inside better off. The 1 percent, whose wealth is distributed globally, simply would move it elsewhere.

Governments failed in not appropriating some of the wealth generated by globalization for the public interest. Perhaps they were afraid of capital flight? Common capacity fees could bring significant revenue to the public purse, provided they are small enough to give a corporate giant no excuse to move elsewhere. Governments failed in allowing their money to become privatized by the banks. They gave up the ability to create their own money, therewith to fund programs to help their people.

These failures can readily be corrected through legislation. For the most part, however, the members of national legislatures do not see this. They see symptoms, such as economic migration, rather than the underlying causes: economic globalization and corporate multinationalism. Citizens take exception to these developments because of the inequalities they generate. They violate the principle of strong reciprocity. As or when populists fail to protect the people from the vagaries of globalization, the people would turn on them, too.

On the other hand, if the people of the fading industrial economies felt assured that their way of life could be preserved then they would be much more accepting of economic globalization, and all that it brings. The era of 'jobs for life' probably is over; however, we are not yet in an era of 'income for life'. People in OECD countries do not have a platform upon which to create their own futures; hence, many remain dependent on 'gig' work to make ends meet, while mourning the loss of their former prosperity. Yet preserving one's way of life is not so much about getting the old job back as maintaining a standard of living. Unless or until legislators understand that they have the ability to offer people an acceptable standard of living—no strings attached—at minimal cost to the private sector, and that they have the ability to take back control of the public money, they will remain the target of popular anger and frustration.

Corporate multinationals, for their part, are not always the greedy behemoths they are sometimes made out to be, even if they remain beyond the control of individual sovereign governments. Many executives worry about the degradation of natural systems, both for reasons of economic uncertainty and for ethical reasons. Many corporations are taking steps to improve energetic and material efficiency. Many, but not all; hence, those at the forefront have called upon governments to create a new, level regulatory playing field for all corporations, one which presumably would benefit the early movers. Cap-and-trade systems and carbon taxes are examples of instruments that have been proposed or adopted. Multinationals need governments to act because governments have the ability to create the rules for all players in a market, thereby formalizing and embedding a culture of strong reciprocity. Absent these rules, a corporation's impulse for self-preservation eventually will override any motivation to reduce its material intensity. In an economic environment that fails to tie profit-making to material efficiency, the market will trend toward the lowest common denominator: the maximization of financial returns without accounting for returns to nature. The long-term consequences would be catastrophic. Not even the 1 percent are immune from ecological collapse, and they know it.

Society is far from the syntopian state outlined above, and many storms lie ahead. But it is possible to see a pathway toward it.

From Exploitative Capitalism to Regenerative Capacitism

From before the industrial revolution to around the middle of the twentieth century, Western society had been ecologically asleep. Among the many things the European colonial powers had spread to the corners of the world was a culture of dominion over nature. A defining feature of modernity, this culture was alien to most of the societies they conquered. Back home, though, the effects of unregulated waste and pollution upon air, water and living systems eventually could no longer be ignored. In the nineteenth century, London installed a sewer system after the Thames became so horribly smelly during the summer months that Parliament, situated on its banks, could no longer meet. Beginning around 1900, industrialized nations passed clean air acts, established national parks, and began to take measures to protect endangered species. Yet all this time, there was little if any official questioning of the structure and functioning of the underlying capitalist paradigm. That was taken as given.

Fifty years after the ecological awakening of the 1960s, it is still taken largely as given, although an emerging school of thought has begun to question it. A number of writers proposed utopian alternatives. Operating far outside the mainstream, they sowed the seeds of concrete proposals that have emerged in the early years of the twenty-first century, such as for carbon taxes, sovereign money, banking reform, basic income and deliberative democracy. Although these proposals have not yet made it into national legislation, they represent the early signs of a trend away from a culture of exploitative capitalism toward what could be called regenerative *capacitism*.

Capitalism, as Dodd and many others have described, is an economic system and a doctrine of the self-expansion of capital, specifically financial capital. Capitalism is a relatively recent phenomenon in human history, having arisen during the Renaissance in Europe. "Before capitalism", Dodd writes, "money was merely superimposed onto pre-existing social formations." Monarchies, in part, were defined by their distinct ability to determine legal tender and to issue it. In this way, they created a standardized taxation medium to fund the affairs of the state. Yet this 'despotic' money did not actually define those social formations, as Dodd explains:

> Money had no 'body' of its own; it was merely appended to other bodies that had been formed independently of it (e.g., nature, the state). For money to have its own body, it had to be capable of creating more money. [...] It therefore had to become capital; or more precisely, it had to be transformed into finance capital.[17]

It was this transformation that led capital eventually to transcend the influence of monarchs and governments to form its own space, a global one. Today's monetary landscape "is not divided neatly into the ... spatial parameters of states", write Gilles Deleuze and Félix Guattari. To the contrary, capitalism "deterritorializes and, through finance, potentially creates its own territorial space ... a space [*in extremis*] without border, regulation or control..."[18]

But for money as finance capital to continue to exist independent of geopolitical control, it must continue to feed its own expansion, that is, to deliver returns. According to David Harvey, capitalism must achieve a compound growth rate of 3 percent per year to maintain its own structural integrity.[19] If Harvey is correct then, in a situation where capital's self-expansion broke down, capitalism as we know it would collapse in upon itself. Finance capital would evaporate, leaving only payment capital circulating within the economy. The self-expansion of capital *has* broken down in the past, at least temporarily, during the various financial crises besetting the financial system during the modern era. The reason these breakdowns were temporary rather than long-lived is because commercial finance made use of public money, placing the onus upon governments to 'bail

[17] Dodd (2014, p. 232).
[18] Dodd (2014, p. 233), citing Deleuze and Guattari (1987).
[19] Harvey (2011, p. 70).

out' financial institutions by issuing new public money. This dependency, argues Mary Mellor, represents a major contradiction within capitalism: "...The privatization of the money supply was based on debt. This provides the 'elastic credit' capitalism needs, but is constantly threatened by implosion when the system can take no more debt." From it, two other contradictions follow: "A second ... is the starvation of monetary resources for the public sector, ... which removes the only source of debt-free money that could sustain demand. A third is linked to the second. Starving the public sector of money undermines the political legitimacy of the system."[20]

The intertwining of public money and private finance might be reason enough for monetary reform, even in the absence of a material coevolution between the economy and nature. That coevolution, however, superimposes a new, more serious challenge onto the self-expansion of capital, which is a limit of absolute scarcity in natural capacity. The problem is particularly wicked because no-one knows where that limit may lie or whether we might have crossed it already.

This is the long-term consequence of capitalism as an exploitative undertaking, which neither its early champions nor its critics foresaw. Capitalism, in its present construction, is parasitic upon the planet. For the economy to coexist with nature, the exploitative must become the regenerative. If it does, it will find nature reciprocally regenerative, thereby creating a positive reinforcement, turning the parasitic into the *sympoietic*.[21] If it does not, it will find nature reciprocally destructive. This is not a moral imperative; it is simply a fact of survival. It is not that capitalism is inherently evil, as Marx and others argued; capitalism as presently configured is merely exploitative. It must evolve toward the regenerative for civilization to persist.

Capacitism, then, is an economic philosophy and a culture of the maintenance of capacity: both the capacity of the economy to provide for its occupants, and the capacity, in turn, of nature to accommodate and provide for the economy.[22] A market-planetarian raft of policies would translate this philosophy into action. Other capacitist economic systems are conceivable, such as the localized, community-based systems proposed by

[20] Mellor (2016, p. 180).

[21] Meaning 'generating together' or 'producing together'.

[22] The term 'capacitism', it should be noted, has alternate uses in other areas. It can refer to discrimination against people who are incapacitated by injuries or disabilities, and it has also been used in philosophy to describe an externalist view of evidence and knowledge (Schellenberg 2018, ch. 10).

the deep ecologists or, hypothetically, centrally planned or authoritarian ones. This volume advocates a market-based form of capacitism as the most effective engine of economic change for today's institutionally dispersed, globally connected society.

Capitalism of the past two centuries has been defined by its celebration of unrelenting growth. Uncritically and with all sincerity, capitalists embraced the exponential. The concept of a growth curve increasing ever more steeply into perpetuity is a preposterous fantasy. It has no grounding in the real world. Faced with actual limits, the typical response of the indoctrinated capitalist was to deny their existence. To do otherwise would be to abandon their worldview.

Capacitism, on the other hand, is defined by its acknowledgment of limits. In actuality, all kinds of growth tend to slow and level off at some point, as long as bounding conditions remain unchanged. The capitalist might retort that the bounding conditions themselves change, and that this is what allows growth to continue. To suggest that bounding conditions continually relax, however, is tantamount to assuming growth itself is unbounded: it is the same fallacy as the denial of limits. In reality, growth curves that begin exponentially tend to turn logistic at some point. This pattern represents, to a first approximation, the growth of most living things. If one removes the capitalist goggles, then it is clear that this is also generally the growth pattern of human social systems. Nature is a free-for-all, but only within limits imposed by resource scarcity. Capacitism simply acknowledges that society—and specifically the economy—resembles nature in this regard. The market provides the free-for-all, and planetarian norms impose limits. These limits must come from a guiding principle or prime directive of some kind, which shapes the controls over market activity. Market planetarianism is a form of capacitist economics.

Where capitalism is an "expropriation of the common", to use Hardt and Negri's term,[23] capacitism would provide a return to the common. "The democratization of money", Mellor writes, "is not about bringing commercial values into public provisioning. Quite the opposite: democratically determined public values would drive the commercial sector."[24] Capitalism no longer would run riot but would be contained. It would no longer dominate the economy but be dominated by it. Market planetarianism is an application of capitalism to good use.

[23] Hardt and Negri (2009, p. 137).

[24] Mellor (2016, p. 181).

American Capacitism

As the largest and one of the more liberal examples of the capitalist system, the economy of the US represents a test-case of the kinds of social, cultural, political and economic changes that are needed to evolve from exploitative capitalism toward regenerative capacitism. "Any ... analysis of the world's current economic problems", wrote Joseph Stiglitz, "must begin with a discussion of America."[25] Whichever direction the US moves, it will pull other countries with it by economic gravitation. The converse is also true: other countries are often loth to make systemic changes of their own unless or until the US does. Being economically and culturally more liberal—or neoliberal—than probably any other large country on the planet, the US might be the last place one would imagine a 'collectivist' or 'planetarian' norm emerging. Yet at the same time, its fervent adherence to the legitimacy of the market represents its most powerful engine of change.

For many decades, the 'American Way' was characterized, and shaped, by America's culture of self-sufficiency. It is a philosophy—if not always a practice—of low taxation, providing the individual with a potential for self-enrichment, while at the same time offering a much smaller social safety net than other comparably wealthy countries. People pride themselves on being able, at least theoretically, to resist government intrusion upon personal freedoms. Aside from the era of the New Deal, America was largely a culture of 'small government', even though the actual scale of America's federal administration is vast. America is rightly described as a 'land of opportunity', a culture in which people have freedom to fail without shame but also without much help back onto their feet. Per-capita consumption is high, as are rates of indebtedness. The financial system, and therefore the economy at large, is prone to volatility, which adds to the financial risks to households. The America of an exploitative capitalism is a high-energy, high-throughput, risk-taking economy.

Contrast this economy with an imaginary future America that is materially stable and broadly prosperous. It would be the 'American Way' of the twenty-second century. In this future America, income taxation is zero or very low for all except for those on incomes substantially above the median. Some of this tax burden might be channeled into elective redistribution. Tax rates on material intensity are high, putting control over material intensity into the hands of the consumer through their choices in the

[25] Stiglitz (2003, p. XLV).

market. Consumers save by becoming materially efficient. The opportunity for self-enrichment still is very much alive in this America, although there is no social safety net. Instead, there is something much stronger: a concrete floor in the form of a basic living program. Work-based welfare programs are a thing of the past. They were onerous, administratively inefficient, inequitable, and undignified. The dignity of a guaranteed living standard instead has ended decades of wage-imprisonment, giving everybody a degree of material opportunity. In the process, it has fostered creativity, innovation and productivity. The old culture of self-reliance and self-sufficiency has evolved into one of shared reliance, which actually has increased people's ability to take risks.

Investment in infrastructure is higher than in the past. Most infrastructure, including utilities, is publicly or co-operatively owned because infrastructure is a public benefit and a natural monopoly. The public, furthermore, controls the public currency, the means through which investment in common assets is made and taxes collected. The government does not need to rely on tax or borrowing for revenue, instead placing new public money into circulation for public investment, and recovering public money through taxes, such as on material intensity. Although the public sector as a whole is larger than in the past, employing more people, the regulatory apparatus of government is smaller because, with the right incentives operating within the economy, market behaviors and natural capacity are in greater alignment. The late-twentieth-century belief that privatizing public assets and services would make them more cost-efficient has been universally rejected. It was always a myth. The mantra in this America is not about making government as small as possible but about giving it full control over the resources it needs to serve the public. At the same time, being focused less on direct regulation and more on public investment, government is no longer vulnerable to the accusation of 'meddling' in business.

Another distinguishing feature of this America is a relative absence of antagonism between 'government' and 'the people'. This antagonism arose originally because people perceived the federal government, particularly as the twentieth century gave way to the twenty-first, as becoming 'too big' and therefore unaccountable. The government made decisions—such as about the use of natural resources—against which affected people found they had little recourse. Conflict, resentment and suspicion of the federal government grew among many sectors of American society. The entrapment of elected officials by well-funded special interests only added to a sense of distance and aloofness of Washington from the interests and needs of common people.

In the future America, antagonism between government and the people has been largely resolved. The people now *are* the government, thanks to the adoption of deeper, more participatory democracy. People no longer cede control every two years to elected representatives who, in turn, have little actual influence over bloated, unaccountable bureaucracies. American governance now is hierarchical, from individuals to local and state governments, up to the federal level. Citizens have direct access to the legislative process through sortition, operating at all levels. In this way, they participate with elected officials in decision-making. Perhaps the most dramatic difference between this America and the old exploitative one is that all this activity now is shaped by a constitutional amendment adopting a prime directive: to maintain and improve the capacity of nature to support the economy, and to maintain and improve the capacity of society to accomplish this.

As a result of these measures, per-capita resource consumption is low and standards of living are high. Rates of indebtedness are low. Credit is available, but not in the public currency, only in private ones, which are not backed by the public currency. As a consequence, savings rates are relatively higher than during the debt-blighted past. Financial volatility is low, not only because debt is less prevalent but also because the financial sector itself is relatively smaller than in the past. The economy is stable.

The America of a regenerative capacitism is a medium-energy, low-throughput economy accommodating both risk-taking and risk aversion. It preserves individual freedoms while also providing for individual engagement and participation in the social process. For these reasons, it is a tolerant society.

The broad prosperity imagined for this future America would have to exist just about everywhere in the world for it to persist domestically in the long term. That same requirement would apply to material efficiency, monetary policy and the relative absence of debt. Within these parameters, however, the economic profiles of different regions could vary according to their societies and cultures.

Getting there will be a matter of recognizing the need for a cultural shift from competition toward co-operation, and for the economic incentives and social support systems to reinforce it. As this book has discussed, social norms are shaped to a very significant degree by the instruments at work within the economy. These norms, in turn, influence the choice of instruments, so that the two evolve in a cycle. This is the mechanism by which today's exploitative capitalism would iterate toward the regenerative capacitism of the future.

Coda: A Chance to Beat the Evolutionary Odds

The writing of this book has covered a span of time of increasing social and political division, particularly within the US and other OECD countries, but also worldwide. This divisiveness has touched, for one, upon the cumulative effects of people's livelihoods upon the natural world, and on the reactionary effects of nature upon society. These influences appear to be growing stronger with each passing decade. The divisiveness also has taken hold in the form of inequality, which in turn has spilled over into migration, issues of national identity and the purpose of statehood. Meanwhile, the global economy has continued to grow, becoming ever more integrated and interdependent.

The issue here is not with protecting nation-states but with protecting people. There remains no part of the world today where people do not have some form of economic or social connection with other parts of the world. Yet the daily struggle for survival, not only among the poor but also among the vast majority of middle earners, places great strain upon the planet's capacity to provide for the economy in the way it is currently configured. This book has explored how the economy could be reconfigured to reduce that strain thereby, over time, reducing the reciprocal threat to the economy from nature. It has examined certain ways we need to think differently about our economic relationship with the rest of the planet, our social relationships with one another, and the ways we study the economic process. Just as with any other science, unless we ask the right questions, we will not find meaningful answers. One of the principal conclusions of this investigation is that the long-term success of any efforts to bring the global economy into alignment with nature critically rests upon our will and ability to resolve the inequality problem. This is a collective problem that can only have a collective solution. The problem is global, so the thinking has to go beyond national borders. Solutions to aligning the economy with nature range from the local to the global, so the institutions to implement them must accommodate this hierarchy.

Is humanity up to it? The answer may depend partly on our consciousness. Just as the brains of squirrels and deer are not adapted to the evolutionarily sudden appearance of roads with vehicle traffic, so it is possible that human brains are not yet adapted to a planetary environment in which their economy has significant and unpredictable effects upon natural systems, and is significantly and unpredictably affected by them in return, these effects playing out over periods longer than a human lifetime. Our consciousness is tuned to the periodicity of social processes, which are much shorter. This is why, in the face of decades of scientific evidence, the social and political response to the coevolution of the economy and nature has been mostly one of denial and inaction. It is very hard to mobilize a large number of people in the face of a threat whose apparent effects are uncertain and not immediate. The global coevolution of the economy and nature is a completely different kind of threat from that of a hungry predator.

Yet this threat, in the long run, will make a mockery of all our cute little social constructs. If you could stand on the surface of Pluto, the sun would appear as little more than a bright star. Earth would be completely invisible. From the vantage-point of Pluto, humanity's great social achievements would be undetectable, as would the rest of life on Earth. The Communist experiment, the capitalist experiment, industrialization, agriculture, civilization, mammals, dinosaurs, flowering plants, insects, fish, marine metazoans, phytoplankton, cyanobacteria, self-replicating organic molecules: they may as well never have existed. And we haven't even left the solar system. Riding as we are today on a wave of early-twenty-first-century technological innovation, it is impossible to really understand how vulnerable civilization is.

The average lifetime of most species of large mammal in the fossil record is about 0.8 million years. The record shows that *Homo sapiens* has existed as a distinct species for about a quarter of a million years. All else being equal, this gives us about another half-million years of existence. Is it likely, on recent performance, that we will reach that average? Or will aliens visiting Earth ten million years hence find only geological evidence of a civilization caught with its pants down? If a civilization takes care of its natural container, then the container will take care of it. Some argue that we will develop technologies enabling us to find other containers—other worlds—just as these hypothetical aliens might have done. This is indeed possible, although those technologies likely will be contingent upon our first developing the technologies, both physical and social, to persist within the present container. The chances are good: civilization has existed for only about one to two percent of the half-million years we might have left. That is plenty of time to get it right first time around, or for many subsequent attempts.

Why do people enjoy getting out for a hike in nature? Nature is wonderful because it exists largely outside our control, even if our actions may inadvertently influence its long-term pathways. Some of us believe that a god controls the outcome, perhaps because we would like to believe we are at the mercy of a sentient being, as opposed to the vagaries of nature, but for all practical purposes they come to the same thing. Whatever one's belief, we do not control the outcome. We can only control our own actions, our own social systems. To paraphrase the biologist Peter Medawar, God, if he or she exists, must be inordinately fond of nonlinearity. This is the property that renders nature fabulously unpredictable and uncontrollable.

To find a way toward a planetary economy and society, the past cannot be a guide. Only the future is a guide, and the future is in our heads. Think about what you want in the future for yourself, your descendants, those you love, those you don't care about, even those you might think you hate. You can be sure they also will be thinking about the same things. What opportunities you would like people to take, so that they follow a different path? These would be opportunities to establish, so as to create a stable society on Earth. The alternative, inevitably, would be societal collapse: not in my lifetime, probably not in your grandchildren's lifetime, but possibly within the next couple of hundred years.

If, on the other hand, the effort were successful, then your descendants would live in a world whose population would be stable, the necessities of life would be provided for all, standards of living would be universally high, and conflict and unrest would be rarely encountered. The reverberations of society's profligate past still being felt, it might also be a warmer, more stormy, less biodiverse world than today's, although the necessary physical and social technologies to cope with those conditions would be well established. Socially, it would be a world of relative constancy, something people were accustomed to for centuries before the age of modernity—a dynamic constancy, one would hope, not a boring one. Yet within this constancy would exist a tranquility people yearn for today.

References

Brannen, P. (2017). *The Ends of the World: Volcanic Apocalypses, Lethal Oceans and Our Quest to Understand Earth's Past Mass Extinctions*. New York: HarperCollins.

Brooks, D. (2018, May 28). The Strange Failure of the Educated Elite. *New York Times*.

Brown, N. O. (1959). *Life Against Death: The Psychoanalytical Meaning of History*. New York: Vintage Books.

Deleuze, G., & Guattari, F. (1987). *A Thousand Plateaus: Capitalism and Schizophrenia*. London and New York: Continuum.

Dodd, N. (2014). *The Social Life of Money*. Princeton University Press.

Fromm, E. (1976). *To Have Or To Be*. New York: Harper & Row.

Hardt, A., & Negri, M. (2009). *Commonwealth*. Harvard University Press.

Harvey, D. (2011). *The Enigma of Capital and the Crises of Capitalism*. London: Profile Books.

Jensen, K., Call, J., & Tomasello, M. (2007). Chimpanzees Are Rational Maximisers in an Ultimatum Game. *Science, 318*(5847), 107–109.

Latif, K., & Sarsour, L. (2016, February 4). We Welcome President Obama's Words in His Mosque Visit. Now We Need Actions. *The Guardian*.

Lovins, L. H., et al. (2018). *A Finer Future: Creating an Economy in Service to Life*. Gabriola Island, BC, Canada: New Society Publishers.

Malhotra, A., & Thorpe, R. S. (1991). Experimental Detection of Rapid Evolutionary Response in Natural Lizard Populations. *Nature, 353*, 347–348.

Marglin, S. (2008). *The Dismal Science: How Thinking Like an Economist Undermines Community*. Harvard University Press.

Mellor, M. (2016). *Debt or Democracy: Public Money for Sustainability and Social Justice*. London: Pluto Press.

Schellenberg, S. (2018). *The Unity of Perception: Content, Consciousness, Evidence*. Oxford University Press.

Stiglitz, J. E. (2003). *The Roaring Nineties*. New York: W.W. Norton.

Warburton, P. (1999). *Debt and Delusion: Central Bank Follies That Threaten Economic Disaster*. London: Allen Lane.

Weigel, M. G. (2019, June 20). Feminist Cyborg Scholar Donna Haraway: 'The Disorder of Our Era Isn't Necessary'. *The Guardian*.

Zahavi, A. (1975). Mate Selection – Selection for a Handicap. *Journal of Theoretical Biology, 53*(1), 205–214.

Index[1]

[1] Note: Page numbers followed by 'n' refer to notes.

F. Murison Smith, *A Planetary Economy*,
https://doi.org/10.1007/978-3-030-49296-0

D

E

H

Q

The manufacturer's authorised representative in the EU is Springer Nature Customer Service Centre GmbH, Europaplatz 3, 69115 Heidelberg, Germany. If you have any concerns regarding our products, please contact ProductSafety@springernature.com

Printed and bound by CPI Group (UK) Ltd, Croydon, CR0 4YY
30/09/2026
02225510-0001